THE GENUS COELOGYNE
A Synopsis

THE GENUS COELOGYNE
A Synopsis

Dudley Clayton

Photographs by

R.S. Beaman, C.L. Chan, C. Clarke, D.A. Clayton,
J.B. Comber, J. Cootes, P.J. Cribb, K. Jayaram, S.K. Jacobson,
P. Jongejan, B. Kieft, C.G. Koops, S. Kumar, A. Lamb,
J. Meijvogel, M. Perry, W.M. Poon, A. Schuiteman,
C. Sussendran, D. Titmus, and A. Vogel

Line illustrations by

Linda Gurr and Judi Stone

Natural History Publications (Borneo)
Kota Kinabalu

in association with

2002

To Jill

Published by

Natural History Publications (Borneo) Sdn. Bhd. (216807-X)
   A913, 9th Floor, Phase 1, Wisma Merdeka
   P.O. Box 15566
   88864 Kota Kinabalu, Sabah, Malaysia
   Tel: 088-233098 Fax: 088-240768
   e-mail: chewlun@tm.net.my
   http://www.nhpborneo.com

in association with

The Royal Botanic Gardens
   Kew, Richmond
   Surrey TW9 3AB, England

Copyright ©2002 The Royal Botanic Gardens, Kew
Photographs copyright ©2002 as credited.
First published 2002

Date of publication: 24th April 2002

All rights reserved. No part of this publication may be reproduced,
stored in a retrieval system, or transmitted in any form or
by any means, electronic, mechanical, photocopying, recording,
or otherwise, without the prior permission of the copyright owners.

The Genus Coelogyne: A Synopsis
   by Dudley Clayton

Scientific Editor: P.J. Cribb

ISBN 983-812-048-0

Design and layout by C.L. Chan

Frontispiece: *Coelogyne cristata* (Photo: D.A. Clayton)

Printed in Malaysia.

CONTENTS

FOREWORD vii

PREFACE ix

INTRODUCTION 1

ECOLOGY AND CLIMATE 7

MORPHOLOGY 11

TAXONOMY 15
    Sectional Treatments 15
    Artificial Key to the Sections 17
    Coelogyne Species and their Synonyms listed by Section 19
    Sectional Characteristics, Artificial Keys to the Species and Descriptions 28
        Section Elatae 28
        Section Proliferae 41
        Section Fuliginosae 47
        Section Micranthae 56
        Section Brachypterae 57
        Section Speciosae 61
        Section Bicellae 83
        Section Moniliformes 85
        Section Longifoliae 95
        Section Cyathogyne 117
        Section Verrucosae 119
        Section Tomentosae 128
        Section Hologyne 154
        Section Rigidiformes 157
        Section Veitchiae 164
        Section Ptychogyne 166
        Section Lawrenceanae 167
        Section Coelogyne 170
        Section Fuscescentes 180
        Section Ocellatae 185
        Section Lentiginosae 192
        Section Flaccidae 207

INDEX OF SYNONYMS AND EXCLUDED SPECIES 217

MOLECULAR PHYLOGENY OF COELOGYNE AND ALLIED GENERA:
AN URGE TO REORGANISE by B. Gravendeel and E.F. de Vogel 227

HYBRIDISATION
    Natural Hybrids 235
    Artificial Hybrids 237

CULTIVATION 241

BIBLIOGRAPHY 245

GLOSSARY OF TERMS 249

ACKNOWLEDGEMENTS 253

COLOUR PLATES 255

INDEX of SCIENTIFIC NAMES 305

FOREWORD

My first meeting with a *Coelogyne* proved to be a momentous one. In 1978 I visited Malaya prior to the World Orchid Conference in Bangkok. The Genting Highlands, one of the first localities I visited, is now one of Malaysia's prime tourist resorts with fine hotels and casinos perched high on a ridge in the Central Highlands of the peninsula. At the time, the first phase of building had just begun and the road-side was littered with felled trees that were laden with epiphytic orchids. Near the ridge-top my eye lighted on a handsome orchid with a pendent inflorescence of several large white flowers marked with yellow on the lip. In my innocence I had imagined that I would be able to identify it using Professor Holttum's fine account of the orchid flora of Malaya. Imagine my surprise when it failed to key out, neither could I find it in the literature of neighbouring countries, nor in the herbarium. My first *Coelogyne*, subsequently described as *Coelogyne kaliana*, was a new species!

I have had many memorable encounters with coelogynes since then. I particularly remember seeing *Coelogyne veitchii* for the first time on a forest track in the mountains of central Guadalcanal. Two glistening sprays hung down from a branch that crossed the track. Although exhausted from carrying a heavy back-pack all day, my spirits soared and I almost danced into camp in the evening. The Solomon Islands also disgorged another new species *Coelogyne susanae*, a fine ally of the more widespread *C. beccarii* and *C. speciosa* that was collected and painted from life by Sue Wickison who illustrated the *Orchids of the Solomon Islands and Bougainville* account.

Coelogyne is an Asiatic genus that can be found from sea level to the tops of the tropical mountains. One of the most surprising sights in Borneo is to find the necklace-like inflorescences of *Coelogyne papillosa* at above 3000 m on Mt Kinabalu. At that elevation it is by far the most showy orchid with brilliant white flowers marked with yellow and brown on the lip. A little lower, coelogynes abound with the brown and cream-flowered *C. radioferens*, and the white flowers of *C. venusta*, *C. swaniana* and *C. clemensii* hanging from the forest trees like so many brilliant necklaces.

Identifying coelogynes has always been problematic. The islands of Borneo and Sumatra are rich in the species of the genus but few keys are currently available to identify them. Dudley Clayton's book is, therefore, especially welcome, providing for the first time an up-to-date account of the genus, keys to the species, black-and-white illustrations and colour photographs of the majority of the species.

When Dudley first came to Kew some six years ago, I imagined that he was a keen grower eager to identify the plants in his collection. I had underestimated him. He has shown incredible determination, dug up many obscure references, chased type specimens, visited continental herbaria and corresponded with other *Coelogyne* enthusiasts around the world. His single-mindedness has resulted in the first comprehensive treatment of the genus since that of Pfitzer and Kraenzlin, published nearly 95 years ago.

Almost every collection, both amateur and commercial, grows at least one coelogyne. Many species make specimen plants: I remember particularly one, the Himalayan *Coelogyne cristata*, that regularly appeared at orchid shows in the UK and carried off best-in-show with monotony. It required a fork-lift truck to move it and a van to transport it from home to a show. I doubt if the owner's greenhouse had room for any other plants once it was in place! Nevertheless, it was a sensational plant and thoroughly deserved its many awards.

This book contains a detailed synopsis of our present state of knowledge of the genus but, as even a cursory reading of the chapter by Barbara Gravendeel and Ed de Vogel of the Rijksherbarium in Leiden will demonstrate, this is only a start. Their work suggests that *Coelogyne* is not even a good genus but contains quite distinct lineages. Furthermore, the rather distinctive *Pholidota* and *Dendrochilum* must be reconsidered before we can get a grip on the future generic delimitation of *Coelogyne*. I am sure that, like me, readers will await with anticipation future work by the team in Leiden.

Phillip Cribb
Curator, Orchid Herbarium,
The Royal Botanic Gardens,
Kew

PREFACE

In 1987, with my retirement from the Royal Air Force after a 35-year Service career, I converted a general interest in orchids to the growing of a variety of species and hybrids. In 1993, the focus of my orchid interest turned towards the genus *Coelogyne* and I have slowly built up a collection of species, a few hybrids and a small sample of the allied genera. I soon realised that it was extremely difficult to obtain information on the plants in my collection and many of these plants were wrongly named. I therefore began to gather information by starting with an examination of the *Encyclopaedia of Cultivated Orchids* by Alex D. Hawkes and the *Manual of Cultivated Orchids* by Helmut Bechtel, Phillip Cribb and Edmund Launert. The information on taxa, so gathered, was compared with the published taxa I found in the *Index Kewensis*. The gap in the number of species and the available information was enormous and I was thankful for my membership of the Royal Horticultural Society as it enabled me to make a series of visits to Vincent Square in London where I explored the resources of the Lindley Library.

Subsequently, I was given access to the Library facilities of the Royal Botanic Gardens at Kew and my studies started to move with more precision and at a more rapid pace. Dr. Phillip Cribb, Jeffrey Wood and Sarah Thomas of the Royal Botanic Gardens, Kew gave me a considerable amount of guidance and Dr. E.F. de Vogel of the National Herbarium, the Netherlands (Rijksherbarium), Leiden was most encouraging when I visited him and he let me have access to his material. He advised me on the approach I should take with my studies.

When other amateur orchid growers with an interest in *Coelogyne* heard about my studies, I was asked many questions about the plants in their collections and I have received various requests for information about the genus as a whole. With the encouragement I received from Phillip Cribb and Jeffrey Wood, I decided to record the results of my investigations. Phillip Cribb very kindly examined my interim manuscript and made many recommendations on the correct methodology to be used in the presentation of the botanical information I had gathered.

The Genus *Coelogyne* is based on an array of the information I have extracted from a wide range of references and my translation, from the Latin, of the type descriptions made by the original authors. It also contains a digest of the latest taxonomic descriptions made by modern authors such as Ed de Vogel and Barbara Gravendeel of the National Herbarium, Leiden, the Netherlands. By necessity, I have amplified some of the original descriptions using the results of my own observations, including a study of the growth habit of my plants during all stages of their development and flowering. I have examined the herbarium specimens at the Royal Botanic Gardens, Kew, Natural History Museum in London, Laboratoire de Phanérogamie, Muséum National d'Histoire in Paris, some specimens from the National Herbarium in Leiden, the Netherlands, and materials loaned from the Orchid Herbarium of Oakes Ames at Harvard University in Massachusetts. A chapter has been included on the molecular phylogeny of *Coelogyne* and allied genera, written by Barbara Gravendeel and Ed de Vogel. Molecular phylogeny provides the potential for a taxonomic revision of the genus and some associated genera but it is clear that there are many stages through which the work associated with the new techniques must pass before the tried and tested methods associated with morphological assessment can be reassessed.

My descriptions are abbreviated compared with those of modern taxonomists but they should meet the needs of growers who want a simple way to identify the species in their collections and hopefully they will help them identify and name their plant(s). I hope the results of this work will also be of use to those in the professional field.

Dudley Clayton
New Alresford
Hampshire

INTRODUCTION

In 1821, John Lindley, in his *Collectanea Botanica*, named and described the genus *Coelogyne* based upon *Coelogyne cristata* Lindl. and *Coelogyne punctulata* Lindl. I have selected *C. cristata* Lindl. (Fig. 1) as the type for the genus.

Coelogyne is a genus of 190 species, two additional subspecies and 12 additional varieties giving a total of 204 taxa. Since records began, over 400 taxa have been included in the genus at various times. Many of these have subsequently proved to be synonyms or taxa that were placed in the wrong genus. Synonyms and other taxa formerly in *Coelogyne* are detailed in the Index of Synonyms and Excluded Species.

CLASSIFICATION

In *The Orchids, Natural History and Classification*, R.L. Dressler (1981) included the genus *Coelogyne* Lindl., in the subtribe *Coelogyninae* Bentham within the tribe *Coelogyneae* Pfitzer which, in turn, is in the subfamily *Epidendroideae* Lindl. He provided descriptions for these taxa as follows:

Subfamily *Epidendroideae* Lindl.—characterised by the anther eventually bending downwards over column apex to become operculate, or operculate at the column apex but not bending downwards. Pollinia 2, 4 or 8, usually hard, waxy, sometimes mealy or sectile. Leaves distichous usually articulated at the base.

Tribe *Coelogyneae* Pfitzer—characterised by 2 or 4 pollinia, superposed or ovoid. The plants usually with pseudobulbs of a single internode. Inflorescence terminal.

Subtribe *Coelogyninae* Bentham—pseudobulbs or corms of a single internode. Leaves convolute or duplicate, plicate or conduplicate, articulate. Inflorescence terminal often produced before the growth of the pseudobulbs, simple, of few to many flowered, flowers either in a spiral or distichous. Flowers small to large, resupinate; base of the lip may be saccate; column short or elongate, apex often petaloid and hooded over the anther; anther terminal or ventral, incumbent; pollinia 2 or 4, superposed or ovoid, with prominent caudicles; stigma entire, often emergent.

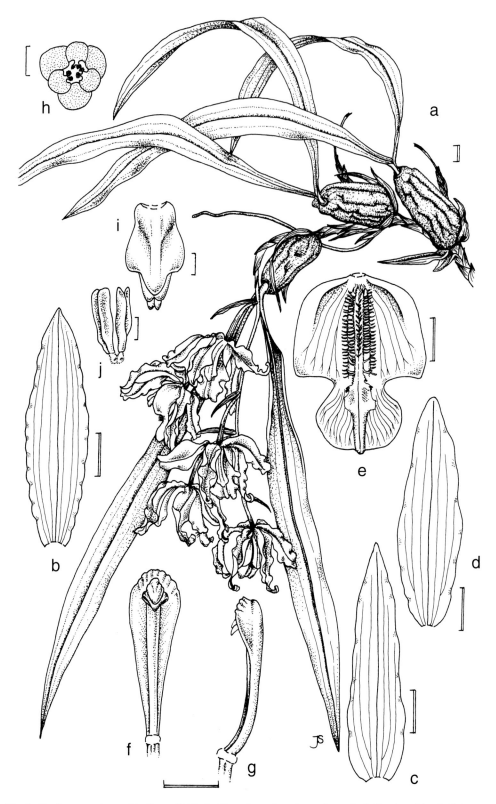

Fig. 1. *Coelogyne cristata* Lindl. **a.** Habit; **b.** Dorsal sepal; **c.** Lateral sepal; **d.** Petal; **e.** Lip; **f.** Column, front view; **g.** Column, side view; **h.** Ovary, transverse section; **i.** Anther cap; **j.** Pollinia. Scale: double bar = 1 cm; single bar = 1 mm. [Cult. D.A. Clayton]. All drawn from living plant by Judi Stone.

INTRODUCTION

GENERIC AFFINITIES

In addition to *Coelogyne* Lindl., the fifteen genera that constitute the subtribe *Coelogyninae* Bentham are:

Bracisepalum J.J. Sm.—two species, affinity with *Dendrochilum* Blume.
Bulleya Schltr.—monotypic, *B. yunnanensis* Schltr. is little known.
Chelonistele Pfitzer—11 species, two varieties, close affinity with *Coelogyne* Lindl.
Dendrochilum Blume—four subgenera, 13 sections, 269 species, 21 varieties; incorporating *Acoridium* Nees & Meyen and *Pseudoacoridium* Ames.
Dickasonia L.O. Williams—monotypic, *D. vernicosa* L.O. Williams.
Entomophobia de Vogel—monotypic, *E. kinabaluensis* (Ames) de Vogel is allied to *Pholidota* (Hook.) Lindl.
Geesinkorchis de Vogel—two species, affinity with *Pholidota* (Hook.) Lindl.
Gynoglottis J. J. Sm.—possibly two species, affinity with *Coelogyne* Lindl.
Ischnogyne Schltr.—monotypic, *I. mandarinorum* (Kraenzl.) Schltr.
Nabaluia Ames—three species, affinity with *Pholidota* (Hook.) Lindl.
Neogyna Rchb.f.—monotypic, *N. gardneriana* (Lindl.) Rchb.f., affinity with *Coelogyne* Lindl.
Otochilus Lindl.—five species, affinity with *Coelogyne* Lindl.
Panisea Lindl.—eight species; incorporating *Sigmatogyne* Pfitzer, affinity with *Coelogyne* Lindl.
Pholidota (Hook.) Lindl.—nine sections, 29 species, seven varieties.
Pleione D. Don—18 species.

Pederson, Wood & Comber (1997) in 'A revised subdivision and bibliographical survey of *Dendrochilum* (Orchidaceae)', published an artificial key to the genera of subtribe *Coelogyninae*.

DERIVATION OF THE NAME *COELOGYNE*

The name *Coelogyne* comes from the Greek **koilos** (hollow, cavity) and **gyne** (female), referring to the deeply set stigmatic cavity at the front of the column. I suppose it should be pronounced 'koi–lo–GUY–nee' but, on the assumption that the traditional English pronunciation is in common use and is to be preferred, then the choice is 'see–lo–GUY–nee', 'see–LODGE–eh–nee' or 'see–law–JI–nee'. I recommend that the last example be ignored and that it is a matter of personal taste whether you chose the first or second. I have settled on pronouncing it 'see–lo–GUY–nee'.

GEOGRAPHIC DISTRIBUTION

Coelogyne species are widely distributed in South and Southeast Asia and across to the Southwest Pacific islands. The distribution is best illustrated by referring to geo-climatic regions that transcend the political boundaries of the nations in the area. In consequence, the Indian subcontinent has been treated as a whole and then divided into north and south. The concept of an Upper and Lower part to Burma (Myanmar) remains appropriate and the name Burma has been retained. The term Indochina has been retained to comprise Cambodia, Laos, and Vietnam. Apart from Tibet (now Xizang), which is separately identified, Hongkong is included in China. The Indonesian islands are divided into the separate regions of Sumatra, Java, Sulawesi and Maluku. The island of Borneo has been treated as a whole and thus includes Brunei and parts of Malaysia (Sabah and Sarawak) and of Indonesia (Kalimantan). South India and Sri Lanka form a natural alliance and similarly Tibet, Nepal, Bhutan, North India, Upper Burma and South China form a Himalayan region. The number of *Coelogyne* species per region is given in Table 1 followed, in brackets by, the number of endemic taxa; the number of taxa in two, usually adjacent, regions and the number of taxa more widely spread throughout the regions. The same information is shown diagrammatically in Fig. 2 and the distribution of the genus is shown in Map 1.

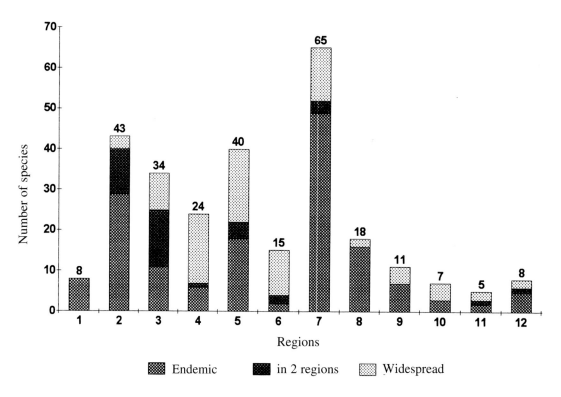

Fig. 2. Distribution of *Coelogyne* species in South and Southeast Asia.

Table 1. Geo-climatic Regions of South and Southeast Asia.

Region 1. South India and Sri Lanka—8 [8, 0, 0]
       2. Tibet, Nepal, Bhutan, North India, Upper Burma and South China—43 [29, 11, 3]
       3. Lower Burma, Thailand and Indochina (Cambodia, Laos, Vietnam)—34 [11, 14, 9]
       4. Thailand Peninsula, Peninsular Malaysia—24 [6, 1, 17]
       5. Sumatra—40 [18, 4, 18]

INTRODUCTION

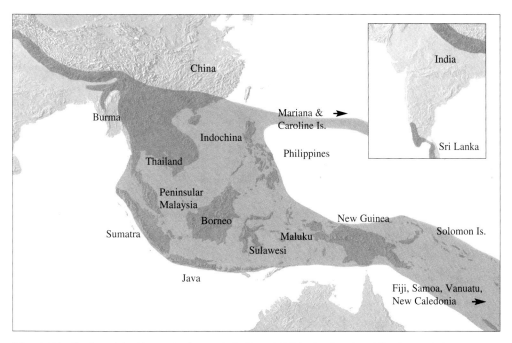

Map 1. Distribution of the Genus *Coelogyne* in India and Sri Lanka (inset) and South-east Asia.

6. Java including Lesser Sunda Islands—15 [2, 2, 11]
7. Borneo (Brunei, Sabah, Sarawak, Kalimantan)—65 [49, 3, 13]
8. The Philippines—18 [16, 0, 2]
9. Sulawesi—11 [7, 0, 4]
10. Maluku—7 [3, 0, 4]
11. New Guinea (Irian Jaya and Papau New Guinea)—5 [2, 1, 2]
12. Mariana, Caroline and Solomon Islands, Vanuatu, New Caledonia, Fiji, Tonga—8 [5, 1, 2]

ECOLOGY AND CLIMATE

Tropical Evergreen Rain Forest (sea level–1500 m) is found in Burma, China, Vietnam, Thailand (eastern coastal region), Peninsular Malaysia, Borneo, Sumatra, Java, the Malay Archipelago, the Philippines and New Guinea. Temperatures are normally between 24°C and 30°C; the temperature rarely falls below 21°C or rises above 32°C. Rainfall is usually more or less constant throughout the year and is usually greater than 2000 mm per year. A distinctive feature is an average daily temperature range which is much greater than the range between hottest and coldest months. The structure of the tropical evergreen rain forest is as follows:

Emergent and Canopy Layer with tall solitary trees (emergents), more than 40 m tall which rise above a closed, dense canopy formed by the crowns of the shorter 20–40 m trees. The canopy cuts off sunlight from the forest below. There is a diversity of fauna in the canopy layer, which may contain some *Coelogyne* species.

Middle Layer below the canopy where the understory crowns do not meet but where a dense layer of growth is still formed from trees 5–20 m tall. There is no clear distinction between the canopy and middle layers. Epiphytes are plentiful in the middle layer.

Shrub Layer which consists of woody shrubs and small trees of varying height between 1–5 m which grow in any space between the boles of larger trees.

Ground Layer comprising decomposing litter (fungi) and shade tolerant herbs and ferns which rarely reach 50 cm in height. Light levels are less than 1% of full daylight level.

The diversity of orchids generally increases with altitude in dipterocarp forest. Epiphytic orchids are quite common in the emergent layer of the rain forest, often in the forks and along the main branches of trees. In the mid-canopy, with more shade, are found the broader leafed epiphytes. Terrestrial orchids are not common on the shaded forest floor, but increase in number near to streams or on ridges. Secondary forest and open grassy areas provide a habitat for orchids that seem to favour the better light conditions.

The term **'Hill Forest'** is often used to define the upper part of the lowland dipterocarp-dominated rain forest between about 600–1200 m.

Soil types and river delta regions modify the ecology in the lowland evergreen rain forests, low-lying coastal areas and along the rivers. Notable variations are:

– Forests over limestone and the changes from ultrabasic soils, oxisols and ultisols to podzolised sands to produce **'Heath Forest (Kerangas)'**.

– Dense narrow bands of trees growing along the river edge in delta areas to produce an emerging canopy and an understory, which is similar to the tropical evergreen rain forests. Also along the riverbanks, in essentially drier areas, the same type of vegetation grows. Both the deltas and river edges are suitable habitats for all types of epiphytic orchids, in particular coelogynes such as *C. asperata*, *C. pandurata* and *C. septemcostata*.

– Where the water tables are high and there is an influence from seawater, beach vegetation, mangrove forests and brackish water forests occur but where it is freshwater, then peat swamp forests and freshwater swamps occur.

Tropical Monsoon Forest (900–1500 m) is widely distributed in Asia (South, Southeast and East), south of Latitude 25°N, in particular, in India (the southern parts of the Western Ghats, eastern Himalayas and Assam), South China (Yunnan), Burma (Tenasserim), Indo-China, Thailand, Sulawesi, East Java, Lesser Sunda Islands and in the south of New Guinea. The climate is seasonally dry with varying degrees of water shortage; rainfall varies from 200 mm per year to 20000 mm per year with the rains occurring in a period varying between 3 and 9 months. In much of India 85% of rainfall occurs during a limited monsoon period. There are pronounced changes in temperature as a result of altitude. The tropical evergreen rain forest alters to tropical monsoon forest, as in East Java, where there is a sharp fall in the total number of plant and animal species which are adapted to endure the seasonal droughts. There is a distinctive period of drought, which may last as much as nine months. The monsoon forests, when compared with the tropical rain forests, have a more open canopy and very dense undergrowth. Many of the trees shed their leaves in the dry season. With the lower levels of monsoon rain, the forests thin into virtual semi-desert vegetation as in Northwest India. The Western Himalayas is variable and dependant on the aspect and altitude of the mountain ranges, distance from the plains and sea, which give rise to the Northeast and Southwest Monsoon-bearing clouds. Rains are at a maximum in the east and least in Kashmir. Altitude is the most important factor controlling temperature. The Himalayas are not a single continuous chain of mountains but a series of more or less parallel converging ranges. The tract is rugged and intersected by valleys and deep gorges cut by the river systems of the Indus, Satluj and Ganges. The outer ranges rise abruptly from the tropical Indo-Gangetic plain. The individual ranges generally present a steep slope towards the plains, and more gentle inclined slopes towards the north. The inner ranges are higher, enclosing the higher and colder valleys, and surmounted still higher by perpetual snow. The main Himalayan range in this region has a number of peaks greater than 7000 m. South-facing slopes near the plains receive most of the rain and mist. This is where the cool-growing coelogynes such as *C. cristata*, *C. flaccida* and *C. nitida* are to be found. The north-facing lower slopes behind high ridges receive very little moisture and orchids are not present. The Western Ghats of India and foothills of the Himalayas in Assam both rise to above 2500 m. Temperatures decrease rapidly at these altitudes with corresponding changes in the vegetation similar to that which occurs in lower and upper montane forests. In southern India, the Nilgiri Hills contain forests of a temperate characteristic and there is an intermingling of temperate and tropical species. The Nilgiris are formed by the conjunction of the Western Ghats, Eastern Ghats and Southern Ghats. The rainfall varies between 900 and 7500 mm per year from both Northeast and Southwest Monsoons. The rains nourish some 2000–2500 sq km of forest and mainly on the western side of this area, there is a wide variety of orchids, including *Coelogyne* species, such as *C. breviscapa*, *C. nervosa* and *C. odoratissima*, between the altitudes 500–2000 m; *C. mossiae* is found at a higher altitudes.

Tropical Lower Montane Rain Forest (900–1800 m) is found in South-east Asia and specifically in Borneo (Kalimantan, Sabah and Sarawak), Java and Sumatra. The climate is similar to that found in the tropical evergreen rain forests but with average temperatures falling with increasing altitude. There are four layers, as with tropical evergreen rain forests, but without the emergent trees. The *Coelogyne* species which commonly grow in this type of forest are in Section *Tomentosae*. In limestone areas, there is a rich abundance of orchid species as lowland forests often surround the outcrops of limestone. Many outcrops are karst formations with steep sheer cliffs and pinnacles. On the higher limestone mountains the high rainfall produces a thick acidic peat to 0.6 m thick which separates the plant roots from the limestone parent rock.

Tropical Upper Montane Forest (1800–2900 m) is found in parts of China, Borneo (Kalimantan and Sabah) and Sumatra. Again the climate is similar to that found in the tropical evergreen rain forests but with average temperatures falling with increasing altitude. The tropical upper montane forests have a dense herb layer. In general vascular epiphytes, mainly orchids, are abundant, particularly in the windier exposed canopy. Many epiphytic orchids are found just above ground level or growing in small clumps of moss at ground level where they are more protected. For orchids, lower light levels and temperatures are limiting factors and hence they are less abundant than in the lower montane forests. On ultramafic soils (Sabah, Mount Kinabalu), the orchids are more distinct and many endemic species are found, including *Coelogyne papillosa*.

Tropical Sub-Alpine Vegetation is found in Borneo (Mount Kinabalu, Sabah) where at the lower altitudes the shallow soils have a thin humus layer and consequently, there is a much thinner herb layer. The forest is drier and, although mosses and liverworts are found on the tree-trunks and ground, it does not compare with the moss forests lower down. The orchid diversity is very much lower. At the higher altitudes, the conditions are similar and the species can be either epiphytic or terrestrial. Above 3700 m, again in Borneo and in the Himalayas, the summit areas of the mountains contain a mosaic of pockets of alpine scrub where a depth of soil has formed but the species diversity is low. Very few orchid species are found at this lower limit of the alpine zone and possibly the only *Coelogyne* is *C. papillosa*.

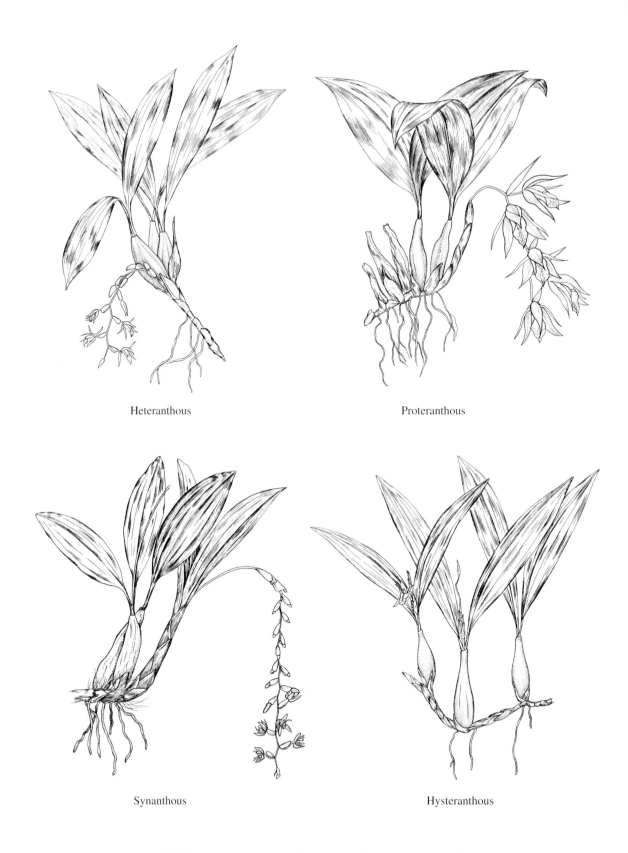

Fig. 3. Four types of inflorescence growth—terminal or lateral to mature pseudobulbs.

MORPHOLOGY

Most *Coelogyne* species are epiphytic but many are sometimes lithophytic or terrestrial herbs. They are short to tall, with crowded pseudobulbs which often form large clumps, to long creeping plants with the pseudobulbs widely spaced on the rhizome.

Roots are slender to a few millimetres thick, of variable length, loosely branched or without branches, either in clusters at the base of closely packed pseudobulbs or isolated at the node adjacent to the pseudobulb or at the nodes along the developing rhizome or near the end.

Rhizomes are slender to stout, terete or more or less flattened, branching or without branches, short or long, with pseudobulbs either clustered or far apart; one to many noded with each internode generally covered with one or more sterile imbricate scales which are long persistent or soon disintegrate; the scales at the node at which the pseudobulb develops extend to enclose the base of the pseudobulb.

Pseudobulbs can be ovoid, conical, pyriform or fusiform to cylindric, the base partly enclosed with scales from the rhizome. Normally they consist of a single internode. In the case of a pseudobulb with one leaf (unifoliate) at the apex, there is one internode but with 2 leaves (bifoliate) at the apex of the pseudobulb, then there is an additional, abbreviated internode.

Leaves are stiffly herbaceous, generally long-lived with subsessile to long channelled petioles, almost orbicular in section. The blade can be small to large, narrow to broad, elliptic to obovate, ovate, oblong to linear-lanceolate or lanceolate, coriaceous or plicate with a few to many main nerves; tip acute, acuminate, mucronate. The base of the blade is generally decurrent onto the petiole.

The **Inflorescence** is terminal or lateral to mature pseudobulbs, Fig. 3, and either heteranthous, proteranthous or synanthous with partially to entirely developed leaves, or hysteranthous, erect or pendulous, 1- to very many-flowered, hairy or glabrous. The peduncle is terete, ovoid or elliptic or flattened elliptic, sometimes broadening, sometimes narrowing to the apex. In some species, during flowering, the peduncle is at the base only and enclosed by the young leaf blade(s) and/or scales of the young shoot. In the latter stages, the peduncle extends and continues to elongate after anthesis. The rachis is erect to arcuate, terete or flattened, zig-zag, with straight or slightly curved internodes or distinctly swollen internodes, either long or short. The inflorescence can be:
– **Heteranthous** when the vegetative shoot from which the inflorescence emerges never develops a leaf or leaves and where the terminal internode of this growth never enlarges to form a pseudobulb. In the old state the inflorescence and residual internode lie lateral to the rhizome, covered in scales or eroded scales.

– **Proteranthous** when it develops on the top of a vegetative shoot of which the leaf or leaves and the terminal internode are not yet developed. In this stage the base of the shoot is covered with scales. After anthesis and during fruit setting the terminal internode develops into a pseudobulb and a leaf or leaves grow from the top. The rhizome is covered with scales, or the scales are eroded and some of these scales enclose the base of the pseudobulb.

– **Synanthous** when it develops on the top of a vegetative shoot of which the terminal internode is not developed, but the leaf or leaves are developing simultaneously. At this stage, scales from which the young leaf or leaves are protruding to a varying extent, cover the base of the shoot. After anthesis and during fruit setting the terminal internode develops into a pseudobulb and, from the top, a leaf or leaves continue to develop. The rhizome is covered with scales, or the scales are eroded and some of these scales enclose the base of the pseudobulb.

– **Hysteranthous** when it develops on the top of a fully developed pseudobulb with a fully-grown leaf or leaves. The rhizome is covered with scales, or the scales are eroded and some of these scales enclose the base of the pseudobulb. For species in sect. *Elatae* Pfitzer and sect. *Proliferae* Lindl. the inflorescence is hysteranthous but the species all have imbricate bracts at the junction between the peduncle and rachis. In the case of the species in sect. *Proliferae*, the rachis extends after anthesis to form a further set of imbricate bracts and the inflorescence flowers again during the next season, Fig. 4.

Fig. 4. A form of hysteranthous inflorescence with imbricate bracts at the junction of the peduncle and rachis (see sect. *Elatae* Pfitzer) and imbricate bracts on an extended rachis, as shown (see sect. *Proliferae* Lindl.).

Flowers open simultaneously or in succession; floral bracts are deciduous or persistent and attached around the base of the pedicel, lanceolate to ovate to oblong, tip acute, acuminate or mucronate. The flowers are small to large, fleshy or membranous, mainly distichous, mostly widely open but in a few cases not so. The pedicel is terete or angular, sometimes slightly twisted, generally glabrous but in some cases, sparsely hairy; ovary usually tapering into pedicel, ribbed, twisted, glabrous or almost glabrous or densely to sparsely hairy.

The **dorsal sepal** is symmetrical, concave, broadly sessile, ovate, ovate-oblong, oblong to lanceolate, elliptic or obovate, narrowly ovate to narrowly elliptic; tip acute to acuminate, obtuse, apiculate; nerves few to many, often with interconnecting veins, midrib more or less prominent, sometimes a low rounded keel, glabrous, sometimes with scattered hairs or stubble.

The lateral sepals are more or less asymmetric, oblique, concave, broadly sessile, ovate, ovate-oblong, oblong to lanceolate, elliptic or obovate, narrowly ovate to narrowly elliptic; tip acute to acuminate, obtuse, apiculate or mucronate; nerves few to many, often with interconnecting veins, midrib more or less prominent, sometimes a low rounded to plate-like keel; glabrous, sometimes with scattered hairs or stubble.

The **petals** are generally symmetrical or more or less falcate, slightly to extremely recurved, sometimes reflexed, narrowly elliptic to linear, sometimes filiform; tip acute to acuminate, obtuse, apiculate or mucronate; generally with a prominent single, centric nerve, nerves one to a few.

The **lip** is sessile to the base of column, often concave, lacking a spur, deeply or obscurely 3-lobed. Generally, the side-lobes are erect on either side of the column and the mid-lobe spreading or reflexed. The lip comprises two portions:
– **Hypochile**, the basal portion of the lip, rather narrowly to broadly attached, not or slightly saccate or not or slightly swollen at the base, when flattened (in outline) orbicular to elliptic, obovate ovate or cordate. Side-lobes erect, more or less broadly rounded in front and either or not recurved; sinus broadly rounded or obtuse, in some cases, slightly converging or diverging, with an entire or erose margin.
– **Epichile** (mid-lobe), the terminal portion of the lip, generally broadly attached, concave or convex, rhombic, obovate, ovate, orbicular, elliptic, transversly elliptic, spathulate or ligulate, sometimes with a distinct, long claw or with or without a short claw; tip acuminate to emarginate, acute or broadly acute, retuse, deeply retuse or obtuse; margin may be fleshy, membranous, porrect, recurved or curved upwards to varying degrees, entire, undulate, crenate, erose, fimbriate or ciliate.
– **Keels** vary from the simple to the complex. In the simple form, more or less elevated plate-like, entire to incised, dentate or swollen sections; more complex forms are widened along the crest, sometimes as entire or interrupted undulating rows, with or without a broad longitudinal groove between, or as a thick callus by fusion of the keels or sometimes fused into irregularly rounded warts and ridges or projections. Keels sometimes with hairs or with minute papillae or dissolved into a mass of papillae or warts.

The **column** can be short to long, generally uniform in each section, sometimes arched to the front, widens to form hood with winged margins, widest below the apex; when flattened, normally spathulate; top margin recurved or not recurved, entire or more or less irregular or dentate, sometimes laterally notched.

The **anther** is ventral at the apex of the column, versatile, bell-shaped in outline, apex with or without an incision at the tip. There are four **pollinia**, attached in pairs, flattened, obliquely elliptic to obliquely orbicular, at the tip attached to a large and irregular shaped caudicle consisting of a mass of sterile pollen. The **stigma** is cup-shaped, semi-elliptic to semi-orbicular, with an elevated, more or less recurved margin; rostellum large, semi-orbicular to more or less triangular, lateral margins incurved, with a truncate, obtuse, broadly rounded or acute apex.

TAXONOMY

SECTIONAL TREATMENTS

John Lindley (1821) established the genus *Coelogyne* in *Collectanea Botanica* (sub t. 33). In his *Genera and Species of Orchidaceous Plants*, 1830–40, he recognised three groups: *Neogyne, Coelogyne* and *Pleione*, the second being further subdivided into five sections; *Flaccidae, Erectae, Proliferae, Filiferae* and *Flexuosae*.

Pfitzer & Kraenzlin (1907), in A. Engler, *Das Pflanzenreich*, revised the genus and recognised 14 sections in two series, *Succedanae* and *Simultanae*. In the former the flowers open in succession, in the latter simultaneously. The breakdown of their sectional treatment of *Coelogyne* Lindl. was as follows:

Series 1. *Succedanae*, Subseries *Nudae*: Sections *Longifoliae, Speciosae*.
Subseries series *Vaginatae*: Sections *Fuliginosae, Ancipites*.

Series 2. *Simultanae*, Subseries *Nudiscapae*: Sections *Fuscescentes, Carinatae, Lentiginosae*.
Subseries *Glumacea*: Sections *Ocellatae, Cristatae, Tomentosae, Verrucosae, Venustae*
Subseries *Imbricatae*: Sections *Elatae, Proliferae*.

They recognised a number of genera within the subtribe as distinct from *Coelogyne* Lindl., including *Ptychogyne* Pfitzer, *Dendrochilum* Blume, *Pleione* D. Don, *Neogyne* Rchb.f., *Gynoglottis* J.J. Sm., *Hologyne* Pfitzer, *Sigmatogyne* Pfitzer, *Crinonia* Blume, *Chelonistele* Pfitzer, *Panisea* Lindl., *Chelonanthera* Blume, *Pholidota* (Hook.) Lindl., *Camelostalix* Pfitzer and *Otochilus* Lindl.

Butzin (1974, 1992) presumably recognised *Coelogyne cristata* Lindl. as the type species for the genus, hence his treatment of Sect. 10 as *Coelogyne*. He recognised 14 sections, as follows:

Sect. 1. *Longifoliae* Pfitzer
Sect. 2. *Speciosae* Lindl.
Sect. 3. *Fuliginosae* Lindl.
Sect. 4. *Ancipites* Pfitzer
Sect. 5. *Bicellae* J.J. Sm.
Sect. 6. *Fuscescentes* Pfitzer
Sect. 7. *Flaccidae* Lindl. (syn. *Carinatae* Pfitzer)
Sect. 8. *Lentiginosae* Pfitzer
Sect. 9. *Ocellatae* Pfitzer

Sect. 10. *Coelogyne* (syn. *Cristatae* Pfitzer)
Sect. 11. *Tomentosae* Pfitzer (including *Venustae* Pfitzer)
Sect. 12. *Verrucosae* Pfitzer
Sect. 13. *Elatae* Pfitzer
Sect. 14. *Proliferae* Lindl.

He recognised four further genera, *Chelonistele* Pfitzer, *Ptychogyne* Pfitzer, *Hologyne* Pfitzer, and *Cyathogyne* Schltr., as distinct from *Coelogyne*.

Seidenfaden (1975) dealt with the species found in Thailand and his key included 12 of Pfitzer & Kraenzlin's 14 sections, those not represented in Thailand being *Ancipites* and *Venustae*. Seidenfaden did not follow Butzin or Pfitzer & Kraenzlin and adopted the sectional name *Coelogyne* in preference to *Ocellatae*, as he had chosen *C. punctulata* Lindl. as the type for the genus. He followed Pfitzer & Kraenzlin and preferred *Carinatae* Pfitzer to *Flaccidae* Lindl. His account of the orchids of Indochina included nine of Pfitzer & Kraenzlin's 14 sections (Seidenfaden 1992). These were *Coelogyne* (syn. *Cristatae*), *Elatae*, *Flaccidae* (syn. *Carinatae*), *Fuscescentes*, *Lentiginosae*, *Ocellatae* (*C. punctulata* no longer considered the type for the genus), and *Verrucosae*. He judged *Coelogyne lawrenceana* Rolfe and *C. eberhardtii* Gagnep. to be in sect. *Speciosae* and *C. dichroantha* Gagnep. to be in sect. *Fuliginosae*.

Pradhan (1979) followed Pfitzer & Kraenzlin and recognised nine of their 14 sections as native to India, and established a tenth and new section, *Micranthae*. His key identifies ten sections and 23 species.

Das and Jain (1980) identified 34 species from nine of Pfitzer & Kraenzlin's sections in India. In their treatment, *Coelogyne punctulata* Lindl. was selected as the type species and, in consequence, their sect. *Coelogyne* = *Ocellatae* Pfitzer and they retained sect. *Cristatae* Lindl.

De Vogel (1993) recommended that *Chelonistele* Pfitzer be kept separate from *Coelogyne*, that sect. *Bicellae* J.J. Sm. should be raised to generic rank, but reduced *Ptychogyne* Pfitzer, *Hologyne* Pfitzer and *Cyathogyne* Schltr. to sectional rank. This resulted in his recognition of 22 sections. In addition to the 14 sections identified by Butzin, he recognised sections *Ptychogyne* Pfitzer, *Hologyne* Pfitzer, *Cyathogyne* Schltr., *Moniliformes* Carr, *Rigidiformes* Carr and *Micranthae* Pradhan. He suggested the need for two new sections, one containing *C. brachyptera* Rchb.f. and similar species and the other for *C. veitchii* Rolfe. He later produced a key for the 24 species in sect. *Tomentosae* Pfitzer (de Vogel 1992).

Gravendeel & de Vogel (1999) published a revision of Section *Speciosae* with a key to 16 taxa in the section and with *Coelogyne speciosa* Lindl. divided into 3 subspecies.

In this synopsis I have reviewed the various infrageneric treatments of *Coelogyne* produced since Lindley's work. I have accepted the views of de Vogel (1993) that sect. *Bicellae* J.J. Sm. should be raised to generic level and his recognition of sections *Ptychogyne* Pfitzer, *Hologyne* Pfitzer and *Cyathogyne* Schltr. However, as no taxonomic revision has been published on the subject, I have retained sect. *Bicellae* and not raised it to generic level but the three subgenera are now formally given the new rank of Section (Sect.). Three new sections have been recognised to accommodate, firstly *C. veitchii* Schltr.; secondly, the three species *C. brachyptera* Rchb.f., *C. parishii* Hook. and *C. virescens* Rolfe which are often placed in sect. *Verrucosae* Pfitzer but deserve to be classified separately; and thirdly, *C. lawrenceana* Rolfe and *C. eberhardtii* Gagnep. which have some similarities to sect. *Speciosae* Lindl. and sect. *Coelogyne* but are judged to be sufficiently different to merit a section of their own. *Coelogyne dichroantha* Gagnep. has caused a fair amount of confusion, because the type is only known from a painting by Eberhardt. Gravendeel and de Vogel in their revision of

Section *Speciosae* described it as a 'dubious species' but retained it in sect. *Speciosae*. During my examination of the Eberhardt drawings and painting of *C. dichroantha*, I came to the conclusion that the characters represented showed a similarity to species assigned to Section *Fuscescentes* and I have therefore included the species in that section. There seems little justification in retaining sect. *Ancipites* Pfitzer (three species)—'flowers opening in succession, one or a few together' as a separate entity from sect. *Elatae* Lindl. (13 species)—'flowers opening simultaneously'. The reasoning behind this judgement is that the manner of flower opening is variable in many of the species with only a tendency towards 'simultaneous'. In other respects the species have common characteristics. Therefore, sect. *Ancipites* is incorporated in sect. *Elatae*. Precedence has been given to the name sect. *Flaccidae* Lindl. over sect. *Carinatae* Pfitzer; this accords with the judgement of both Butzin and de Vogel. The assessment results in the recognition of 22 Sections for which an artificial key is provided.

No account has been taken of the molecular phylogeny of *Coelogyne* produced by Gravendeel and de Vogel and outlined at the end of this synopsis. The consequences of this work will undoubtedly be profound and will soon be published by the authors. The essence of their work currently in progress at the National Herbarium of the Netherlands is to be found in the chapter '*Molecular Phylogeny of Coelogyne and Allied Genera: An Urge to Reorganise*'.

Artificial key to the sections of Genus Coelogyne

1. Scape with imbricating sterile bracts 2
   Scape bare to first flower, or rarely with one or a few sterile bracts 6

2. Scape with imbricating sterile bracts at the junction of peduncle and rachis 3
   Scape with imbricating sterile bracts mainly at base of the peduncle 4

3. Rachis producing single set of flowers .... **Sect. 1. *Elatae* Pfitzer**
   Rachis extending with new imbricate bracts to produce further annual sets of flowers ....
   .... **Sect. 2. *Proliferae* Lindl.**

4. Lip with 2 or 3 keels, mid-lobe margins fimbriate .... **Sect. 3. *Fuliginosae* Lindl.**
   Lip with 3 or 4 keels, mid-lobe margins entire or undulate 5

5. Lip with 3 keels, mid-lobe margins entire .... **Sect. 4. *Micranthae* Pradhan**
   Lip with 3 or 4 papillose keels, mid-lobe margins undulate ....
   .... **Sect. 5. *Brachypterae* D.A. Clayton**

6. Flowers opening in succession 7
   Flowers opening simultaneously 10

7. Flowers with sepals mainly 4–6 cm long; lip with double cristate and verrucose keels, with or without hairs, or ridges which form a papillae patch .... **Sect. 6. *Speciosae* Lindl.**
   Flowers smaller; lip with entire or irregular cristate keels 8

8. Lip adnate to the base of column .... **Sect. 7. *Bicellae* J.J. Sm.**
   Lip not adnate to the base of column 9

9. Rachis with swollen and short internodes; leaves papyraceous, thin-textured .... **Sect. 8. *Moniliformes* Carr**
   Rachis with slender and long internodes; leaves coriaceous .... **Sect. 9. *Longifoliae* Pfitzer**

10. Ovary with hairs or papillae on surface **11**
    Ovary glabrous **13**

11. Flowers densely packed, spirally arranged; nodes on rachis not pronounced and the internodes not zig-zag .... **Sect. 10. *Cyathogyne* (Schltr.) D.A. Clayton**
    Flowers well spaced in 2 opposite rows; nodes on rachis pronounced (zig-zag) **12**

12. Mid-lobe equal to or greater than half the length of lip, with papillae on mid-lobe .... **Sect. 11. *Verrucosae* Pfitzer**
    Mid-lobe equal to or less than one-third length of lip; if longer than one-third length of lip, with no papillae on mid-lobe .... **Sect. 12. *Tomentosae* Pfitzer**

13. Flowers fleshy; lip more or less entire without distinct side-lobes, basal part embracing column **14**
    Flowers membraneous; lip with distinct side-lobes, basal part not embracing column ... **16**

14. Inflorescence heteranthous **15**
    Inflorescence proteranthous to synanthous ..... **Sect. 13. *Hologyne* (Pfitzer) D.A. Clayton**

15. Pseudobulbs with 1 leaf; inflorescence erect .... **Sect. 14. *Rigidiformes* Carr**
    Pseudobulbs with 2 leaves; inflorescence pendulous ..... **Sect. 15. *Veitchiae* D.A. Clayton**

16. Lip with transverse fold at base .... **Sect. 16. *Ptychogyne* (Pfitzer) D.A. Clayton**
    Lip lacking transverse fold at base **17**

17. Sepals, petals and lip margin generally undulate; lip keels fimbriate, deeply serrate to deeply crenulate or lacinate **18**
    Sepals, petals and lip margins not undulate; lip keels not fimbriate, deeply serrate, deeply crenulate or lacinate but entire or erose, with papillae or verrucose, dentate or crenulate to irregularly cristate **19**

18. Inflorescence hysteranthous, sterile bracts generally at interface between peduncle and rachis; flowers with sepals at least 7 cm long; lip with lacinate keels .... **Sect. 17. *Lawrenceanae* D.A. Clayton**
    Inflorescence generally synanthous or heteranthous, exceptionally hysteranthous, no sterile bracts between interface of peduncle and rachis; flower with sepals 2–4.5 cm long; lip with deeply serrate or deeply lacinate keels .... **Sect. 18. *Coelogyne***

19. Dorsal sepal erect, away from the column; lateral sepals and petals wide spread away from the column; sepals and petals of about equal length **20**
    Dorsal sepal forming a hood over the column; dorsal sepal larger than lateral sepals and petals; lateral sepals and petals not wide spread away from the column .... **Sect. 19. *Fuscescentes* Pfitzer**

20. Dorsal sepal and lateral sepals of about equal width, petals narrower 21
    Dorsal sepal, lateral sepals and petals of about equal width; sepals, petals and lip tend towards being fleshy .... **Sect. 20. *Ocellatae* Pfitzer**

21. Lip with mid-lobe large in relation to the overall size of the flower; mid-lobe may be clawed; lip with margins which tend towards being membraneous .... **Sect. 21. *Lentiginosae* Pfitzer**
    Lip with mid-lobe not large in relation to overall size of the flower; lip without evident claw; lip with margins which tend towards being fleshy .... **Sect. 22. *Flaccidae* Lindl.**

COELOGYNE SPECIES AND THEIR SYNONYMS LISTED BY SECTION
[Synonyms listed in order of published date]

1. Section **Elatae** Pfitzer
    C. anceps Hook.f.
    C. barbata Griff.
    C. calcicola Kerr
    C. filipeda Gagnep.
        C. prolifera auct. non Lindl.
    C. ghatakii T.K. Paul, S.K. Basu & M.C. Biswas
    C. griffithii Hook.f.
    C. holochila P.F. Hunt & Summerh.
        C. elata auct. non Lindl.
    C. leucantha W.W. Sm.
    C. lockii Aver.
    C. pendula Summerh. ex Parry
    C. pulchella Rolfe
    C. rigida C.S.P. Parish & Rchb.f.
        C. tricarinata Ridl.
    C. sanderae Kraenzl. ex O'Brien
        C. annamensis Ridl.
        C. ridleyi Gagnep.
        C. darlacensis Gagnep.
    C. stricta (D. Don) Schltr.
        Cymbidium strictum D. Don.
        Coelogyne elata Lindl.
    C. tenasserimensis Seidenf.
    C. zhenkangensis S.C. Chen & K.Y. Lang

2. Section **Proliferae** Lindl.
    C. ecarinata C. Schweinf.
    C. longipes Lindl.
    C. prolifera Lindl.
        C. flavida Lindl.
    C. raizadae S.K. Jain & S. Das
        C. longipes auct. non Lindl.

C. schultesii S.K. Jain & S. Das
   C. prolifera auct. non Lindl.
   C. longipes var. *verruculata* (Lindl.) S.C. Chen
   C. flavida auct. non Hook.f. ex Lindl.
C. ustulata C.S.P. Parish & Rchb.f.

3. Section **Fuliginosae** Lindl.
   C. arunachalensis H.J. Chowdhery & G.D. Pal
   C. chrysotropis Schltr.
   C. fimbriata Lindl.
      Pleione chinense Kraenzl.
      P. fimbriata (Lindl.) Kraenzl.
      Coelogyne ovalis auct. non Lindl.
      C. loatica Gagnep.
      C. xerophyta Hand.-Mazz.
      C. leungiana S.Y. Hu
      C. primulina Barretto
   C. fuliginosa Lodd. ex Hook.
      Pleione fuliginosa (Hook.f.) Kuntze
   C. longeciliata Teijsm. & Binn.
   C. ovalis Lindl.
      Broughtonia linearis Wall. ex Lindl.
      Coelogyne decora Wall. ex F. Voigt
      C. pilosissima Planch.
   C. padangensis J.J. Sm. & Schltr.
   C. pallens Ridl.
      C. fimbriata var. *annamica* Gagnep.
   C. triplicatula Rchb.f.

4. Section **Micranthae** Pradhan
   C. micrantha Lindl.
      C. papagena Rchb.f.
      C. clarkei Kraenzl.

5. Section **Brachypterae** D.A. Clayton
   C. brachyptera Rchb.f.
   C. parishii Hook.
   C. virescens Rolfe

6. Section **Speciosae** Lindl.
   C. beccarii Rchb.f.
      C. micholitziana Kraenzl.
      C. beccarii Rchb.f. var. *micholitziana* Kraenzl.
      C. beccarii Rchb.f. var. *tropidophora* Schltr.
   C. caloglossa Schltr.
   C. carinata Rolfe
      C. sarrasinorum Kraenzl.
      C. truncicola Schltr.

C. oligantha Schltr.
C. alata A. Millar *nom. nud.*
C. celebensis J.J. Sm.
C. platyphylla Schltr.
C. formosa Schltr.
C. fragrans Schltr.
C. fuerstenbergiana Schltr.
C. guamensis Ames
C. palawensis Tuyama
C. lycastoides F. Muell. & Kraenzl.
C. whitmeei Schltr.
C. macdonaldii F. Muell. & Kraenzl.
C. lamellata Rolfe
C. rumphii Lindl.
Angraecum nervosum Rumphius
Coelogyne psittacina Rchb.f.
C. psittacina var. *huttonii* (Rchb.f.) Rchb.f.
C. salmonicolor Rchb.f.
C. speciosa (Blume) Lindl. var. *salmonicolor* Schltr.
C. bella Schltr.
C. salmonicolor Rchb.f. var. *virescentibus* J.J. Sm. ex Dakkus
C. septemcostata J.J. Sm.
C. speciosa auct. non (Blume) Lindl.
C. membranifolia Carr
C. speciosa (Blume) Lindl. **subsp. speciosa**
Chelonanthera speciosa Blume
Pleione speciosa (Blume) Lindl.) Kuntze
Coelogyne speciosa (Blume) Lindl. var. *albicans* H.J. Veitch
C. speciosa (Blume) Lindl. var. *alba* Hort.
C. speciosa (Blume) Lindl. var. *rubiginosa* Hort.
C. speciosa (Blume) Lindl. **subsp. fimbriata** (J.J. Sm.) Gravendeel
C. speciosa (Blume) Lindl. var. *fimbriata* J.J. Sm.
C. speciosa (Blume) Lindl. **subsp. incarnata** Gravendeel
C. susanae P.J. Cribb & B.A. Lewis
C. tiomanensis M.R. Henderson
C. tommii Gravendeel & P. O'Byrne
C. tomiensis O'Byrne *nom. illeg.*
C. usitana Röth & Gruss
C. xyrekes Ridl.
C. xanthoglossa Ridl.

7. Section **Bicellae** J.J. Sm.
C. bicamerata J.J. Sm.
C. calcarata J.J. Sm.

8. Section **Moniliformes** Carr
C. crassiloba J.J. Sm.
C. gibbifera J.J. Sm.

C. macroloba J.J. Sm.
C. harana J.J. Sm.
C. incrassata (Blume) Lindl. var. **incrassata**
   Chelonanthera incrassata Blume
   Pleione incrassata (Blume) Kuntze
C. incrassata (Blume) Lindl. var. **sumatrana** J.J. Sm.
C. incrassata (Blume) Lindl. var. **valida** J.J. Sm.
C. kelamensis J.J. Sm.
C. longpasiaensis J.J. Wood & C.L. Chan
C. monilirachis Carr
C. naja J.J. Sm.
C. tenuis Rolfe
   C. bihamata J.J. Sm.
C. vermicularis J.J. Sm.
   Chelonistele vermicularis (J.J. Sm.) Kraenzl.

9. Section **Longifoliae** Pfitzer
   C. bilamellata Lindl.
      Panisea bilamellata (Lindl.) Rchb.f.
      Pleione bilamellata (Lindl.) Kuntze
   C. borneensis Rolfe
   C. brachygyne J.J. Sm.
   C. candoonensis Ames
   C. compressicaulis Ames & C. Schweinf.
   C. contractipetala J.J. Sm.
   C. cuprea H. Wendl. & Kraenzl. var. **cuprea**
   C. cuprea H. Wendl. & Kraenzl. var. **planiscapa** J.J. Wood & C.L. Chan
   C. dulitensis Carr
   C. elmeri Ames
   C. endertii J.J. Sm.
   C. integra Schltr.
   C. kinabaluensis Ames & C. Schweinf.
   C. longifolia (Blume) Lindl.
      Chelonanthera longifolia Blume
      Cymbidium stenopetalum Reinw. ex Lindl.
      Pleione longifolia (Blume) Kuntze
   C. longirachis Ames
   C. motleyi Rolfe ex J.J. Wood, D.A. Clayton & C.L. Chan
   C. planiscapa Carr var. **planiscapa**
   C. planiscapa Carr var. **grandis** Carr
   C. prasina Ridl.
      C. rhizomotosa J.J. Sm.
      C. modesta J.J. Sm.
      C. vagans Schltr.
      C. rhizomatosa J.J. Sm. var. *quinquelabolata* J.J. Sm.
   C. quinquelamellata Ames
   C. radicosa Ridl.
      C. carnea Hook.f., *non* (Blume) Rchb.f.

C. stipitibulbum Holttum
C. remediosae Ames & Quisumb.
C. steenisii J.J. Sm.
C. stenobulbon Schltr.
C. stenochila Hook.f.
C. tenompokensis Carr
C. trilobulata J.J. Sm.
C. tumida J.J. Sm.

10. Section **Cyathogyne** (Schltr.) D.A. Clayton
C. multiflora Schltr.

11. Section **Verrucosae** Pfitzer
C. asperata Lindl.
C. lowii Paxt.
?*C. macrophylla* Teijsm. & Binn. *nom. nud.*
C. pustulosa Ridl.
C. edelfeldtii F. Muell. & Kraenzl.
Pleione asperata (Lindl.) Kuntze
C. imbricans J.J. Sm.
C. marthae S.E.C. Sierra
C. mayeriana Rchb.f.
C. pandurata Lindl.
Pleione pandurata (Lindl.) Kuntze
C. papillosa Ridl. ex Stapf
C. peltastes Rchb.f.
C. verrucosa S.E.C. Sierra
C. zurowetzii Carr

12. Section **Tomentosae** Pfitzer
C. acutilabium de Vogel
C. bruneiensis de Vogel
C. buennemeyeri J.J. Sm.
C. distans J.J. Sm.
C. echinolabium de Vogel
C. genuflexa Ames & C. Schweinf.
C. reflexa J.J. Wood & C.L. Chan
C. hirtella J.J. Sm.
C. radiosa J.J. Sm.
C. judithiae P. Taylor
C. kaliana P.J. Cribb
C. latiloba de Vogel
C. longibulbosa Ames & C. Schweinf.
C. moultonii J.J. Sm.
C. muluensis J.J. Wood
C. odoardi Schltr.
C. palawanensis Ames
C. pholidotoides J.J. Sm.

C. pulverula Teijsm & Binn.
    C. dayana Rchb.f.
C. radioferens Ames & C. Schweinf.
C. rhabdobulbon Schltr.
    C. pulverula auct. non Teijsm. & Binn.
C. rochussenii de Vriese
    C. plantaginea Lindl.
    C. macrobulbon Hook.f.
    C. stellaris Rchb.f.
    C. steffensii Schltr.
    Pleione rochussenii (De Vries) Kuntze
    P. plantaginea (Lindl.) Kuntze
    P. macrobulbon (Hook.f.) Kuntze
C. rupicola Carr
C. squamulosa J.J. Sm.
C. swaniana Rolfe
    C. quadrangularis Ridl.
C. testacea Lindl.
    C. sumatrana J.J. Sm.
C. tomentosa Lindl.
    C. massangeana Rchb.f.
    Pleione tomentosa (Lindl.) Kuntze
    Coelogyne densiflora Ridl.
    C. dayana Rchb.f. var. *massangeana* Ridl.
    C. tomentosa Lindl. var. *massangeana* Ridl.
    C. cymbidioides auct. non Rchb.f.
C. velutina de Vogel
    C. tomentosa Lindl. var. *penangensis* Hook.f.
C. venusta Rolfe

13. Section **Hologyne** (Pfitzer) D.A. Clayton
    C. malipoensis Z.H. Tsi
    C. miniata (Blume) Lindl.
        Chelonanthera miniata Blume
        Coelogyne simplex Lindl.
        Pleione miniata (Blume) Kuntze
        Coelogyne lauterbachiana Kraenzl.
        Hologyne miniata (Blume) Pfitzer
        H. lauterbachiana Pfitzer
    C. obtusifolia Carr

14. Section **Rigidiformes** Carr
    C. albobrunnea J.J. Sm.
    C. clemensii Ames & C. Schweinf. var. **clemensii**
    C. clemensii Ames & C. Schweinf. var. **angustifolia** Carr
    C. clemensii Ames & C. Schweinf. var. **longiscapa** Ames & C. Schweinf.
    C. craticulaelabris Carr
    C. exalata Ridl.

C. subintegra J.J. Sm.
C. plicatissima Ames & C. Schweinf.
C. rigidiformis Ames & C. Schweinf.

15. Section **Veitchiae** D.A. Clayton
    C. veitchii Rolfe

16. Section **Ptychogyne** (Pfitzer) D.A. Clayton
    C. flexuosa Rolfe
        C. bimaculata Ridl.
        Ptychogyne flexuosa (Rolfe) Pfitzer
        P. bimaculata (Ridl.) Pfitzer

17. Section **Lawrenceanae** D.A. Clayton
    C. eberhardtii Gagnep.
    C. lawrenceana Rolfe
        C. fleuryi Gagnep.

18. Section **Coelogyne**
    C. concinna Ridl.
    C. cristata Lindl.
        Cymbidium speciosissimum D. Don
    C. cumingii Lindl.
        C. longebracteata Hook.f.
        C. casta Ridl.
        Pleione cumingii (Lindl.) Kuntze
    C. foerstermannii Rchb.f.
        C. maingayi Hook.f.
        Pleione foerstermannii (Rchb.f.) Kuntze
        P. maingayi (Hook.f.) Kuntze
        C. kingii Hook.f.
    C. glandulosa Lindl. var. **glandulosa**
    C. glandulosa Lindl. var. **bournei** S. Das & S.K. Jain
    C. glandulosa Lindl. var. **sathyanarayanae** S. Das & S.K. Jain
        C. mossiae Rolfe *pro parte*
    C. kemiriensis J.J. Sm.
    C. malintangensis J.J. Sm.
    C. mooreana Sand. ex Rolfe
        C. psectrantha Gagnep.
    C. nervosa A. Rich
        C. corrugata Wight
    C. sanderiana Rchb.f.

19. Section **Fuscescentes** Pfitzer
    C. assamica Linden & Rchb.f.
        C. fuscescens Lindl. var. *assamica* (Linden & Rchb.f.) Pfitzer & Kraenzl.
        C. annamensis Rolfe
        C. siamensis Rolfe

C. dalatensis Gagnep.
Cymbidium evardii Guill.
Coelogyne saigonensis Gagnep.
C. dichroantha Gagnep.
C. fuscescens Lindl. var. **fuscescens**
C. fuscescens Lindl. var. **brunnea** Lindl.
C. brunnea Lindl.
C. cynoches C.S.P. Parish & Rchb.f.
C. fuscescens Lindl. var. **integrilabia** (Pfitzer) Schltr.
C. integrilabia Pfitzer
C. fuscescens Lindl. var. **viridiflora** U.C. Pradhan
C. picta Schltr.

20. Section **Ocellatae** Pfitzer
C. corymbosa Lindl.
C. gongshanensis H. Li
C. hitendrae S. Das & S.K. Jain
C. nitida (Wall. ex D. Don) Lindl.
Cymbidium nitidum (Lindl.) Wall. ex D. Don, *non* Roxb.
Coelogyne conferta Hort
C. ochracea Lindl.
C. occulata Hook.f. var. **occulata**
C. occulata Hook.f. var. **uniflora** N.P. Balakr.
C. punctulata Lindl.
Cymbidium nitidum Roxb.
Coelogyne ocellata Lindl.
C. brevifolia Lindl.
C. goweri Rchb.f.
C. ocellata Lindl. var. *maxima* Rchb.f.
C. ocellata Lindl. var. *boddaertiana* Rchb.f.
C. nitida (Roxb.) Hook.f., *non* (Wall. ex D. Don) Lindl.
C. punctulata Lindl. var. *hysterantha* Tang & Wang

21. Section **Lentiginosae** Pfitzer
C. breviscapa Lindl.
C. angustifolia Wight
C. chlorophaea Schltr.
C. chloroptera Rchb.f.
C. confusa Ames
C. lacinulosa J.J. Sm.
C. lentiginosa Lindl.
C. loheri Rolfe
C. marmorata Rchb.f.
C. zahlbrucknerae Kraenzl.
C. merrillii Ames
C. monticola J.J. Sm.
C. mossiae Rolfe
C. odoratissima Lindl.

   C. trifida Rchb.f.
   C. angustifolia A. Rich.
  C. schilleriana Rchb.f. & C. Koch
   Pleione schilleriana (Rchb.f. & C. Koch) B.S. Williams
  C. sparsa Rchb.f.
   Pleione sparsa (Rchb.f.) Kuntze
  C. suaveolens (Lindl.) Hook.f.
   Pholidota suaveolens Lindl.
  C. taronensis Hand.-Mazz.
  C. undatialata J.J. Sm.
  C. vanoverberghii Ames
  C. zeylanica Hook.f.

22. Section **Flaccidae** Lindl.
  C. albolutea Rolfe
  C. esquirolei Schltr.
  C. flaccida Lindl.
   C. lactea Rchb.f.
  C. hajrae Phukan
  C. huettneriana Rchb f.
  C. integerrima Ames
  C. quadratiloba Gagnep.
   C. thailandica Seidenf.
  C. trinervis Lindl.
   C. cinnamomea Teijsm & Binn.
   C. rhodeana Rchb.f.
   C. rossiana Rchb.f.
   C. angustifolia Ridl
   C. pachybulbon Ridl.
   C. wettsteiniana Schltr.
   C. stenophylla Ridl.
  C. viscosa Rchb.f.
   C. graminifolia C.S.P. Parish & Rchb.f.

SECTIONAL CHARACTERISTICS, ARTIFICIAL KEYS TO THE SPECIES AND DESCRIPTIONS

The keys to the species within sections have been presented in the order in which the sections were identified and numbered in the Artificial Key to the Sections. A general description is given at the beginning of each section. To aid identification of the species, the key descriptions are amplified with additional information in [square brackets]. There follows, in alphabetical order, a description of the species in the section.

SECTION ELATAE

Section 1. **Elatae** Pfitzer & Kraenzlin in Engler, *Pflanzenr. Orch.-Mon.-Coelog.* 78 (1907). Type: *Coelogyne elata* Lindl. [= *Coelogyne stricta* (D. Don) Schltr.].
Sect. *Ancipites* Lindl. ex Pfitzer & Kraenzlin, *loc. cit.* 39. Type not selected.

Generally moderate to large plants but two species diminutive in comparison. Pseudobulbs close together or some distance apart on a sheathed, stout to slender rhizome, more or less ovoid, ellipsoid or oblong. Inflorescence hysteranthous with the exception of 2 species, peduncle bare at the base; leathery imbricate bracts at the junction between the peduncle and the rachis; rachis produces a single set of flowers; flowers opening either simultaneously, a few together or in succession. Distribution map 2.

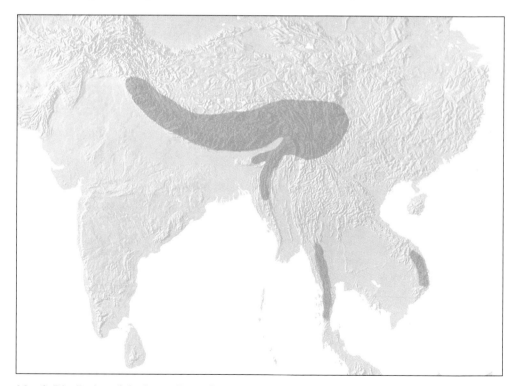

Map 2. Distribution of *Coelogyne* Sect. *Elatae*.

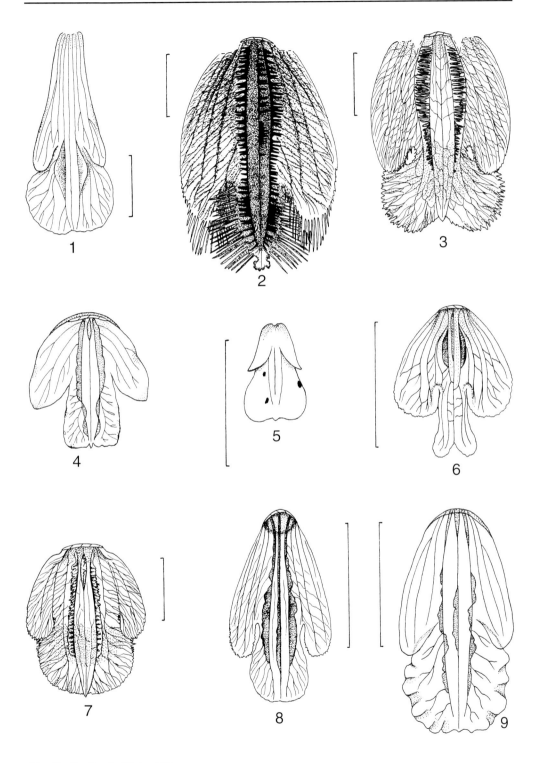

Fig. 5a. **Section 1. Elatae** Pfitzer.
1. Coelogyne anceps Hook.f. [after Hook.f.]; **2. C. barbata** Griff. [Cult. D.A. Clayton DAC 46]; **3. C. calcicola** Kerr [Kew Spirit Coll. 21038]; **4. C. filipeda** Gagnep. [after Averyanov]; **5. C. ghatakii** T.K. Paul, S.K. Basu & M.C. Biswas [after Paul (1989)]; **6. C. griffithii** Hook.f. [*G. Watt* 67880]; **7. C. holochila** P.F. Hunt & Summerh. [Cult. D.A. Clayton DAC 49]; **8. C. leucantha** W.W. Sm. [*Forrest* 27098]; **9. C. lockii** Aver. [after Averyanov]. Drawn by Linda Gurr. Scale bar = 1 cm.

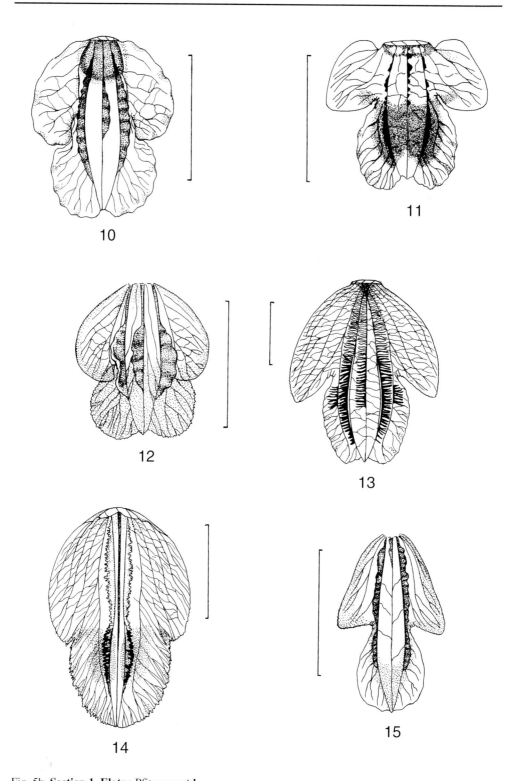

Fig. 5b. **Section 1. Elatae** Pfitzer **contd.**
10. C. pendula Summerh. ex Parry [*Parry* 241]; **11. C. pulchella** Rolfe [*Moore* 16470]; **12. C. rigida** C.S.P. Parish & Rchb.f. [*Parish* 42]; **13. C. sanderae** Kraenzl. ex O'Brien [Cult. Kew Coll.]; **14. C. stricta** (D.Don) Schltr. [Cult. D.A. Clayton DAC 28]; **15. C. tenasserimensis** Seidenf. [*Swinhoe* 103]. Drawn by Linda Gurr. Scale bar = 1 cm.

Key to species in section Elatae

1. Lip with 2 or 3 keels **3**
   Lip with 5 keels **2**

2. Lip with 5 prominent keels, outer keels broader. [Flowers greenish-yellow] .... ***C. griffithii***
   Lip with 5 keels, median keel fleshy on the mid-lobe of the lip .... ***C. zhenkangensis***

3. Dorsal sepal more than 1.5 cm long **4**
   Dorsal sepal less than 1.5 cm long **10**

4. Pseudobulbs to more than 5 cm long **5**
   Pseudobulbs less than 4 cm long. [Lip mid-lobe orbicular, margins entire, 2 keels, smooth, terminating just beyond middle of the mid-lobe. Flowers white.] .... ***C. anceps***

5. Lip with entire or laminate, sometimes undulate keels **6**
   Lip with fimbriate, crenulate or dentate keels **7**

6. Lip with entire keels. [Sepals and petals yellow-green, lip with dark purple veins, keels white.] .... ***C. filipeda***
   Lip with laminate, sometimes undulate keels. [Sepals and petals white, lip white with a yellow or yellow-brownish spot on centre of the midlobe.] .... ***C. lockii***

7. Lip with fimbriate keels **8**
   Lip with crenulate or dentate keels **9**

8. Dorsal sepal more or less 4 cm long. [Lip mid-lobe triangular with fimbriate margins and 3 fimbriate keels terminating at the tip of the mid-lobe. Flowers white, lip with sepia fimbriae.] .... ***C. barbata***
   Dorsal sepal nearly 3 cm long. [Lip mid-lobe quadrangular with fimbriate margins and 2 fimbriate keels terminating at the middle of the mid-lobe, keels become a warty projection towards the tip. Flowers (creamy)-white, some yellow on the lip.] .... ***C. calcicola***

9. Lip mid-lobe broadly ovate, distinctly 3-lobed, 3 minute-dentate to entire and undulate keels, near base of the lip, keels become irregular papillae .... ***C. holochila***
   Lip mid-lobe suborbicular or cordate, margin erose, 2 crenulate keels terminating at the base of the mid-lobe .... ***C. stricta***

10. Pseudobulbs less than 4 cm long **11**
    Pseudobulbs more than 5 cm long **12**

11. Lip mid-lobe nearly orbicular-obcordate, bi-lobed, 3 simple, straight keels terminating just beyond middle of the mid-lobe. [Flowers white with some yellow.] .... ***C. leucantha***
    Lip mid-lobe broad, rounded, nearly truncate at the tip, 3 undulate keels terminating two thirds onto the mid-lobe, median keel near the base only. [Sepals and petals greenish-yellow, side-lobes green with purple near the base, mid-lobe with some brown especially at the median, margins lighter colour, keels dark brownish-purple.] .... ***C. tenasserimensis***

12. Lip with fimbriate or crenulate keels **13**
    Lip with entire or undulate keels **14**

13. Lip with 3 crenulate and fleshy keels terminating at the base of the mid-lobe, median keel shorter. [Flowers white with large sienna-brown blotch on the hypochile of the lip, which becomes darker on the keels.] .... *C. pulchella*
    Lip with 3 fimbriate keels. [Flowers white, lip deep orange-yellow, brown keels.] .... *C. sanderae*

14. Inflorescence slender and arching. Lip mid-lobe orbicular or ovate **15**
    Inflorescence slender. Lip mid-lobe quadrate. [Flowers greenish-yellow overall with some brownish patches on the side-lobes.] .... *C. ghatakii*

15. Inflorescence slender and arching. Lip mid-lobe ovate, bi-lobed, margins undulate, 2 keels terminating at the base of the mid-lobe. [Flowers dull yellow and cream coloured.] .... *C. pendula*
    Inflorescence slender and arching. Lip mid-lobe suborbicular, 3 crenulate or fimbriate keels terminating at the base of the mid-lobe, median keel shorter. [Flowers yellow with red keels.] .... *C. rigida*

1. COELOGYNE ANCEPS

Coelogyne anceps Hook.f., *Fl. Brit. India* 5: 840 (1890). Type: Peninsular Malaysia, Perak, *Scortechini* 1333 (holo. K).

DESCRIPTION. *Pseudobulbs* on a stout, creeping, sheathed rhizome, 2.5 cm apart, ellipsoid, grooved with age, sheathed at base, 2.5–4 × 1cm. *Leaves* 2, elliptic, subacute, coriaceous, with 5 prominent nerves, 6.5–10 × 2–4 cm with 3 cm long, grooved petiole. *Inflorescence* hysteranthous, peduncle stiff, flat, 0.3 cm wide with thin edges, 15–17 cm long, imbricating bracts at interface between peduncle and rachis, rachis erect, zig-zag, 5–8 cm long, 4-flowered, flowers opening in succession, floral bracts deciduous. *Flowers* white. *Sepals* linear-oblong, acute, 2.5–3 × 0.6 cm. *Petals* narrowly linear, obtuse, 2.5–3 × 0.3 cm. *Lip* long, 3-lobed, side-lobes narrow, elongate, with short free tips, mid-lobe orbicular, callus of 2 short glabrous keels extending from just below base to just beyond centre of mid-lobe. *Column* 1.25 cm long, apex narrows to acute tip.

DISTRIBUTION. Peninsular Malaysia (Perak).
HABITAT. Not known.
ALTITUDE. Not known.
FLOWERING. Not known.
NOTES. The epithet refers to the narrow, elongated side-lobes with short, free tips. Only known from two collections made somewhere in Peninsular Malaysia (Perak). Originally a sect. *Ancipites* species.

2. COELOGYNE BARBATA

Coelogyne barbata Lindl. ex Griff. in *Itin. Notes* 72 (1848). Type: Northeast India, Khasia Hills, Moosmai, *Griffith* 5104 (holo. K; iso. P). **Plate 1A**.

DESCRIPTION. *Pseudobulbs* on a stout, 1 cm dia., heavily sheathed rhizome, 1.6–2.7 cm apart, pyriform, angular, curved, longitudinally grooved, particularly with age, 7.5–10 × 2.5 cm, enclosed completely with 15–16 cm long, overlapping, persistent, papery bracts. *Leaves* 2, lanceolate, acuminate, margins undulate, with 9 main nerves, to 45 × 5–5.5 cm with 14 cm long, channelled petiole. *Inflorescence* hysteranthous, peduncle a long time developing. Peduncle bare to first node, 0.3 cm dia., to 40 cm long. Then a section about 3 cm. long between nodes with single enclosing bract, further imbricate bracts at interface with the rachis extending for about 7 cm; rachis 7–9 cm long, zig-zag, 4- to 14-flowered, flowers opening in sequence but remaining open simultaneously, floral bracts deciduous. *Flowers* 5–8 cm across, lasting well, musk-scented. Sepals and petals pure white, lip on outside of side-lobes whitish, on inside brown veining, lip with sepia-brown hypochile, mid-lobe, keels and fimbriate margins, column white with some yellow beneath hood. *Dorsal sepal* ovate-oblong, acute, 7 nerves, 4.5 × 1.7 cm. *Lateral sepals* ovate-oblong, acute, oblique, 7 nerves, 4.6 × 1.5 cm. *Petals* linear, acute, 3-nerved, 4.3 × 0.3 cm. *Lip* 3-lobes, 3.3 cm long, 2.5 cm across side-lobes, side-lobes erect, embracing the column, ovate, front triangular, back margin becoming fimbriate toward the front, front margin with deep fimbriate, mid-lobe narrowly triangular, tip acute, incurved, margin with 0.5 cm long fimbriate projections, callus of 3 keels extending from base of the lip to tip of the mid-lobe, initially parallel, then converging at the tip of the mid-lobe, lateral keels more elevated than median keel, all keels with fimbriate projections, median keel less so. *Column* stiff, slightly arcuate, 2.5 cm long, expanding gently into slightly winged hood, tip nearly straight, erose with notches at each end.

DISTRIBUTION. Nepal, Bhutan, Northeast India (Arunachal Pradesh, Khasia Hills, Manipur, Meghalaya, Nagaland), Upper Burma, China.

HABITAT. Epiphytic on trees, sometimes lithophytic on rocks in lower montane forest.

ALTITUDE. 1000–1800 m.

FLOWERING. September–December.

NOTES. The epithet refers to the bearded callus of the lip of the flower.

3. COELOGYNE CALCICOLA

Coelogyne calcicola Kerr in *J. Siam Soc., Nat. Hist. Suppl.* 9: 233 (1933). Type: Laos, Muang Cha, Xieng Khouang, *Kerr* 978 (holo. K; iso. C). **Plate 1B**.

DESCRIPTION. *Pseudobulbs* on a stout, creeping, sheathed rhizome, 3–4 cm apart, ovoid, angled, 5–10 cm long, enclosed with persistent bracts at base. *Leaves* 2, oblong-lanceolate or oblong-obovate, somewhat acute, many nerves, dorsal prominent beneath, lower surface minutely verrucose, 13–20 × 4–5.5 cm with 3.8–6 cm long, narrow petiole. *Inflorescence* hysteranthous, erect, rigid, peduncle bare, 7–15 cm long, 6- to 7-densely imbricate bracts at interface between peduncle and rachis, rachis 5–7 cm long, to 6-flowered, flowers opening simultaneously. *Flowers* spreading, sepals and petals creamy white. *Dorsal sepal* oblong-lanceolate, somewhat acute, 7–9-nerved, 2.8 × 1.4 cm. *Lateral sepals* oblong-lanceolate, somewhat acute, 7–9 nerves, 2.7 × 0.9 cm. *Petals* linear, obtuse, 3-nerved, 2.6 × 0.35 cm. *Lip* 3-lobed, 2.7 × 2.5 cm, side-lobes small compared with remainder of lip, rounded, front margin fimbriate, mid-lobe somewhat quadrangular, 1.3 × 1.4 cm, margin with long fimbriate except around tip, callus of 2 keels extending from base of lip and reaching middle of mid-lobe, fimbriate, towards tip keels broaden into warty protrusion. *Column* slightly arched, 1.7 cm long, hood winged.

DISTRIBUTION. Upper Burma, China (Yunnan), Thailand (Chieng Kwang), Laos (Sinlumkaba (Kachin)), Vietnam

HABITAT. Epiphytic on trees and lithophytic on outcrops of limestone in lower montane forest.

ALTITUDE. 900–1500 m.

FLOWERING. April–May.

NOTES. The epiphet refers to the limestone rocks on which the species grows.

4. COELOGYNE FILIPEDA

Coelogyne filipeda Gagnep. in *Bull. Mus. Hist. Nat. Paris* sér. 2, 22: 506 (1950). Type: Vietnam, Langbian, icon. *Eberhardt* 715 (holo. P).

Coelogyne prolifera sensu Gagnep. in Lecompte, *Fl. Gen. Indoch.* 6: 316 (1934); Seidenf., *Contr. Orch. Fl. Cambodia, Laos & Vietnam*: 35 (1975), *non* Lindl.

DESCRIPTION. ***Pseudobulbs*** on creeping rhizome covered with closely overlapping scales, ovoid, tip attenuated, ribbed, 7 × 4 cm at base. ***Leaf*** 1, linear, acute-acuminate, sessile, 15 × 1.3–1.5 cm. ***Inflorescence*** hysteranthous, peduncle thread-like, base bare, 1.8–2 cm long, imbricating bracts at interface between peduncle and rachis, rachis arched, 4-flowered. ***Flowers*** about 4 cm across, yellow-green, lip dark purple veined, keels white. ***Dorsal sepal*** ovate, acute, 2 × 0.9 cm. ***Lateral sepals*** ovate, acute, 2 × 0.9 cm. ***Petals linear***, thread-like, 2 cm long. ***Lip*** 3-lobed, lobes fixed at base, ovate, obtuse, erect, 1.3 × 0.7 cm., mid-lobe quadrangular, tip minutely notched, callus of 2 keels, not erose. ***Column*** notched.

DISTRIBUTION. Vietnam (Da Lat).

HABITAT. Not known.

ALTITUDE. Not known.

FLOWERING. Not known.

NOTES. The epithet presumably refers to the thread-like nature of the inflorescence. The painting by Eberhardt shows a single leaf to each pseudobulb. There is the possibility that the species has 2 leaves to each pseudobulb, a feature in common with all the other sect. *Elatae* species.

5. COELOGYNE GHATAKII

Coelogyne ghatakii T.K. Paul, S.K. Basu & M.C. Biswas in *J. Bombay Nat. Hist. Soc.* 86 (3): 425 (1989). Type: Northeast India, Manipur, Imphal Valley, *Ghatak* 2213a (holo. CAL).

DESCRIPTION. ***Pseudobulbs*** ovoid-oblong, 4-angled with 4 grooves, 5.5 × 2.2 cm, and dark green. ***Leaves*** 2, elliptic-lanceolate, acute-acuminate, entire, coriaceous, with 6–7 nerves, 10–15 × 2.5–3 cm, dark green, petiole grooved, 2–3 cm long. ***Inflorescence*** peduncle slender, glabrous, green, bare, 6.5 cm long, rachis 5.5 cm long, 6- to 8-flowered, flowers opening in succession. ***Flowers*** 1–1.5 cm across, greenish-yellow overall with 2–3 brownish patches on side-lobes and a brown dot at base of mid-lobe and a further 3–4 brown dots on the mid-lobe, column light yellow. ***Sepals*** spreading, oblong-lanceolate, acute, entire, glabrous, 3 nerves, 0.6–0.7 × 0.3 cm. ***Petals*** spreading, linear, obtuse, entire, glabrous, 1-nerved, 0.6–0.7 × 0.5 cm.

Lip deeply 3-lobed, glabrous, side-lobes ovate-oblong, entire, mid-lobe subquadrate, recurved, retuse, narrowed at base, entire, undulate, callus of 2 keels prominent near base of mid-lobe and extending to apex. *Column* 0.5 cm long, hood broadly winged with wings serrate.

DISTRIBUTION. Northeast India (Manipur, Imphal Valley).
HABITAT. Epiphytic on tall trees and sometimes lithophytic on moss-covered rocks.
ALTITUDE. 1500 m.
FLOWERING. April–June.
NOTES. The species named after Ghatak, the collector of the type specimen. Originally a sect. *Ancipites* species.

6. COELOGYNE GRIFFITHII

Coelogyne griffithii Hook.f., *Fl. Brit. India* 5: 638 (1890). Type: Northeast India, Manipur, Khongui valley, *G. Watt* 6780 (holo. K; syn. CAL).

DESCRIPTION. *Pseudobulbs* on a sheathed rhizome, linear-oblong, compressed, sulcate, 4–10 × 1.4–3 cm, enclosed with bracts at base. *Leaves* 2, elliptic-lanceolate, acuminate, entire, coriaceous, with 6–7 nerves, 10–26 × 5–8 cm with 5 cm long petiole. *Inflorescence* hysteranthous, peduncle slender, glabrous, green, bare, 5–15 cm long, rachis zig-zag, 10–20 cm long, 6- to 18-flowered, flowers opening in succession. *Flowers* 2.5 cm across, light-brownish overall. *Sepals* broadly oblong, acute, entire, glabrous, 3 nerves, 1–1.3 × 0.6 cm. *Petals* very slender, obtuse, entire, glabrous, 1-nerved, 1–1.3 × 0.1 cm. *Lip* 3-lobed, glabrous, side-lobes rounded with broad crenulations, mid-lobe narrow truncate at base, recurved, retuse, entire, undulate, callus of 5 short, prominent keels with outer ones broader. *Column* 0.5 cm long, hood narrowly winged with wings serrate.

DISTRIBUTION. India (Himalayas), Upper Burma.
HABITAT. Epiphytic on tall trees and sometimes lithophytic on moss-covered rocks.
ALTITUDE. 1300–1600 m.
FLOWERING. April–June.
NOTES. The species dedicated to William Griffith (1810–45) who was born at Ham, Surrey, England and who went to Madras as an assistant surgeon in 1832. He succeeded N. Wallich as the Superintendent of the Calcutta Botanic Gardens in 1842. Originally a sect. *Ancipites* species.

7. COELOGYNE HOLOCHILA

Coelogyne holochila P.F. Hunt & Summerh. in *Kew Bull.* 20: 52 (1966). Type: Burma, Chin Hills, Mt. Victoria; cult. Glasnevin B.G., *Wheeler Cuffe* s.n. (holo. K). **Plate 1C.**

Coelogyne elata sensu Hook. in *Bot. Mag.* 83: t. 5001 (1857), *non* Lindl.

DESCRIPTION. *Pseudobulbs* on a sheathed, creeping, 1 cm thick rhizome, 4–7 cm apart, erect, oblong-cylindric, obtusely angled, narrowly grooved, 6–12 × 2–4 cm, when young, chalk like sheaths enclose base. *Leaves* 2, oblanceolate, acute or slightly acuminate, coriaceous, margins slightly undulate, with 7 nerves, slightly raised, dorsal prominent beneath, 25–45 × 3–7 cm with 4–8 cm long, slender, grooved petiole. *Inflorescence* hysteranthous, peduncle bare, erect, 19.5–30 cm long, imbricate bracts at interface between peduncle and rachis, rachis bare, zig-zag, 10 cm long, 6- to 10-flowered, flowers opening simultaneously, floral bracts deciduous. *Flowers* more than 5 cm across, fragrant, lasting well, waxy, sepals and petals white or creamy-white, the lip pure white and marked with orange or yellow. *Dorsal sepal* lanceolate, acute, dorsal keel prominent beneath, 2.5–3 cm × 1–1.2 cm. *Lateral sepals* spreading, lanceolate, acute, dorsal keel prominent beneath, 2.5–3 × 0.9–1.1 cm. *Petals* spreading, or slightly reflexed, lanceolate or tongue-shaped, acute, 2.4–2.7 × 0.4–0.6 cm. *Lip* porrect, strongly concave, flattened, broadly ovate, tip rounded, indistinctly 3-lobed, callus of 3 keels, median keel low, with very minute dentate-to nearly entire margin, about 0.4 cm long, lateral keels extending from base of lip, terminating near tip of mid-lobe, undulate, at base of lip, irregular papillose and breaking up. *Column* slender, semi-terete, slightly arcuate, 1.7 cm long, hooded, winged, tip obscurely 3-lobed.

DISTRIBUTION. Nepal, Bhutan, Northeast India (Himalayas, Sikkim, Khasia Hills, Lushai Hills), Upper Burma, China (Yunnan).

HABITAT. Epiphytic on trees and lithophytic but raised above mossy rocks, by stout roots, in lower and upper montane forest.

ALTITUDE. 1000–2500 m.

FLOWERING. March–April (May–June), depending on altitude.

NOTES. The epithet refers to the almost entire lip. The species with which it is commonly confused, *C. stricta*, has a markedly 3-lobed lip with fimbriate margins.

8. COELOGYNE LEUCANTHA

Coelogyne leucantha W.W. Sm. in *Notes Roy. Bot. Gard. Edinburgh* 13: 198 (1921). Type: Burma, Hpimaw, *Kingdon-Ward* s.n. (holo. E).

Coelogyne leucantha W.W. Sm. var. *heterophylla* Tang & Wang in *Acta Phytotax. Sinica* 1 (1): 78 (1951).

DESCRIPTION. *Pseudobulbs* on a moderately thick, sheathed rhizome, close together, ovate-oblong, 1.5–2 × 0.8–1 cm, enclosed with bracts at base. *Leaves* 2, narrowly oblong-lanceolate, acute-apiculate, coriaceous, with 9 main nerves, 12–13 × 2 cm with 4–5 cm long, grooved petiole. *Inflorescence* hysteranthous, peduncle straight, slender, 10–12 cm long, imbricate bracts at interface between peduncle and rachis, rachis slightly zig-zag, 8.4–10.4cm long (including imbricate bracts), 10-flowered, flowers opening simultaneously, floral bracts deciduous. *Flowers* white. *Dorsal sepal* oblong-ovate, acute, 7-nerved, 1.4 × 0.5 cm. *Lateral sepals* oblique, ovate, acute, 5-nerved, 1.3 × 0.4 cm. *Petals* linear, acute, 1-nerved, 1.2–1.4 × 0.2 cm. *Lip* 3-lobed, about 1.4 cm long, side-lobes small, elliptic, front rounded, margins entire, mid-lobe nearly orbicular-obcordate, bi-lobed, margin undulate-crenulate, callus of 3 keels extending from base of lip to just past the middle of mid-lobe, simple, straight, slightly elevated. *Column* slender, straight, 1.4 cm long, slightly expanded into a hood, no evident wings, tip indeterminate.

DISTRIBUTION. China, Burma (Upper (Hpimow)).

HABITAT. Epiphytic on trees and lithophytic on rocks on grassy hillside on granite substrate.

ALTITUDE. 2450 m.

FLOWERING. June.

NOTES. The epithet refers to the pure white flowers. W.W. Smith suggested that the growth habit was similar to *C. prolifera* Lindl. and *C. flavida* Lindl. (sect. *Proliferae* Lindl.). Also he stated that it is a near ally of *C. pulchella* Rolfe (sect. *Elatae* Pfitzer). *Coelogyne heterophylla* Tang & Wang (?type K, not in Index Kewensis) was subsequently identified as *C. leucantha* W.W. Sm. var. *heterophylla*. However, in my judgement, based on the description available, it is identical to *C. leucantha* W.W. Sm.

9. COELOGYNE LOCKII

Coelogyne lockii Aver. in *Lindleyana* 15 (2): 73–76, f.1 (2000). Type: Northern Vietnam, Ha Quang Prov., Dong Van Distr., Ho Quang Phin municipality, vicinity of Ta Xa village, 14–16 km. SSW of Dong Van town, *P.K Loc, P.H. Hoang & L. Averyanov* (holo. LE; iso AAU, HN, K, MO, P).

DESCRIPTION. ***Pseudobulbs*** on a sheathed, creeping, rigid, 0.4–0.6 cm thick rhizome, 1–3 cm apart, ovoid to cylindrical-ovoid, becoming irregularly wrinkled with age, 2–4(5) cm long, 0.8–1.5 cm wide near base. ***Leaves*** 2, narrowly elliptic, slightly broadening towards an acute apex, coriaceous, many nerved, (6) 8 –12 (16) cm with a 3–12 cm long petiole. ***Inflorescence*** hysteranthous, peduncle rigid, erect, naked at base, (11) 15–27 cm long, imbricate bracts between the peduncle and the rachis, rachis erect, zig-zag, (4) 6–11 cm long, (5) 6- to 11-flowered, flowers opening more or less simultaneously, floral bracts deciduous. ***Flowers*** with sepals and petals white, widely spreading, lip white with bright yellow or yellow-brownish spot on the centre of the mid-lobe, column and anther light brown. ***Sepals*** slightly concave, narrowly ovate, acute, 7-nerved, 1.7–1.9 × 0.6 cm. ***Petals*** narrowly lanceolate to linear, acute, 3-nerved, 1.7–1.9 × 0.2 cm. ***Lip*** 3-lobed, 1.7–1.9 × 0.9 cm, side-lobes erect, oblong with broadly rounded front, mid-lobe broadly ovate or orbicular, sometimes with a small central mucro, 0.7–0.8 × 0.7–0.8 cm, margins thin, slightly undulate, callus of 3 keels extending from base of lip to about halfway onto mid-lobe, median keel shorter than lateral keels, keels laminate, sometimes undulate. ***Column*** narrow, elongate, about 1.4 cm long, narrowly winged.

DISTRIBUTION. Vietnam (northern part), possibly in south China near the Vietnamese border.

HABITAT. Epiphytic or lithophytic. On trees and on granite or karst limestone ridges in very wet mossy primary forest. Tend to be on trees at the higher elevations.

ALTITUDE. 1100–2200 m.

FLOWERING. April–May.

NOTES. The species is dedicated to the distinguished Vietnamese botanist and plant geographer, Professor Phan Ke Loc of Hanoi University, who first found the species on Mount Phan Tsi Pan (Fan Si Pan) in April 1997. The species is said to differ from the closely related taxa, *Coelogyne tenasserimensis* Seidenf. and *C. filipeda* Gagnep., in having a more oblong lip with 3 laminate keels which are sometimes undulate (wavy). Also the flowers are a pure white apart from the yellow spot on the mid-lobe and light brown column and anther.

10. COELOGYNE PENDULA

Coelogyne pendula Summerh. ex Parry, *The Lakhers*, 606 (1932). Type: Northeast India, Lushai Hills, *Parry* 241 (holo. K).

DESCRIPTION. *Pseudobulbs* on a stout, creeping, sheathed rhizome, 4 cm apart, cylindrical to narrowly ovoid, angular, 8–9 × 2 cm, enclosed with bracts at base. *Leaves* 2, lanceolate, acuminate, with 5 prominent nerves, 25–26 × 5 cm with 5.5 cm long, grooved petiole. *Inflorescence* hysteranthous, peduncle slender, arcuate, bare, 31 cm long, imbricate bracts at interface between peduncle and rachis, rachis slender, pendulous, slightly zig-zag, 35 cm long, 23-flowered, flowers opening simultaneously, floral bracts deciduous. *Flowers* dull yellow and cream coloured. *Dorsal sepal* lanceolate, acuminate, with 1 prominent nerve beneath, 1.5 × 0.4 cm. *Lateral sepals* lanceolate, acuminate, with 1 prominent nerve beneath, 1.5 × 0.3 cm. *Petals* linear, 1.2 cm long. *Lip* 3-lobed, side-lobes erect, ovate, front rounded, erose, mid-lobe ovate, bi-lobed, callus of 2 keels extending from base of lip to base of mid-lobe, undulate. *Column* slender, straight, 1 cm long, hooded.

DISTRIBUTION. Northeast India (Lushai Hills).
HABITAT. Epiphytic lower montane forest.
ALTITUDE. 610 m.
FLOWERING. Not known.
NOTES. The epithet refers to the pendulous nature of the inflorescence. I prepared the description based on a single specimen (K).

11. COELOGYNE PULCHELLA

Coelogyne pulchella Rolfe in *Bull. Misc. Inf., Kew*: 194 (1898). Type: Burma, cult. Eldon Place Nursery, Bradford, *J.W. Moore* s.n. (holo. K).

DESCRIPTION. *Pseudobulbs* on a stout, sheathed, creeping rhizome, 2–3 cm apart, ovoid to ovoid-oblong, rather drawn out in upper part with several obscure angles which become pronounced with age, dark green, 4–7 × 2–3 cm. *Leaves* 2, oblong-lanceolate or elliptic, acute, subcoriaceous, with 7 nerves, dorsal prominent, 13–17 × 2.5–5 cm with 4 cm long, grooved petiole. *Inflorescence* hysteranthous peduncle bare, erect, stiff, 10–11 cm long, imbricate bracts at interface of peduncle and rachis, rachis slender, 14–15 cm long, 4- to 12-flowered, flowers closely spaced, flowers opening simultaneously, floral bracts deciduous. *Flowers* 2.5–3 cm across, pure white with large sienna-brown blotch on disc that becomes darker on keels and a smaller blotch at extreme base of lip. *Sepals* ovoid-oblong, acute, 1.25–1.5 × 0.75 cm. *Petals* linear, subacute, 1.25–1.5 × 0.2–0.3 cm. *Lip* 3-lobed, 1.25–1.5 cm long, side-lobes erect, semi-ovate, rounded at tip, front margins nearly 0.5 cm long, mid-lobe much larger with two rounded, crenulate, undulating lobes at apex, callus of 3 keels, fleshy, crenulate, extending from base of lip to base of mid-lobe, median keel short. *Column* incurved, flattened on the front, less than 0.5 cm long, tip denticulate.

DISTRIBUTION. Lower Burma (Tenasserim), China (Yunnan).
HABITAT. Epiphytic.
ALTITUDE. Not known.
FLOWERING. March, September–October.
NOTES. The epithet refers to the beautiful form of the flowers.

12. COELOGYNE RIGIDA

Coelogyne rigida C.S.P. Parish & Rchb.f. in *Trans. Linn. Soc. London* 30: 146 (1874). Type: Burma, Moulmein, *Parish* 42 (holo. K, iso. HN, LE).

Coelogyne tricarinata Ridl. in *J. Fed. Mal. States Mus.* 5: 156 (1915). Type: Peninsular Malaysia, Kounong, *H.C. Robinson* s.n. (holo. K).

DESCRIPTION. ***Pseudobulbs*** on a stout, creeping, heavily sheathed rhizome, 3–5 cm apart, large, narrowly oblong, 7.5–13 × 2.5–7 cm, enclosed with bracts at base. ***Leaves*** 2, elliptic-lanceolate, acuminate, with 7 nerves, 10–15 × 2.5–6 cm with 2.5–6 cm long, grooved petiole. ***Inflorescence*** hysteranthous, peduncle slender, arching, 8.5–20 cm long, imbricate bracts at interface between peduncle and rachis, rachis slender, arching, zig-zag, 10–11 cm long, 5- to 12-flowered, flowers opening simultaneously, floral bracts deciduous. ***Flowers*** small, yellow with red, crenulate keels on lip. ***Dorsal sepal*** ovate, acuminate, 3-nerved, 1 × 0.35 cm. ***Lateral sepals*** lanceolate-ovate, acuminate, 3-nerved, 1 × 0.25 cm. ***Petals*** narrowly linear, 1-nerved, 1 cm long. ***Lip*** 3-lobed, saccate at base, side-lobes large and rounded, margins entire, mid-lobe suborbicular, retuse, callus of 3 crenulate keels, extending from base of lip to base of mid-lobe, median keel shorter. ***Column*** 0.7–0.9 cm long, narrowly winged, serrulate at tip.

DISTRIBUTION. ?Northeast India (Himalayas), China, Lower Burma (Tenasserim), Thailand (Khao Nong, Nakhon Sawan, Mae Tanum) and Peninsular Malaysia.

HABITAT. Epiphytic on trees in evergreen and deciduous forest.

ALTITUDE. 700–1220 m.

FLOWERING. June–July.

NOTES. The epiphet refers to the rigid form of the species.

13. COELOGYNE SANDERAE

Coelogyne sanderae Kraenzl. ex O'Brien in *Gard. Chron.* ser. 3, 13: 360 & 547 (1893). Type: Upper Burma, collected for F. Sander (holo. B†).

Coelogyne annamensis Ridl. in *J. Nat. Hist. Siam. Soc.* 4: 117 (1921). Type: Vietnam, Langbian, *B. Kloss* s.n. (holo. BM).

C. ridleyi Gagnep. in Lecompte, *Fl. Gen. Indo-Chine*, 6: 320 (1933). Type: Vietnam, Langbian, *B. Kloss* 191 (holo. BM)

C. darlacensis Gagnep. in *Bull. Mus. Hist. Nat. Paris* sér. 2, 22: 505 (1950). Type: Vietnam, *Poilane* 32588 (holo. P)

DESCRIPTION. ***Pseudobulbs*** on a creeping rhizome, 2 cm apart, narrowly ovoid-oblong, 3–7.5 × 1.5 cm. ***Leaves*** 2, narrowly elliptic-lanceolate, acute, glossy, undulate, many nerved, to 16 × 5 cm. ***Inflorescence*** hysteranthous, peduncle bare, erect, 20 cm long, prominent imbricate bracts at interface between peduncle and rachis, rachis slender, zig-zag, to 8 cm long, loosely 4- to 7- flowered, flowers opening simultaneously, floral bracts deciduous. ***Flowers*** about 7.5 cm across fragrant, rather waxen in texture, white, lip deep orange-yellow, keels brown, fimbriate. ***Sepals*** oblong-lanceolate, acute, 1.5 cm long. ***Petals*** linear, acute, widely spreading, 1.5 cm long. ***Lip*** 3-lobed with short, erect, almost round, side lobes with fimbriate, dentate margins, mid-lobe nearly round-elliptical, slightly reflexed, shallow notched, fimbriate, dentate margins, callus of 3 keels, fimbriate, hairy fringe. ***Column*** slender, arcuate, 2 cm long, winged, lip obtuse.

DISTRIBUTION. Upper Burma, China (Yunnan), Vietnam (Da Lat, Mount Chu-yang-sinh).

HABITAT. Epiphytic on trees.

ALTITUDE. 1700–?3000 m.

FLOWERING. May–September.
NOTES. The species dedicated to Fred Sander of F. Sander & Sons of St. Albans.

14. COELOGYNE STRICTA

Coelogyne stricta (D. Don) Schltr. in *Repert. Sp. Nov. Regni Veg.* 4: 184 (1919). Type: Nepal, *Wallich* s.n. (Wall. Cat. no. 1959) (holo. BM; iso. K, LINN). **Plate 1D.**

Cymbidium strictum D. Don. in *Prodr. Fl. Nepal.* 35 (1825).
Coelogyne elata Lindl., *Gen. Sp. Orch. Pl.* 40 (1830). Types: Nepal & Sylhet, *Wallich* (syn. K).

DESCRIPTION. ***Pseudobulbs*** on a stout sheathed rhizome, to 5 cm apart, oblong-cylindric, glossy green, 7.5–15 × 2.5–6.5 cm, enclosed with bracts at base. ***Leaves*** 2, elliptic-oblong, acute, coriaceous, 22–35 × 5–7 cm narrowing at base into 3–7.5 cm long petiole. ***Inflorescence*** hysteranthous, peduncle erect, 26 cm long, imbricate bracts at interface between peduncle and rachis, rachis stiff, about 15 cm long, 10- to 15-flowered, flowers opening simultaneously. ***Flowers*** 3 cm across, white, yellow-orange patch near base of mid-lobe, keels orange. ***Sepals*** oblong, somewhat acute, spreading, 1.5–2.4 × 0.4–1 cm. ***Petals*** narrowly oblong, acute, 1.5–2.4 × 0.4–1 cm. ***Lip*** 1.4–2.3 × 1–1.2 cm, 3-lobed, side-lobes erect, narrow, entire, mid-lobe somewhat orbicular or cordate, erose, callus of 2 keels, crenulate or fimbriate, extending from base of lip to near middle of mid-lobe. ***Column*** slender, straight to slightly arcuate, 1.4 cm long, slightly expanded into hood, small wings, notched at sides, front nominally tri-lobed and margin irregular.
DISTRIBUTION. Nepal, Bhutan, Northeast India (Himalayas, Sikkim, Assam, Lushai Hills), Upper Burma, China.
HABITAT. Epiphytic on broad-leaved trees on steep riverbanks in lower and upper montane forest.
ALTITUDE. 1100–2000 m.
FLOWERING. April–June, October.
NOTES. The epithet refers to the upright growth habit of the species.

15. COELOGYNE TENASSERIMENSIS

Coelogyne tenasserimensis Seidenf. in *Dansk Bot. Ark.* 29(4): 73, f. 31 (1975). Type: Thailand, Ban Mussoe, *Seidenfaden & Smitinand* 7379 (holo. C). **Plate 1E.**

DESCRIPTION. ***Pseudobulbs*** on a thick rhizome enclosed with light brown scales, about 3 cm apart, ovoid, glossy, 3–5 × 1.8 cm (when dried), enclosed with bracts at base. ***Leaves*** 2, fleshy, prominent mid nerve, with 5 nerves, dorsal prominent, 9–15 × 1.5–2 cm with 1.5–2.3 cm long petiole. ***Inflorescence*** hysteranthous, peduncle erect, slender, bare at base, 3–7 cm long, imbricate bracts at interface between peduncle and rachis, rachis slender, straight, 3–5 cm long, 4- to 6-flowered, flowers opening simultaneously, floral bracts deciduous. ***Flowers*** sepals and petals greenish-yellow, lip with some brown, especially at centre, margins of mid-lobe lighter, keels dark brownish purple, side-lobes green, purple near base. ***Sepals*** 1–1.3 × 0.3–0.4 cm, ***Petals*** linear, thin-textured, 1–1.3 × 0.3–0.4 cm. ***Lip*** 0.9–1.4 cm long, side-lobes erect, front margin rounded, mid-lobe broad, rounded, nearly truncate at tip, as broad as long,

0.5–0.8 cm, margins undulate, callus of 3 keels, 2 slightly undulate, extending from base of lip to two thirds the length of mid-lobe, elevated and less undulate on mid-lobe, abruptly cut-off at tip, median keel only near base of lip. *Column* 0.7–0.8 cm long, broadly winged.

DISTRIBUTION. Upper Burma (Mandalay), Lower Burma (Tenasserim), Thailand (Ban Phang Mapho, Doi Chiang Dao Kanburi, Ban Mussoe).

HABITAT. Epiphytic on trees, not high, in shady areas of dry evergreen hill forest.

ALTITUDE. 600–850 m.

FLOWERING. March.

NOTES. The epiphet refers to the region of Tenasserim where the type specimen was collected.

16. COELOGYNE ZHENKANGENSIS

Coelogyne zhenkangensis S.C. Chen & K.Y. Lang in *Acta Phytotax. Sin.* 21 (3): 345 (1983). Type: China, Yunnan, Zhenkang, *C.W. Wang* 72347 (holo. PE).

DESCRIPTION. *Pseudobulbs* on a creeping, sheathed rhizome, 0.3–0.5 cm dia., 3–5.5 cm apart, cylindric, apex somewhat attenuated, 5–7 × 0.5–0.8 cm, enclosed with oblong-lanceolate bracts at base, 2–4.5 cm long, outer surface of bracts with wart-like growths. *Leaves* 2, persistent, narrowly ovate-lanceolate or oblong-lanceolate, acuminate or shortly mucronate, 8–10.8cm × 2–3.2 cm with 1.4–2.5 cm long, wedge shaped petiole. *Inflorescence* hysteranthous, overall length 8–11 cm, peduncle bare, imbricate bracts at interface between peduncle and rachis, rachis with slight zig-zag, 3- to 4-flowered. *Flowers* colour not known. *Dorsal sepal* linear, acuminate, fleshy, 5-nerved, 1.1 × 0.2 cm. *Lateral sepals* linear, acuminate, 5-nerved, 1.1 × 0.2 cm. *Petals* narrowly linear, 3-nerved, 1.1 × 0.07 cm. *Lip* small, anchor-shaped, base concave, 1.1 cm long, 3-lobed, side-lobes and base of lip oblong, 0.2 × 0.12 cm, tip irregularly dentate, mid-lobe narrowly oblong-oblanceolate, 0.9 × 0.23 cm, tip acute, callus of 5 keels from base of lip, only thick and fleshy median keel extending to tip of mid-lobe. *Column* not known.

DISTRIBUTION. China (Yunnan).

HABITAT. Epiphytic on trees in upper montane forest.

ALTITUDE. 2500 m.

FLOWERING. March.

NOTES. The epiphet refers to the region, Zhenkang, Yunnan, China where the type specimen was collected.

SECTION *PROLIFERAE*

Section 2. *Proliferae* Lindl., *Fol. Orch.-Coelog.* 10 (1854); Pfitzer & Kraenzlin in Engler, *Pflanzenr. Orch.-Mon.-Coelog.* 82 (1907); P. Ormerod in *Australian Orchid Review*, 62,1: 19–25 (1997). Type: *Coelogyne prolifera* Lindl.

Pseudobulbs generally apart on a slender sheathed rhizome, more or less ovoid or tapered to oblong-cylindric. Inflorescence hysteranthous with imbricate bracts at the junction between the peduncle and the rachis. Flowers open simultaneously. Rachis extends and produces further imbricate bracts and then produces a further annual set of flowers. Distribution map 3.

Map 3. Distribution of *Coelogyne* Sect. *Proliferae*.

KEY TO SPECIES IN SECTION PROLIFERAE

1. Pseudobulbs ovoid or ellipsoid **2**
    Pseudobulbs cylindric or narrowly ovoid **4**

2. Lip mid-lobe orbicular to quadrate **3**
    Lip mid-lobe small, spathulate. [Lip bi-lobed with crenulate margin, 2 smooth keels terminating on the mid-lobe. Flowers yellowish-brown.] .... ***C. ustulata***

3. Lip mid-lobe orbicular or subquadrate, 2 keels faint near base of lip, elevated and prominent on the mid-lobe. [Flowers white to brown, side-lobes of lip with brown.] .... ***C. shultesii***

    Lip mid-lobe orbicular to subquadrate, 2 keels terminating two thirds onto the mid-lobe. [Flowers mostly yellow to yellowish-green.] .... ***C. prolifera***

4. Lip with 2 or 3 keels but no papillae, terminating well onto the mid-lobe **5**
    Lip with keels in the form of minute papillae on the mid-lobe and ciliate. [Flowers rich mahogany red with black lip.] .... ***C. ecarinata***

5. Lip mid-lobe oblong-obcordate, retuse, undulate, crenulate, 3 keels, median keel short, lateral keels parallel, terminating two thirds onto mid-lobe, initially low, then elevated, end abruptly. [Flowers white or creamy, sometimes with yellow.] .... ***C. raizadae***
    Lip mid-lobe cordate, distinctly bi-lobed and mucronate tip, 2 slender keels terminating at the middle of mid-lobe, prominent on mid-lobe only. [Flowers yellow.] ...... ***C. longipes***

SECTION PROLIFERAE

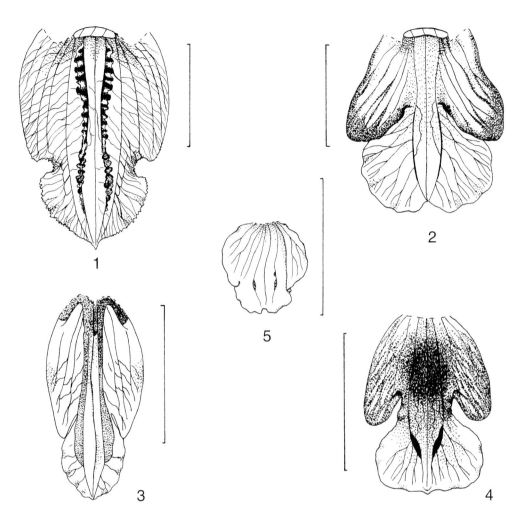

Fig. 6. **Section 2. Proliferae** Lindl.
1. Coelogyne longipes Lindl. [Kew Spirit Coll. 19755]; **2. C. prolifera** Lindl. [Kew Spirit Coll. 14524]; **3. C. raizadae** S.K. Jain & S. Das [*Jeff* s.n.]; **4. C. schultesii** S.K. Jain & S. Das [Cult. D.A. Clayton DAC 20]; **5. C. ustulata** C.S.P. Parish & Rchb.f. [after Seidenfaden (1975)]. Drawn by Linda Gurr. Scale bar = 1 cm.

17. COELOGYNE ECARINATA

Coelogyne ecarinata C. Schweinf. in *Brittonia* 4: 33 (1941). Type: Burma, Hpare, *Kingdon-Ward* 434 (holo. AMES; iso. E, NY). **Plate 1F.**

DESCRIPTION. ***Pseudobulbs*** apparently flattened-cylindric, greenish-yellow and dull in dried specimen. Rhizome elongate, stout. ***Leaves*** 2, elliptic-oblanceolate, sharply acute, coriaceous, with 7 or more prominent nerves beneath, to 18.5 × 3.5 cm, gradually narrowed at base into short, sulcate petiole. ***Inflorescence*** hysteranthous, peduncle bare, cluster of densely imbricating bracts, 2.7–4.5 cm long, at base of rachis and thereafter at intervals along the rachis as it extends with each succeeding annual season, rachis slightly zig-zag, densely and many-flowered, flowers opening simultaneously, floral bracts deciduous. *Flowers* rich mahogany red with black tip to lip. *Dorsal sepal* oblong-ovate, acute, 5-nerved, 1.1 × 0.6 cm. *Lateral sepals* oblong-ovate, acuminate, lightly cymbiform, concave at base, 5-nerved, 1.2 × 0.55–0.59 cm. *Petals* linear or lanceolate-linear, acute, 1-nerved, 1.1–1.3 cm long. *Lip* tubular, concave and saccate at base, embracing the column, 3-lobed, side-lobes erect, 0.86–0.9 cm across, obliquely semi-orbicular, free part very short, porrect, hypochile ovate in outline, fleshy, mid-lobe elliptic-ovate, 0.5 × 0.38–0.4 cm, somewhat acute, minutely papillose and ciliate, tip reflexed. *Column* stout, slightly broadened, arcuate, 0.67 cm long, tip lightly 3-lobed with lobes irregularly denticulate.

DISTRIBUTION. Upper Burma (Hpare), China (Yunnan).
HABITAT. Epiphytic in subtropical forest.
ALTITUDE. 900–1800 m.
FLOWERING. March–May.
NOTES. The epithet refers to the lack of keels on the lip to the flower. P. Ormerod (in *Australian Orchid Review*, Vol. 62, Pt. 1 (1997)) considered *C. ecarinata* C. Schweinf. a synonym of *C. ustulata* C.S.P. Parish & Rchb.f.

18. COELOGYNE LONGIPES

Coelogyne longipes Lindl., *Fol. Orch.-Coelog.* 10 (1854). Type: India, Khasia Hills, *J.D. Hooker & Thomson* 129 (holo. K-LINDL.; iso. BM, C, CAL, P). **Plate 2A & B.**

DESCRIPTION. *Pseudobulbs* on a slender, sheathed rhizome, 4–5 cm apart, long, slender, cylindric or narrowly ovoid, 3.5–10 × 1–1.8 cm. *Leaves* 2, lanceolate, acuminate, with 5–7 nerves, dorsal prominent, 6–18 × 1.5–5 cm with 1.5–5 cm long, grooved petiole. *Inflorescence* hysteranthous, peduncle bare, slender, 7–15 cm long, imbricating bracts between peduncle and rachis, rachis slightly zig-zag, 5 cm long, 3- to 5-flowered, flowers opening in succession, floral bracts deciduous. Extending each year with further imbricating bracts between old and new growths. *Flowers* small, 1–2 cm across, yellow. *Dorsal sepal* ovate-oblong, acute, 7-nerved, 1.2–2 × 0.6 cm. *Lateral sepals* ovate-oblong, acuminate, 5-nerved, 1.2 × 0.4 cm. *Petals* very slender, straight, 1 cm long. *Lip* contracted at saccate base, side-lobes erect, triangular, front rounded, entire, mid-lobe heart-shaped to bi-lobed, mucronate tip, margins erose, slightly undulate, callus of 2 slender keels extending from base of lip to near middle of mid-lobe, prominent on mid-lobe only. *Column* slender, arcuate, 1.1 cm long, expanding abruptly into broadly winged hood, tip rounded.

DISTRIBUTION. Bhutan, Northeast India (Himalayas, Sikkim, Khasia Hills), Burma, Southwest China, Thailand (Doi Suthep), Laos.
HABITAT. Epiphytic on mossy branches of trees in upper montane forest.
ALTITUDE. 1300–2300 m.
FLOWERING. December, February, April–June, August–September.
NOTES. The epithet refers to the long column-foot.

19. COELOGYNE PROLIFERA

Coelogyne prolifera Lindl., *Gen. Sp. Orch. Pl.* 40 (1830). Type: Nepal, near Noakote & Toka, *Wallich* s.n. (Wall. Cat. no. 1956) (holo. K-LINDL; iso. BM, K, CAL). **Plate 2C.**

Coelogyne flavida Lindl., *Fol. Orch.-Coelog.* 10 (1854). Type: India, Khasia Hill, *Lobb* s.n. (syn. K-LINDL; isosyn. K).

DESCRIPTION. *Pseudobulbs* on a stout, creeping, sheathed rhizome, 3–5 cm apart, ovoid or oblong, compressed, 2.5–6 × 1.5 cm, enclosed with bracts at base. *Leaves* 2, narrowly lanceolate, acuminate, with 7 nerves, dorsal prominent below, 6–18 × 1.5–2 cm with 3–4 cm long, grooved petiole. *Inflorescence* hysteranthous, peduncle slender, bare, 10–15 cm long, imbricating bracts at interface between peduncle and rachis, rachis erect, slender, zig-zag, 3.5–6 cm long, 3- to 10-flowered, flowers opening simultaneously, floral bracts deciduous. Extending each year to flower again with further imbricating bracts between old and new growth. *Flowers* small 1–1.5 cm across, sepals and petals greenish-yellow, side-lobes brown veins, mid-lobe yellow, and column yellow. *Dorsal sepal* triangular to oblong, obtuse, 1-nerved, 0.9 × 0.3 cm at base. *Lateral sepals* triangular to oblong, obtuse, 1-nerved, 0.8 × 0.25 cm at base. *Petals* linear, acute, reflexed, 0.6 cm long. *Lip* 3-lobed, side-lobes erect, embracing the column, small, obtuse, mid-lobe extending forward and below side-lobes, hypochile with parallel sides, cuneate obcordate, mid-lobe 0.7 cm long, 0.1 cm across at base, orbicular to somewhat quadrate, retuse, margins undulate, dentate, tip rounded, callus of 2 keels extending from base of lip to near base of mid-lobe, then diminishing. *Column* straight, 0.5 cm long, winged, lip bi-lobed.

DISTRIBUTION. Nepal, Bhutan, Northeast India (Sikkim, Khasia Hills, Manipur), Upper Burma, China (Yunnan (Tengyueh)), ?Thailand.

HABITAT. Epiphytic on trunks and lower branches of trees in hill, lower and upper montane forest.

ALTITUDE. (300)1000–2300 m.

FLOWERING. March–June (May–July).

NOTES. The epithet refers to the flowering in subsequent years on the same inflorescence.

20. COELOGYNE RAIZADAE

Coelogyne raizadae S.K. Jain & S. Das in *Proc. Ind. Acad. Sci. Bombay* B87(5): 119 (1978). Type: Northeast India, Khasia Hills, Mawsmai, *Das* 55419 (holo. CAL; iso. ASSAM). **Plate 2D.**

Coelogyne longipes sensu Hook.f., *Fl. Brit. India* 5: 839 (1890), *non* Lindl. (1854).

DESCRIPTION. *Pseudobulbs* on an ascending, stout rhizome, 2–8 cm apart, cylindric or narrowly oblong, tapering above to narrowly ovoid, 3–9.5 × 0.6–1.2 cm, slightly ridged when old, sheathed with large bracts. *Leaves* 2, narrowly elliptic-oblong, acute-acuminate, entire, 5–7 nerved, mid nerve prominent, 6–11 × 1.6–2.5 cm, narrowing gently at base into 1.3–5 cm long, grooved petiole. *Inflorescence* hysteranthous, peduncle erect, slender, smooth, greenish, 3–15 cm long, imbricating bracts at interface between peduncle and rachis, rachis slightly zig-zag, 2- to 6-flowered. Extending each year with further imbricating bracts between old and new growths. *Flowers* small, 1–1.2 cm across. Sepals and petals white or creamy, lip white or creamy, sometime tinged with light yellow or light brown at front of side-lobes, keels yellow. *Dorsal sepal* oblong-lanceolate, acute, 3–5-nerved, mid nerve prominent, 1.2–1.5 × 0.3–0.5 cm. *Lateral sepals* oblong-lanceolate, somewhat acute, 3–5-nerved, mid nerve prominent, 1.2–1.5 × 0.3–0.4 cm. *Petals* filiform, reflexed, entire, acute, 1-nerved. *Lip* 1.1–1.5 cm long, 0.6–0.7 cm across, smooth, 3-lobed, grooved at base, side-lobes 0.8–1.1 × 0.2–0.3 cm, narrowly oblong, straight on sides, obtuse or somewhat acute, entire, clasping the column, mid-lobe 0.3–0.4 cm long, oblong-obcordate, retuse, undulate, crenulate, callus of 3 keels, median keel short, lateral keels parallel, extending from base of lip to two thirds on to mid-lobe, low initially then rising, abrupt ends. **Column** slightly arcuate, 0.9–1.3 cm long, narrowly winged at tip, entire.

DISTRIBUTION. East Tibet, Nepal, Bhutan, Northeast India (Sikkim, Arunachal Pradesh, Meghalaya, Nagaland), Southwest China, Laos.

HABITAT. Epiphytic or lithophytic in lower montane forest.

ALTITUDE. 1800–2140 m.

FLOWERING. March–June.

NOTES. The species is named to honour Prof. M.B. Raizada, formerly head of the Botany Division in the Forest Research Institute, Dehra Dun, and a leading plant taxonomist in India.

21. COELOGYNE SCHULTESII

Coelogyne schultesii S.K. Jain & S. Das in *Proc. Ind. Acad. Sci. Bombay* 87(5): 121 (1978). Type: Northeast India, Khasia Hills, Cherrapunji, *Das* 60256 A (holo. CAL; iso. ASSAM).
Plate 2E.

Coelogyne prolifera sensu Lindl., *Fol. Orch. Coelog.* 10 (1854), *non* Lindl. (1830).
C. longipes var. *verruculata* (Lindl.) S.C. Chen in *Acta Phytotax. Sin.* 21, 3: 346 (1983). Type: China, Yunnan, Salwin valley, Sichientong, *T.T. Yu* 19264 (holo. PE).
C. flavida sensu Seidenf in *Dansk Bot. Ark.* 29, 4:82, f. 37 & 38 (1975), *non* Hook.f. ex Lindl.

DESCRIPTION. *Pseudobulbs* on a stout, 0.5–1 cm thick, creeping, sheathed rhizome, 1–3.5 cm apart, ovoid to ellipsoid, rarely smooth, irregularly ridged and furrowed, 1.5–5 × 1.7–2 cm, green, sheathed at base. *Leaves* 2, oblong to elliptic-lanceolate, acute or acuminate, entire, with 7 nerves, 6.3–18.6 × 1–1.3 cm, gradually tapering at base into smooth, 1–1.3 cm long, grooved petiole. *Inflorescence* hysteranthous, peduncle erect or arched, slender, smooth, green to greenish-brown, 5–25 cm, rachis slender, imbricating bracts at interface between peduncle and rachis, slightly zig-zag, 7–8 cm long, 4- to 9-flowered. Extending each year with further imbricating bracts between old and new growths. *Flowers* 1.5–3 cm across, sepals and petals brownish-yellow to dark brown, sometimes light greenish, lip brownish-yellow to dark brown, base creamy, mid-lobe and front part of side-lobes dark brown, column light yellow. *Dorsal sepal* oblong-lanceolate, acute, entire, smooth, membraneous, 5–7-nerved, 1.2–1.8 × 0.6–0.9 cm. *Lateral sepals* spreading, oblong-lanceolate, acute, entire, smooth, membraneous, 5–7-nerved, 1.2–1.8 × 0.4–0.6 cm. *Petals* spreading or reflexed, filiform, acuminate, entire, smooth, membraneous, 1-nerved, 1.2–1.8 × 0.1–0.2 cm. *Lip* 1–1.8 × 0.4–0.6 cm, deeply 3-lobed, side-lobes 0.8–1.2 × 0.3–0.5cm, oblong, obtuse, entire, tips clasping column, with 2–4 nerves prominent on outer surface, mid-lobe 0.7–1 × 0.8–1.1 cm, orbicular or somewhat quadrate, reflexed, retuse, narrow at base, entire or finely dentate, undulate, callus of 2 keels, faint near base of lip high and prominent on mid-lobe, extending from base of lip to about two thirds onto mid-lobe. *Column* slightly arcuate, 0.3–0.5 cm long, entire, broadly winged towards tip.

DISTRIBUTION. Nepal, Bhutan, Northeast India (Sikkim, Arunachal Pradesh, Assam, Khasia Hills, Manipur, Meghalaya, Nagaland), Upper Burma, Lower Burma (Tenasserim)), Southwest China, Thailand (Chiang Mai, Doi Suthep).

HABITAT. Epiphytic on tall rhododendrons, *Schima* and other deciduous trees, or lithophytic on moss-covered rocks in lower and upper montane forest.

ALTITUDE. 1000–2000 m.

FLOWERING. Sometimes January, March–June.

NOTES. The species is named in honour of Prof. Richard Evans Schultes of Harvard University, whose research on orchids and economic botany were a source of inspiration to Jain and Das.

22. COELOGYNE USTULATA

Coelogyne ustulata C.S.P. Parish & Rchb.f. in *Trans. Linn. Soc. London* 30: 146 (1874). Type: Burma, Moulmein, *Parish* 174A (holo. W (herb. Rchb.f. 43808-9); iso. K).

DESCRIPTION. *Pseudobulbs* on a stout rhizome, close together, ovoid, 2.5–4 × 1.5 cm. *Leaves* 2, elliptic-lanceolate, acuminate, rigid, 5-nerved, dorsal prominent, 4–5 × 2 cm with 1.5 cm long, grooved petiole. *Inflorescence* hysteranthous, peduncle slender 4 cm long, imbricating bracts at interface between peduncle and rachis, rachis slender, erect, zig-zag, 4 cm long, 5-flowered, flowers opening simultaneously, floral bracts deciduous. *Flowers* small yellow and brown. *Dorsal sepal* oblong, acute, 3-nerved, 0.75 × 0.4 cm. *Lateral sepals* oblong, acute, 3-nerved, 0.75 × 0.4 cm. *Petals* linear-lanceolate, acute, 1-nerved, 0.8 × 0.2 cm. *Lip* 3-lobed, side-lobes very broad, obtuse to rounded, mid-lobe small, spathulate, retuse, bi-lobed, crenulate margin, callus of 2 keels extending from base of lip to mid-lobe, curved, smooth. *Column* (no detail).

DISTRIBUTION. Lower Burma (Tenasserim).

HABITAT. Epiphytic in lower montane forest.

ALTITUDE. 900–1500 m.

FLOWERING. Not known.

NOTES. The epithet refers to the dark coloured parts, especially the lip of the flowers.

SECTION *FULIGINOSAE*

Section 3. *Fuliginosae* Lindl., *Fol. Orch.-Coelog.* 10 (1854). P.B. Pelser, B. Gravendeel & E.F. de Vogel in *Blumea* 45(2): 253–273 (2000). Type: *Coelogyne fuliginosa* Lodd. ex Hook.

Pseudobulbs generally on a creeping, branching, sheathed rhizome, sometimes close together but mostly some distance apart. Leaves are smallish, more or less elliptic or lanceolate. Inflorescence is hysteranthous or proteranthous to synanthous. Scape with imbricate bracts closely packed at the base of the peduncle. Margins of the lip generally fimbriate or ciliate and there are 2 or 3 undulate, crenulate or dentate keels on the lip. Flowers are usually yellow and brown.

Pelser *et al.* (2000) indicated that this section contains only two species, one widespread, *Coelogyne fimbriata* Lindl. and the other localised, *C. triplicatula* Rchb.f. I have not been convinced that the reduction to synonymy of all but two of the existing species to be entirely justified. Distribution map 4.

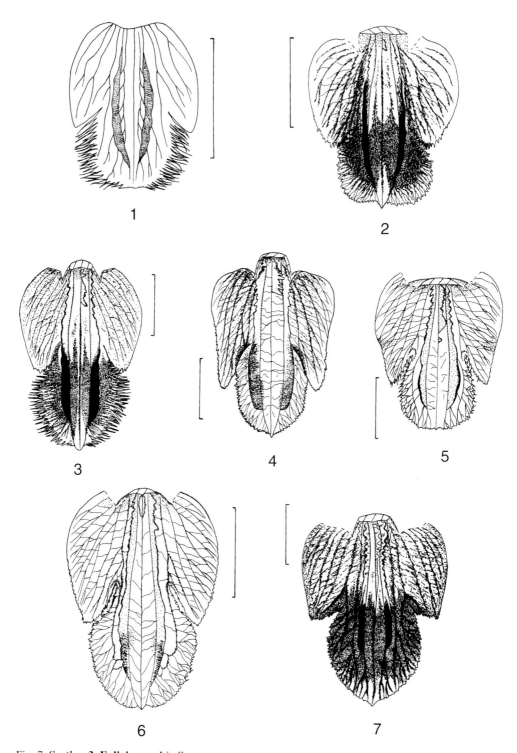

Fig. 7. **Section 3. Fuliginosae** Lindl.
1. Coelogyne arunachalensis H.J. Chowdhery & G.D. Pal [after Nord. J. Bot. 17(4), f.1 (1997)]; **2. C. fimbriata** Lindl. [Cult. D.A. Clayton DAC 36]; **3. C. fuliginosa** Lodd. ex Hook. [Cult. D.A. Clayton DAC 54]; **4. C. ovalis** Lindl. [Cult. D.A. Clayton DAC 10]; **5. C. padangensis** J.J. Sm. & Schltr. [*Wood* 956]; **6. C. pallens** Ridl. [*Wood* 862]; **7. C. triplicatula** Rchb.f. [Cult. D.A. Clayton DAC 63]. Drawn by Linda Gurr. Scale bar = 1 cm.

SECTION FULIGINOSAE

Map 4. Distribution of *Coelogyne* Sect. *Fuliginosae*.

KEY TO SPECIES IN SECTION FULIGINOSAE

1. Pseudobulbs more than 4 cm long 2
   Pseudobulbs 2.4–4 cm long, ovoid to ellipsoid. [Inflorescence with slender and erect peduncle. Lip mid-lobe orbicular, fimbriate, callus of 3 keels, median keel virtually a brown nerve, lateral keels terminating at the tip of the mid-lobe, initially diverge then converge at the tip of the mid-lobe. Sepals and petal pale yellow-green, lip whitish or pale yellow marked with brown.] .... *C. fimbriata*

2. Inflorescence synanthous 3
   Inflorescence hysteranthous 4

3. Lip mid-lobe oblong, lower part of margin ciliate-fimbriate, 2 keels terminating at the middle of the mid-lobe, thickening on the mid-lobe .... *C. chrysotropis*
   Lip mid-lobe elongated, obtuse, tip notched, margins fimbriate, fleshy, 2 keels terminating at the tip of the mid-lobe, crenulate, irregular on the mid-lobe .... *C. padangensis*

4. Lip with undulate or entire keels 5
   Lip with erose, dentate keels. [Lip margin with long ciliate. Flowers dark purple, lip spotted tobacco brown.] .... *C. longeciliata*

5. Lip mid-lobe oblong or rectangular 6
   Lip mid-lobe orbicular or ovoid 7

6. Lip mid-lobe obovate-oblong, tip obtuse. [3 keels terminating two thirds onto the mid-lobe, median keel shorter and not so pronounced, keels elevated, undulate. Flowers pale greenish or buff, lip pale buff.] .... *C. pallens*

Lip mid-lobe ovate-rectangular, tip ovate and slightly acute. [3 keels extending from the base of the lip, median keel quickly vanishes, lateral keels undulate between side-lobes, elevated on mid-lobe, entire and thickened, converging. Sepals and petals ochre-yellow, lip somewhat white with branched veins.] .... *C. triplicatula*

7. Lip with undulate keels 8
Lip with base truncated, with 3 keels, lateral keels extending from the base of the lip to 0.5 cm from the tip of the mid-lobe; median keel near base of the lip only. [Flowers pale yellow, keels light brownish turning colourless, lip base and column base brownish.] .... *C. arunachalensis*

8. Lip mid-lobe ovate, rounded, mucronate tip; margins fimbriate, with 3 keels, median keel indistinct, lateral keels terminating near the tip of the mid-lobe, undulate. [Flowers pale yellowish-green, lip-marked brown, keels darker.] .... *C. ovalis*
Lip mid-lobe orbicular, tip obtuse, margins fimbriate with long hairs, with 3 keels, median keel short and indistinct, lateral keels terminating near tip of the mid-lobe, undulate, sometimes 2 further short keels on the mid-lobe only, outside lateral keels. [Flowers light brown with greenish tinge, lip with deep red-brown margins.] .... *C. fuliginosa*

23. COELOGYNE ARUNACHALENSIS

Coelogyne arunachalensis H.J. Chowdhery & G.D. Pal in *Nord. J. Bot.* 17(4): 369–371 (1997). Type: Northeast India, Arunachal Pradesh, Lower Subansiri, Doimukhl, *G.D. Pal 1790* (holo. CAL).

DESCRIPTION. *Pseudobulbs* on a creeping, woody, 0.3–0.4 cm thick, sheathed rhizome, 2–3 cm apart, oblong-cylindric, 4.5–6.5 × 1–2 cm, 2–4 angled, glossy, green, sheathed at base; sheaths ovate, caducous at maturity. *Leaves* 2, elliptic-oblong, acute, cuneate at base, coriaceous, 10–14 × 2.5–3.2 cm with 1 cm. long petiole. *Inflorescence* hysteranthous, peduncle erect with imbricate bracts at the base, bracts ovate, ovate-lanceolate or elliptic, 1–2.5 × 0.4–1 cm, acute or obtuse, convolute, greyish, membraneous, rachis 6–8 cm long, 1- to 3-flowered, flowers opening in succession, floral bracts ovate-lanceolate, acuminate, 3–4 × 1–1.4 cm, deciduous, yellowish. *Flowers* not widely spreading, 2.5–3 cm across, sepals, petals and lip yellowish, lateral keels initially light brown turning somewhat colourless on the hypochile, median keel faintly brownish, lip base and column base brownish tinge. *Dorsal sepal* erect, oblong to elliptic-oblong, acute, convolute at base, tip and margins slightly reflexed, 9-nerved, dorsal prominent, 2.5–2.6 × 1.1–1.2 cm. *Lateral sepals* erect, oblique, oblong-lanceolate, acuminate, slightly reflexed at tip, 6–7-nerved, 2.6–2.7 × 1–1.1 cm. *Petals* linear, slightly narrow in the middle, obtuse, reflexed, 3-nerved, 2.4–2.5 × 0.2–0.3 cm. *Lip* oblong, about 2.5 cm long, 3-lobed, side-lobes ovate-triangular, 1.6–1.7 cm long, erect, acute or obtuse, front free and more or less triangular, triangle as long as broad basal part, margins entire on basal section, finely fimbriate on triangular section, mid-lobe orbicular, 1.5 × 1.8 cm, truncate at tip, margins fimbriate, callus of 3 keels, lateral keels extending from base of the lip to 0.5 cm from the tip of mid-lobe, elevated about 1.5 mm, median keel very short and at the base of the lip only. *Column* arcuate, 1.5 cm long, broadly winged.

DISTRIBUTION. Northeast India (Arunachal Pradesh).
HABITAT. Epiphytic on upper branches of trees in open evergreen or mixed forest.
ALTITUDE. 500–800 m.

FLOWERING. October–December.

NOTES. The epithet refers to the province, Arunachal Pradesh, India where the type was found. The species is allied to *C. ovalis* but can be distinguished by the presence of light—brownish keels, the absence of dark brown veins on the lip and brown patches on either side of the keels on the hypochile and epichile. The front (free) part of the side-lobes or lip somewhat triangular and as long as broad as the basal part; flowers smaller, not fully opening and dorsal sepal smaller.

24. COELOGYNE CHRYSOTROPIS

Coelogyne chrysotropis Schltr. in *Orchis* 5: 58 (1911). Type: Sumatra, Belawan, cult. Glasnevin B.G. (holo. B†).

DESCRIPTION. *Pseudobulbs* on a creeping, strongly branched rhizome, some distance apart, cylindric-conical, lightly sulcate, 4.5 × 1.2 cm. *Leaves* 2, erect-spreading, elliptic, shortly acuminate, glabrous, papery, 15 × 3–4 cm, base gently narrows into petiole. *Inflorescence* synanthous, sheathing on peduncle, 2- to 3-flowered, flowers opening in succession, floral bracts deciduous. *Flowers* golden in colour. *Sepals* oblong, acute, deeply keeled, 2.5 cm long with oblique lateral sepals. *Petals* narrowly linear, acute, 2.5 cm long. *Lip* at base concave, 3-lobed, side-lobes erect, ovate-triangular, obtuse, margins ciliate-fimbriate, mid-lobe 1.2 cm long oblong, obtuse with margins in the lower half sparsely ciliate-fimbriate, rest entire, callus of 2 keels extending from base of lip to tip of mid-lobe and somewhat enlarged halfway along mid-lobe. *Column* with clearly bi-lobed tip, lobes minutely crenulate.

DISTRIBUTION. Sumatra (Belawan).
HABITAT. Not known.
ALTITUDE. 600 m.
FLOWERING. Not known.
NOTES. The epithet refers to the golden colour of the flowers.

25. COELOGYNE FIMBRIATA

Coelogyne fimbriata Lindl., *Bot. Reg.* t. 868 (1825); Wall. Cat. no. 1957 (1829). Type: China, J.D. Parks s.n. (holo. K-LINDL; iso. P, C). **Plate 3A.**

Pleione chinense Kraenzl. in *Rev. Gen. Bot.* 2: 680 (1891).
Pleione fimbriata (Lindl.) Kraenzl. in *Rev. Gen. Bot.* 2: 680 (1891).
Coelogyne ovalis sensu Pfitzer & Kraenzl. in Engler, *Pflanzenr., Orch.-Mon.-Coelog.* 53 (1907), *non* Lindl.
C. laotica Gagnep. in *Bull. Mus. Hist. Nat. Paris*, sér. 2, 2: 425 (1930). Type: Indochina, Mekong river, *Thorel* s.n. (holo. P).
C. xerophyta Hand-Mazz. in *Symb. Sin.* 7: 1346 (1936). Type: China, NW Yunnan, Mekong river, *Felsen & Boeumen* 8457 (holo. W).
C. leungiana S.Y.Hu in *Quart. J. Taiwan Mus.* 25 (3–4): 223 (1972). Type: China, New Territories, *Hu* 9089 (holo. TAI).
C. primulina Barretto in *Orchid Rev.* 98 (1156): 37–43 (1990). Type: China, Hong Kong, Victoria peak, *Barretto* 315 (holo. K; iso. HK).

DESCRIPTION. *Pseudobulbs* on a creeping, sheathed rhizome, 3–4 cm apart, ovoid to ellipsoid, 2.5–4 × 1.2 cm. *Leaves* 2, oblong-elliptic, acute, with 5 nerves, 7.5–9 × 1.2–1.4 cm with 0.5 cm long petiole. *Inflorescence* hysteranthous, peduncle sheathed with imbricating bracts at base, slender, erect, 2.5 cm long, rachis zig-zag, 2.5 cm long, 1- to few-flowered, flowers opening in succession, floral bracts deciduous. *Flowers* 3–3.5 cm across, sepals and petals pale yellow-green, lip whitish or pale yellow marked with dark brown. *Dorsal sepal* lanceolate, acute, 5-nerved, 2 × 0.8 cm. *Lateral sepals* oblique, lanceolate, acute, 5-nerved, 2 × 0.6 cm. *Petals* linear-filiform, 2 cm long. *Lip* 3-lobed, side-lobes oblong-elliptic, front obtuse, front margins fimbriate, mid-lobe orbicular, margins fimbriate with long hairs, callus of 3 keels extending from base of lip, median keel quickly becomes brown nerve line, 2 lateral undulate keels continue but diverge and then converge and nearly meet near tip of mid-lobe. *Column* slender, arcuate, 1.8 cm long, expanding into hood, tip crenate.

DISTRIBUTION. Nepal, Bhutan, Northeast India (Subtropical Himalayas), Burma, China (Yunnan (R. Mekong), Guizhou, Guangxi, Guangdong, Hongkong), Thailand, Cambodia, Laos, Vietnam, Peninsular Malaysia.

HABITAT. Epiphytic on trees by streams.

ALTITUDE. 640–1300 m. (2290 m. in Manipur).

FLOWERING. July–November.

NOTES. The epithet refers to the fimbriate margins of the lip of the flower. Two varieties have been named, *C. fimbriata* var. *annamica* Gagnep. (1933) which is said to be a small, delicate plant, different to type in that the side-lobes project more and the mid-lobe is more pointed, see synonyms to *C. pallens* Ridl. below. There is also *C. fimbriata* var. *quadricristata* Tang & Wang but it has no recorded description; there is a specimen in the Kew Herbarium with this name.

26. COELOGYNE FULIGINOSA

Coelogyne fuliginosa Lodd. ex Hook. in *Bot. Mag.* 75: t. 4440 (1849). Type: India, Loddiges' collector 1838 (holo. K). **Plate 3B**.

Pleione fuliginosa (Hook.f.) Kuntze in *Rev. Gen. Pl.* 2: 680 (1891).

DESCRIPTION. *Pseudobulbs* on a slender, creeping, branched, sheathed rhizome, 2.5–3 cm apart, narrowly oblong and angular, 4–6 × 1.4 cm, bare at base. *Leaves* 2, lanceolate, acute, with 9 nerves, dorsal nerve prominent, 12–15 × 2.5–3.3 cm with 1.4 cm long, grooved petiole. *Inflorescence* hysteranthous, peduncle distinct with imbricating bracts at base, 3.5 cm long, rachis slender, zig-zag, 3.5 cm long, 2- to 3-flowered, flowers open in succession, floral bracts deciduous. *Flowers* light brown with a greenish tinge, lip with deep red-brown fringed margins. *Dorsal sepal* lanceolate, acute, 7 nerves, 3 × 1 cm. *Lateral sepals* lanceolate, acute, 5-nerved, 2.5 × 0.5 cm. *Petals* linear, 3–3.5 cm long. *Lip* 3-lobed, side-lobes erect, rounded, inner margins fimbriate, mid-lobe orbicular, tip obtuse, margins fimbriate with long hairs, callus of 3 keels, median keel short and indistinct, near base of lip only, 2 lateral, undulate keels extending from base of lip to near tip of mid-lobe, diverging then converging, 2 further short keels sometimes on outside of other keels on mid-lobe only. *Column* slender, 1.8 cm long, expanding slightly to form winged hood, tip crenulate.

DISTRIBUTION. Northeast India (Sikkim, Khasia Hills), Lower Burma (Tenasserim), Sumatra, Java.

HABITAT. Epiphytic on trees and lithophytic on rocks in valley.
ALTITUDE. 900–1500 m.
FLOWERING. December–January.
NOTES. The epithet refers to the sooty colour of the flowers.

27. COELOGYNE LONGECILIATA

Coelogyne longeciliata Teijsm. & Binn. in *Tijdschr. Ned.-Indie* 27: 16 (1867). Type: India, possibly Assam, *Lobb* s.n. (holo. not located).

DESCRIPTION. *Pseudobulbs* on a rhizome (no detail), remote, oblong, base narrows to point, angular, furrowed, 4.5 cm long. ***Leaves*** 2, oblong-lanceolate, acuminate, 3.5 × 1.3 cm, (no detail on petiole). ***Inflorescence*** peduncle and rachis erect, zig-zag, (no detail on number of flowers, flowers opening and floral bracts). ***Flowers*** dark violet, lip spotted tobacco-brown. ***Sepals*** oblong, acute, 3 × 1.2 cm. ***Petals*** linear, reflexed, 3 cm long. ***Lip*** 3-lobed, side-lobes and mid-lobe acute, margins with long ciliate, callus of 3 keels extending from base of lip to base of mid-lobe where median keel terminates, lateral keels continue onto mid-lobe, erose-dentate. ***Column*** (no detail).
DISTRIBUTION. Northeast India, probably Himalayas.
HABITAT. Not known.
ALTITUDE. Not known.
FLOWERING. Not known.
NOTES. The epithet refers to the flowers which have long hair-like fringed margin to the lip.

28. COELOGYNE OVALIS

Coelogyne ovalis Lindl. in *Bot. Reg.* 24: misc. 91 (1838). Type: Nepal, *Wallich* s.n. (Wall. Cat. no. 1957 *pro parte*) (holo. K-LINDL). **Plate 3C.**

Broughtonia linearis Wall. ex Lindl., *Gen. Sp. Orch. Pl.* 41 (1830) *nomen*.
Coelogyne decora Wall. ex F. Voigt, *Hort. Suburb. Calc.* 621 (1845). Type: Provenance unknown, icon. K (holo. K-LINDL.)
C. pilosissima Planch., *Hort. Donat* 144 (1858). Type: Provenance unknown, cult. Donat 144 (holo. not located).

DESCRIPTION. *Pseudobulbs* on a creeping, branching, sheathed rhizome, 5–7 cm apart, ovoid-fusiform or fusiform, smooth, 5–8 × 1.5 cm., ridged with age, enclosed with bracts at base. ***Leaves*** 2, narrowly elliptic, acute to acuminate, 9–17 × 2.5–4 cm, narrowing into 2 cm long, grooved petiole. ***Inflorescence*** hysteranthous, peduncle slender, lax, enclosed at base by 3 scales, 5 cm long, rachis, slender, 7 cm long, few-flowered, flowers opening in succession, floral bracts deciduous. ***Flowers*** about 3–4 cm across, pale yellowish-green, lip marked brown, 2 keels darker brown, column yellowish-green. ***Dorsal sepal*** ovate-lanceolate, acute, 2.7 × 1.3 cm. ***Lateral sepals*** lanceolate, acuminate, reflexed, 3 × 0.7 cm. ***Petals*** linear, acute, reflexed, 2.7 × 0.1 cm. ***Lip*** saccate at base, 3-lobed, side-lobes erect, partly around column, oblong to triangular, lower part of margin ciliate, mid-lobe ovate, tip rounded, mucronate, margin ciliate, callus of 3 keels at base of lip, 2 extending to near tip of mid-lobe, initially close together, diverging then converging, undulate, median keel indistinct. **Column** arcuate, 1.5 cm long, expanding to form a hood 0.7 cm across near the tip, tip rounded, notched at each side.

DISTRIBUTION. Tibet, Nepal, Bhutan, Northeast India (Himalayas, Sikkim, Assam, Manipur, Khasia Hills), Burma, China (Yunnan (Tengyueh)), Thailand.

HABITAT. Epiphytic on branches of trees and lithophytic on humus covered rocks in hill forest.

ALTITUDE. 800–1700 m.

FLOWERING. July–December.

NOTES. The epithet refers to the oval shape of the mid-lobe of the lip of the flowers.

29. COELOGYNE PADANGENSIS

Coelogyne padangensis J.J. Sm. & Schltr. in *Bot. Jahrb. Syst.* 45, : 104, 6 (1911). Type: Sumatra, Kampaeng Tengah, *Schlechter* 15950 (holo. B†; iso. BO). **Plate 3D.**

DESCRIPTION. **Pseudobulbs** on an elongated, sheathed rhizome, distant, erect, narrowly oblong, 4-angled, apex slightly attenuated, 3.5–6 × 0.7–0.9 cm. **Leaves** 2, erect-spreading, elliptic-lanceolate, somewhat acute, 9–18 × 2.3–3.2 cm including a slightly grooved petiole. **Inflorescence** synanthous, (no detail of the peduncle or rachis), 3- to 5-flowered. **Flowers** colour not known. **Sepals** ovate-lanceolate, acute, fleshy, lateral sepals slightly narrower than dorsal. **Petals** narrowly linear, somewhat acute, a little shorter than sepals. **Lip** curved inwards at base, wedge-shaped, 3-lobed, side-lobes erect, triangular, obtuse, tiny fimbriate, 2.1 × 1.7 cm, mid-lobe elongated, obtuse, tip shallowly notched, margins totally fimbriate, fleshy, 1.2 cm long, callus of 2 keels extending from base of lip to tip of mid-lobe, crenulate, irregular on mid-lobe. **Column** semi-terete, 1.5 cm long, tip bi-lobed with small teeth between.

DISTRIBUTION. Sumatra (West Coast (Pariaman, Kampaeng Tengah, Ayam, Danau Maninjau, Padang, Bukit Gampang Estate)).

HABITAT. Epiphytic in lower montane forest.

ALTITUDE. 1200 m.

FLOWERING. January.

NOTES. The epithet refers to the generalised region where the type specimen was collected.

30. COELOGYNE PALLENS

Coelogyne pallens Ridl. in *J. Roy. Asiat. Soc. Straits Branch* 39: 81 (1903). Types: Peninsular Malaysia, Perak, Taiping Hills, (Bujang Malacca), *Curtis* s.n., *Ridley* s.n. (syn. SING, isosyn. C). **Plate 3E.**

Coelogyne fimbriata var. *annamica* Gagnep. in Lecompte, *Fl. Gen.* 1, C6: 309 (1933). Type: Vietnam, Langbian, *Finet* 216 (holo. P).

DESCRIPTION. **Pseudobulbs** on a rising slender rhizome covered with overlapping sheaths, 4–7 cm apart, 5.5–8 × 1–1.5 cm, enclosed with bracts at base. **Leaves** 2, elliptic, acuminate, with 5 nerves, 18–33 × 2 cm with 1.3 cm long, grooved petiole. *Inflorescence* hysteranthous, peduncle enclosed in imbricating bracts, 2–4 cm long, rachis slender, bare, 2 cm but could extend, 1-flowered (probably more). *Flowers* pale greenish or buff with pale buff lip. ***Dorsal sepal*** lanceolate, acuminate, many nerved, 2.4 × 0.8 cm. ***Lateral sepals*** linear-lanceolate, acuminate, many nerved, 2.3 × 0.4 cm. *Petals* thread-like, 2.3 cm long. *Lip* 3-lobed, side lobes erect, elliptic, front triangular, porrect, mid lobe about 1.5 × 0.8 cm, oblong, tip triangular, margins fimbriate with dark hairs up to 1 mm long, callus of 3 elevated, undulate keels extending from base of lip to three quarters the length of mid-lobe, median keel slightly shorter and not so pronounced. *Column* slender, 1.3 cm long, expanding to form hood.

DISTRIBUTION. Burma, Vietnam, Peninsular Malaysia (Perak), Peninsular Thailand, Borneo (Sarawak, Sabah).

HABITAT. Epiphytic in hill forest on ridges.

ALTITUDE. 1000–1600 m.

FLOWERING. December–March.

NOTES. The epithet refers to the pale colour of the flowers.

31. COELOGYNE TRIPLICATULA

Coelogyne triplicatula Rchb.f. in *Bot. Zeitung (Leipzig)* 22: 415 (1864); *Xen Orch.* 2:159, t. 166 (1870). Type: Burma, Moulmein, *Parish* s.n. (holo. W; syn. K). **Plate 3F.**

DESCRIPTION. *Pseudobulbs* on a slender, many branched, rising, elongated rhizome with greenish and blackish markings on imbricate bracts. Not very far apart, pyriform or oblong, 4-angled, many longitudinal grooves, 4 cm × 2 cm. *Leaves* 2, lanceolate, slightly acute, margins undulate, 10 × 2 cm with short petiole. *Inflorescence* hysteranthous, base of peduncle bare, overall about 10 cm long, 1- to 2-flowered, flowers opening in succession, floral bracts lanceolate, acuminate, convolute, deciduous. *Flowers* with ochre-yellow sepals and petals, lip somewhat white with branched veins on the mid-lobe. *Sepals* oblong, acute, slightly reflexed, 2.8 cm long. *Petals* linear, acute, reflexed, about 2.8cm long. *Lip* sub-saccate, 3-lobed, side-lobes small, trapezoid, angular on lower edge, front rounded, erect, ciliate margin, mid-lobe semi ovate-rectangular, tip ovate, slightly acute, margins minutely ciliate, callus of 3 keels extending from the base of the lip, undulate between side-lobes, and elevated on mid-lobe, entire and thickened, converging, median keel quickly vanishes. *Column* slender, expanding into a winged hood, crenulate margins.

DISTRIBUTION. Lower Burma (Moulmein).

HABITAT. Not known.

ALTITUDE. Not known.

FLOWERING. Not known.

NOTES. The epithet refers to three longitudinal folds but, in relation to this species, the meaning is not clear. Type specimen was obtained from Mr. John Day of Tottenham, London. Many authors place this species as a synonym of *C. fuliginosa* Lindl. or *C. ovalis* Lindl. I concur with the judgement made by Pelser of the National Herbarium, the Netherlands, Leiden when he recognised the distinctiveness of the species.

SECTION *MICRANTHAE*

Section 4. **Micranthae** Pradhan, *Indian Orchids: Guide to Identification and Culture* 2: (1979). Type: *Coelogyne micrantha* Lindl.

Pseudobulbs a short distance apart, variable in shape. Inflorescence is hysteranthous with imbricate bracts at the base of the peduncle. Lip with the mid-lobe margin entire, sometimes undulate, at least 3 keels on the mid-lobe, sometime smooth, sometimes with papillae.

A monotypic section. Distribution map 5.

Map 5. Distribution of *Coelogyne* Sect. *Micranthae*.

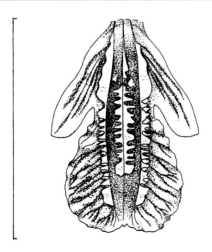

Fig. 8. **Section 4. Micranthae** Pradhan
1. Coelogyne micrantha Lindl. [*Dick* s.n.]. Drawn by Linda Gurr. Scale bar = 1 cm.

32. COELOGYNE MICRANTHA

Coelogyne micrantha Lindl. in *Gard. Chron.* 173 (1855). Type: India, *Dick* s.n. (holo. K).

Coelogyne papagena Rchb.f. in *Bot. Zeitung* (*Berlin*) 214 (1862). Type: Provenance unknown, *Low* s.n. (holo. W).
C. clarkei Kraenzl. in *Gard. Chron.* ser. 3, 13: 741 (1893). Type: Origin unknown, cult. Sander s.n. (holo. B†)

DESCRIPTION. ***Pseudobulbs*** on a branching, sheathed rhizome, 3 cm apart, oblong, compressed, 1.5–6 × 0.6–1 cm, sheathed at base. ***Leaves*** 2, oblong or linear-lanceolate, acuminate, 7-nerved, 6.5–20 × 1–2 cm with 0.4 cm long, grooved petiole. ***Inflorescence*** hysteranthous, peduncle sheathed with a few imbricating bracts at base, erect, 2 cm long, rachis erect, slightly zig-zag, 2.5–3.5 cm long, 5- to 7-flowered, flowers opening simultaneously, floral bracts deciduous. ***Flowers*** small, 1.25 cm across, green with black-spotted lip. ***Sepals*** oblong, acute, 1 × 0.5cm. ***Petals*** filiform, 0.9 × 0.1 cm. ***Lip*** 3-lobed, side-lobes small, oblong, obtuse, margins entire, mid-lobe large, broadly oblong or rounded, retuse, margins entire but undulate, callus of 3 elevated keels extending from base of lip along hypochile then becoming interlocked with a further 3 to 5 papillose keels to form an oblong warty mass extending halfway onto mid-lobe, 2 elevated lateral keels then reappear towards front of mid-lobe. ***Column*** 0.7 cm long, slightly winged.

DISTRIBUTION. Northeast India (Himalayas, Khasia Hills, Lushai Hills), Lower Burma (Tenasserim).

HABITAT. Epiphytic on trees in hill forest.

ALTITUDE. 900–1500 m.

FLOWERING. January–March.

NOTES. The epithet refers to the small flowers. The synonym, *C. clarkei* Kraenzl., is only known from a short description and it was previously judged to be 'a botanical species in the way of *C. griffithii* Hook.f. and *C. anceps* Hook.f., with close affinities to the latter' and was thus included in sect. *Ancipites* by earlier authors. However, P. Ormerod (pers. comm.) has suggested, and I agree, it is a synonym of *C. micrantha*. Furthermore, previous authors have grouped *C. micrantha* with *C. treutleri* Hook.f. in either sect. *Fuliginosae* or sect. *Micranthae*. Again, Ormerod (pers. comm.) has suggested that *C. treutleri* may not be a *Coelogyne* but an *Epigeneium*. A dissection of the only known specimen (Sikkim, *Treutler* s.n. (holo. K)) by J.J. Wood has confirmed it to be an *Epigeneium,* conspecific with *E. yunnanense* Tang & Z.H. Tsi in *Acta Phytotax Sin.* 22 (6): 484 (1984). Thus, *C. treutleri* Hook.f. becomes *Epigeneium treutleri* (Hook.f.) Ormd.

SECTION *BRACHYPTERAE*

Section 5: *Brachypterae* **D.A. Clayton sect. nov.**

Pseudobulbi elongato-cylindrici vel anguste conici aetate provecta sulcati in rhizomate crasso vaginato arcte paullum distantes siti. Folii duo, elliptici vel lanceolati, plicati, undulati. Inflorescentia hysterantha, pedunculo basi in bracteis imbricantibus incluso, rachidi leviter fractiflexa. Labellum cum tribus vel quattuor carinis undulatis instructum, lobi medii marginibus etiam undulatis. Typus: *Coelogyne brachyptera* Rchb.f.

THE GENUS COELOGYNE

Pseudobulbs on a stout, sheathed rhizome, close together or slightly apart, elongated cylindric or narrowly conical, grooved with age. Leaves 2, elliptic to lanceolate, plicate, undulate. Inflorescence hysteranthous, peduncle enclosed in imbricating bracts at the base, rachis slightly zig-zag. Lip with 3 or 4 undulate keels and mid-lobe margins undulate. Distribution map 6.

A section of three species.

KEY TO SPECIES IN SECTION BRACHYPTERAE

1. Inflorescence with 3 keels 2
   Inflorescence with 4 keels. [Lip mid-lobe pandurate, margin undulate, with 4 pectinate (combed) keels terminate on the mid-lobe and become verrucose. Flowers yellow-green, lip bluish green with blotch of dark purplish-black.] .... *C. parishii*

2. Lip mid-lobe pandurate, orbicular, margin undulate, with 3 slender undulate, slightly verrucose keels terminating at the tip of the mid-lobe. [Flowers green, lip white with a brown spot and a transverse orange bar on the mid-lobe.] .... *C. brachyptera*
   Lip mid-lobe pandurate, ovate, margin strongly undulate, with 3 almost entire keels. [Flowers pale green with yellow, lip darkly marked with spots.] .... *C. virescens*

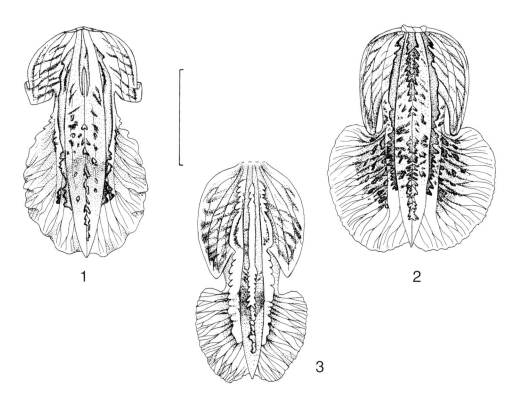

Fig. 9. **Section 5. Brachypterae** D.A. Clayton
1. Coelogyne brachyptera Rchb.f. [after *Bot. Mag.* t.8582 (1914)]; **2. C. parishii** Hook. [Kew Spirit Coll. 13568]; **3. C. virescens** Rolfe [Cult. D.A. Clayton DAC 30]. Drawn by Linda Gurr. Scale bar = 1 cm.

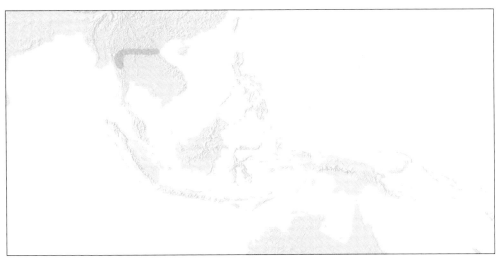

Map 6. Distribution of *Coelogyne* Sect. *Brachypterae*.

33. COELOGYNE BRACHYPTERA

Coelogyne brachyptera Rchb.f. in *Gard. Chron.* ser. 2, 16: 6 (1881). Type: Lower Burma, Tenasserim, Cult. Low s.n. (holo. W). **Plate 4A.**

DESCRIPTION. *Pseudobulbs* on a short, stout, sheathed rhizome, 1.5 cm apart, elongated, cylindric, somewhat 4-angled and longitudinally grooved and bent when old, 6–15 × 1 cm, enclosed with persistent bracts at base. *Leaves* 2, elliptic-lanceolate, subacute, plicate, 8-nerved, 12.5–21 × 3.2–5 cm with 1 cm long petiole. *Inflorescence* hysteranthous, peduncle enclosed at base with loosely imbricating bracts, erect, slender, 6–7 cm long, rachis erect, zig-zag, 4–4.5 cm long, (sometimes overall 15–19 cm), 2- to 7-flowered, flowers opening simultaneously, floral bracts persistent. *Flowers* about 6–7 cm across, fragrant, green, lip white with brown spots and orange transverse band. *Dorsal sepal* spreading, ovate-lanceolate, subacute, 7-nerved, 3–3.5 × 0.5 cm *Lateral sepals* oblong-lanceolate, acute, 7-nerved, 3–3.5 × 0.5 cm. *Petals* linear-lanceolate, acute, with a prominent nerve, 2.5–3 × 0.3 cm. *Lip* 3-lobed, about 2.5 cm long, 1 cm across side-lobes, side-lobes erect, front obtuse, suborbicular, margin undulate, mid-lobe pandurate, orbicular, margin undulate, tip rounded, 1.5 cm wide, callus of 3 slender, flexuous, slightly verrucose keels extending from base of lip to tip of mid-lobe. *Column* slender, arcuate, 1.5 cm long, expanding into slightly winged hood.

DISTRIBUTION. Lower Burma (Tenasserim).
HABITAT. Epiphytic.
ALTITUDE. Not known.
FLOWERING. April.
NOTES. The epiphet refers to the short wings on the short side-lobes of the lip. In discussion with Dr E.F. de Vogel (pers. comm.), we agreed to place this species with *C. parishii* Hook.f. and *C. virescens* Rolfe in a new section. Other authors place them in sect. *Verrucosae* Pfitzer whilst Rchb.f. thought that *C. brachyptera* was close to *C. lentiginosa*. From a study of the herbarium specimens of the three species and observation of growing material, I am in accord with the views expressed in 1975 to Seidenfaden by Hunt & Summerhayes who felt that the name of the Indochina species *C. virescens* has been

overlooked and that cultivated plants of it were invariably grown mistakenly under the name *C. parishii* Hook.f. (*Bot. Mag.* 88: t. 5323 (1862). There is a third entity, *C. brachyptera* Rchb.f. (*Gard. Chron.* ser. 2, 16: 6 (1881) and *Bot. Mag.* 140: t. 8582 (1914)), that is closely related to the other two and which could be confused with them. The three species may be distinguished thus:

– keels of labellum broken up into numerous series of papillae—*C. parishii*.
– keels of the labellum almost entire: Labellum with orange transverse band—*C. brachyptera*.
– labellum with numerous dark spots but no orange band—*C. virescens*.

I have placed the three species in a new section that has been called *Brachypterae* to identify the first named species in the group.

34. COELOGYNE PARISHII

Coelogyne parishii Hook. in *Bot. Mag.* 88: t. 5323 (1846). Type: Lower Burma, Tenasserim, Moulmein, *Parish* (holo. K). **Plate 4B.**

DESCRIPTION. *Pseudobulbs* on a creeping, sheathed rhizome, close together, cylindric to narrowly conical, 4-angled, 11–15 × 0.7 cm, yellowish with age. *Leaves* 2, elliptic or lanceolate, obtuse, 7-nerved, 18 × 3.5 cm with, not very evident, 1 cm long petiole. *Inflorescence* hysteranthous, peduncle enclosed with loosely imbricating bracts at base, drooping or erect, 4 cm long, rachis slender, to 15 cm long, 3- to 5-flowered, flowers opening virtually simultaneously, floral bracts deciduous. *Flowers* large, 7–7.5 cm across, green to yellow-green, lip bluish-green, blotched with dark purplish-black. *Sepals* lanceolate, acute, 3 × 0.8 cm. *Petals* linear-lanceolate, acute, 2.7 × 0.3 cm. *Lip* 2.3 × 1.6 cm, side-lobes auriculate, mid-lobe pandurate, truncate at base, broadly clawed, broader than long, margins undulate, callus of 4 pectinate (combed) keels from base of lip onto mid-lobe, verrucose on median line. *Column* slender, arcuate, green, 1.5 cm long.

DISTRIBUTION. Lower Burma (Tenasserim).
HABITAT. Not known.
ALTITUDE. 120 m.
FLOWERING. April–June.
NOTES. The species dedicated to the Rev. Charles Samuel Pollock Parish (1822–1897) who collected in Burma where he was Chaplain to the British Army at Moulmein. See notes under *Coelogyne brachyptera*.

35. COELOGYNE VIRESCENS

Coelogyne virescens Rolfe in *Bull. Misc. Inf., Kew*: 70 (1908). Type: Vietnam, *Micholitz* s.n. (holo. K). **Plate 4C.**

DESCRIPTION. *Pseudobulbs* on a creeping, sheathed rhizome, 2.5 cm apart, cylindric to very narrowly conical, 4-angled and lightly grooved, pale green, 9–12 × 1.5 cm, enclosed with bracts at base. *Leaves* 2, elliptic, subacute, margin undulate, with 7 fairly prominent nerves, 14–19 × 3.2–5.5 cm with, not very evident, 1.6 cm long petiole. *Inflorescence* hysteranthous, peduncle enclosed with loosely imbricating bracts, erect, 6 cm long, rachis erect, slender, slightly zig-zag, 7–15 cm long, up to 5-flowered, flowers opening simultaneously, floral bracts persistent. *Flowers* 5.5 cm across, 6.5 cm long, pale green with yellow sepals and petals, lip darkly marked with spots *Sepals* spreading widely, lanceolate, acuminate, keeled, 3.5 × 1 cm. ***Petals*** linear-lanceolate, acuminate, tip slightly reflexed. ***Lip*** 3-lobed, slightly reflexed, side-lobes oblong-obtuse, erect, mid-lobe oval with strongly undulate tip, mid-lobe raised and truncated towards the base, callus of 3 almost entire keels, crenulate at the sides. ***Column*** slender, expanding into hood, tip with incisions at each end.

DISTRIBUTION. North Thailand, Vietnam.
HABITAT. Epiphytic on trees in deciduous forest.
ALTITUDE. 200–300 m.
FLOWERING. March–April.
NOTES. The epiphet refers to the green sepals and petals and predominantly green lip. See notes under *Coelogyne brachyptera*.

SECTION *SPECIOSAE*

Section 6. ***Speciosae*** Lindl., *Fol. Orch.-Coelog.* 10 (1854). B. Gravendeel & E.F. de Vogel in *Blumea* 44(2): 253–320 (1999). Type: *Coelogyne speciosa* (Blume) Lindl.

The pseudobulbs are borne on a generally stout, woody, sheathed rhizome, close together or a short distance apart, ovoid, cylindric-ovoid or cylindric, smooth and 4-angled, 1 or 2 leaves at the apex. Inflorescence mainly synanthous but in one case proteranthous, peduncle generally bare although it is initially enveloped in the developing petiole, leaves and scales of the young shoot. Flowers medium to large. The flowers generally open in succession although there are cases where the flowers tend to open simultaneously; the floral bracts are deciduous or persistent. The number of keels is variable and the keels may form a thick callus, be undulating with an entire or interrupted margin, fused into an irregular shape, have rounded warts or ridges, have elongate projections with an longitudinal groove between or have tapering projections with or without hairs. Distribution map 7.

Map 7. Distribution of *Coelogyne* Sect. *Speciosae*.

KEY TO SPECIES IN SECTION SPECIOSAE
(after B. Gravendeel & E.F. de Vogel)

1. Lip with hairs or elongate papillae on the keels; flowers opening in succession 2
   Lip with minute papillae on the keels; flowers opening in succession or (nearly) simultaneously 7

2. Lip with hairs on the keels; keels with 2 projections over the width of the keels, separated by a longitudinal groove or with up to 5 projections over the width of the keel 3
   Lip with elongate papillae on the keels; keels plate-like, undulating. [Sepals and petals brownish-salmon, lip pale salmon with dark brown inside, keels blackish-brown.] .... *C. xyrekes*

3. Lip with 2–3 keels, hairs on the keels 0.1–0.5 mm long 4
   Lip with 5–8 keels, hairs on keels 0.7–1.5 cm long. [Sepals and petals green to very pale green, lip whitish to medium cream, side lobes orange to dull orange-brown, keels orange to orange-brown] .... *C. septemcostata*

4. Lip with 5 elongate projections over the width of the keel; hairs more or less stellate at the apex of the elongated projections of the keels; lip 3.3–6.1 cm long 5
   Lip with 2 projections over the width of the keel; hairs implanted on the rims of the longitudinal groove of each keel; lip 2.7–3.4 cm long. [Sepals and petals pale salmon-pink or creamy-white, lip pale salmon-pink or creamy-white, side-lobes orange-brownish veins, keels and central median streak on the lip of similar colour.] .... *C. salmonicolor*

5. Sepals and petals yellowish-green; lateral sepals less than 5 cm long 6
   Sepals and petals salmon coloured; lateral sepals longer than 5.6 cm. [Lip outside salmon with a brownish tinge, side-lobes brown to red-brown with creamy-white spots, lateral keels red-brown, median keel yellow.] .... *C. speciosa* subsp. *incarnata*

6. Lip with margin of the mid-lobe irregularly erose; locality Java and Flores. [Sepals and petals light green to yellowish-green, transparent. Lip white to cream with dense red to orange-brown lines on the base.] .... *C. speciosa* subsp. *speciosa*
   Lip with margin of the mid-lobe fimbriate with projections up to 0.35 cm long; locality Sumatra. [Sepals and petals ochrish-yellow, lip cream with brown lines on inside of side-lobes, inside median orange, keels initially orange, then brown.] ....
   .... *C. speciosa* subsp. *fimbriata*

7. Lip with 5–13 keels 8
   Lip with 2–4 keels 12

8. Pseudobulbs with 1 or 2 leaves; flowers opening simultaneously 9
   Pseudobulbs with 2 leaves; flowers opening in succession 10

9. Pseudobulb with 1 leaf; lip with side-lobes not projecting in front. [Sepals and petals pale green, lip whitish and apart from the margins and the apical part of the mid-lobe, mainly tinged orange-brown, keels orange-brown.] .... *C. lycastoides*

Pseudobulb with 2 leaves; lip with side-lobes clearly projecting in front. [Sepals and petals pale green to whitish, lip greenish-white, inside light orange-brown with lighter spots, warts on keels cream but on mid-lobe darker brown.] .... *C. macdonaldii*

10. Lip with keels consisting of low callus patches and many warts; hypochile much shorter than epichile **11**
    Lip with plate-like keels; hypochile about as long as the epichile. [Flower colour not known.] .... *C. guamensis*

11. Leaves with 3–5 main nerves; lip with keels consisting of broad and large callus patches. [Sepals and petals pale green to yellowish-white, lip white, side-lobes and base of mid-lobe blackish-brown, callus cream coloured to light brown to reddish-orange to purple-brown.] .... *C. beccarii*

    Leaves with 5–7 main nerves; lip with keels consisting of many rows of small warts. [Sepals and petals pale yellow-green to creamy-green with orange or red-brown lip, pale yellow or white apex.] .... *C. susanae*

12. Inflorescence synanthous **13**
    Inflorescence proteranthous. [Sepals and petals light green to creamy-yellow, transparent, lip light brown to rusty brown, side-lobes darker red-brown veins on outside, forward margins almost black, mid-lobe light brown with creamy-white tip, keels brown tipped.] .... *C. tommii*

13. Lip with keels or keels having interrupted rows **14**
    Lip with plate-like keels, not interrupted **16**

14. Dorsal sepal and lateral sepals less than 3.3 cm long, petals less than 2 cm long. [Sepals and petals pale green, lip pale green to green, side-lobes and mid-lobe with orange-brown markings, keels with light orange-brown sides, crest paler.] .... *C. carinata*
    Doral sepal and lateral sepals more than 3.3 cm long, petals more than 4 cm long **15**

15. Inflorescence usually bearing 2–6 flowers in succession [Flowers fragrant, sepals and petals light green to golden yellow, lip greenish-white or cream coloured, brownish to orange to yellow markings on mid-lobe and side-lobes, keels brown, warts on keels brown to light orange.] .... *C. fragrans*
    Inflorescence bearing up to 20 flowers in succession. Flowers creamy green with dark brown column, lip and a creamy anther .... *C. usitana*

16. Lip with keels containing 5 elongate projections over the width of the keel or plate-like **17**
    Lip with keels containing 1 or 2 projections over the width of the keel, separated by a longitudinal groove **19**

17. Lip with plate-like undulating keels **18**
    Lip with keels containing 5 elongate projections over the width of the keel. [Sepals and petals yellowish-green to yellowish-cream, lip cream to whitish, at base orange-yellow, side-lobes tinged red to orange-brown, inside veined red-brown, sinus and keels red-brown.] .... *C. rumphii*

18. Lip with mid-lobe kidney-shaped; 2–3 undulate, plate-like, entire keels. [Flower colour not known.] .... *C. formosa*
    Lip with mid-lobe ob-rhomboid to obovate; 2–3 undulate, plate-like keels with papillae. [Sepals yellow to pale salmon, petals pale greenish to pale salmon, lip mid-lobe margin white with brown veins, centre dark brown, edges light brown, side-lobes paler with brown veins.] .... *C. tiomanensis*

19. Lip with mid-lobe elliptic; scattered warts near the tip. [Sepals and petals pale green, mid-lobe with dark coloured, yellow-brown or orange-yellow markings in middle, keels dark-brown or chestnut coloured.] .... *C. celebensis*
    Lip with mid-lobe orbicular or ovate; no scattered warts near tip **19**

20. Lip with mid-lobe orbicular; keels crested. [Flower colour not known.] .... *C. caloglossa*
    Lip with mid-lobe ovate; keels with varying elevation. [Flower colour not known.] .... *C. fuerstenbergiana*

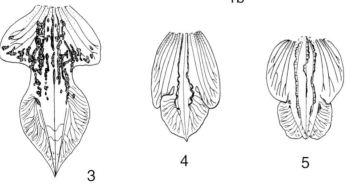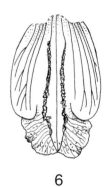

Fig. 10a. **Section 6. Speciosae** Lindl. **contd.**
1a–b. C. speciosa (Blume) Lindl. **subsp. fimbriata** (J.J. Sm.) Gravendeel; **2. C. speciosa** (Blume) Lindl. **subsp. incarnata** Gravendeel. **3. C. susanae** P.J. Cribb & B.A. Lewis; **4. C. tiomanensis** M.R. Henderson; **5. C. tommii** Gravendeel & P. O'Byrne; **6. C. xyrekes** Ridl. All after Gravendeel & de Vogel (1999). Drawn by Linda Gurr. Scale bar = 1 cm.

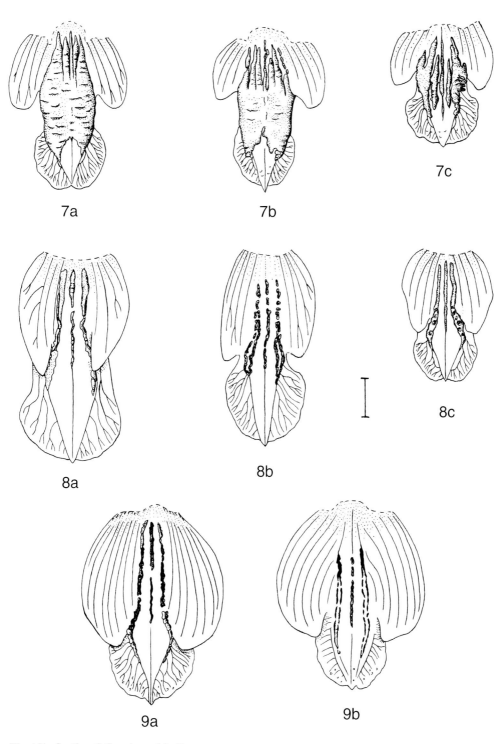

Fig. 10b. **Section 6. Speciosae** Lindl.
7a–c. Coelogyne beccarii Rchb.f.; **8a–c. C. carinata** Rolfe; **9a–b. C. celebensis** J.J. Sm. All after Gravendeel & de Vogel (1999). Drawn by Linda Gurr. Scale bar = 1 cm.

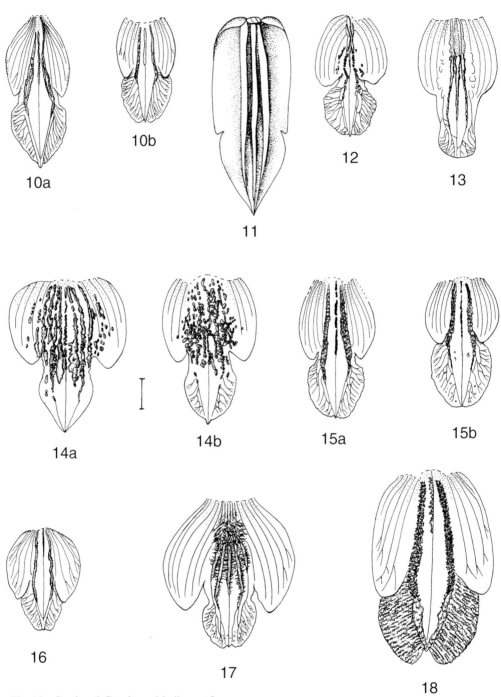

Fig. 10c. **Section 6. Speciosae** Lindl. **contd.**
10a–b. C. **fragrans** Schltr.; **11.** C. **fuerstenbergiana** Schltr. [after Orchis 8: t 3 (1914)]; **12.** C. **guamensis** Ames; **13.** C. **lycastoides** F. Muell. & Kraenzl.; **14a–b.** C. **macdonaldii** F. Muell. & Kraenzl.; **15a–b.** C. **rumphii** Lindl.; **16.** C. **salmonicolor** Rchb.f.; **17.** C. **septemcostata** J.J. Sm.; **18.** C. **speciosa** (Blume) Lindl. **subsp. speciosa**. All after Gravendeel & de Vogel (1999) except 5. Drawn by Linda Gurr. Scale bar = 1 cm.

36. COELOGYNE BECCARII

Coelogyne beccarii Rchb.f. in *Bot. Centralbl.* 28: 345 (1886). Type: New Guinea, Arfak Mts., *O. Beccari* 888 (holo. FI; iso. K). **Plate 4D.**

Coelogyne micholitziana Kraenzl. in *Gard. Chron.* ser. 3, 8: 300 (1890). Type: New Guinea, *Micholitz* s.n. (holo. K).
C. beccarii Rchb.f. var. *micholitziana* Kraenzl. in *Gard. Chron.* ser. 3, 10: 300 (1891).
C. beccarii Rchb.f. var. *tropidophora* Schltr. in *Repert. Sp. Nov. Regni Veg. Beih.* 1: 103 (1914). Type: New Guinea, Finisterre Mts., *Schlechter* 19110 (holo. B†; iso. AMES, E, G, K, L).

DESCRIPTION. ***Pseudobulbs*** on a thick, woody rhizome, erect, 0.9 cm apart, cylindrical-ovoid, green, smooth, develops 4 deep grooves and 4-angles, 4.8–8.5 cm long. ***Leaves*** 2, linear to linear-lanceolate, acuminate, 7–9-nerved, 22–35.4 × 1.9–6 cm which narrows gradually into 2.3–8.5 cm long petiole. ***Inflorescence*** synanthous with partially to entirely developed leaves, peduncle during flowering enclosed at base by petiole and scales of young shoot, peduncle erect to suberect, 18–44 cm long, rachis slender, erect to suberect, zig-zag, 1.9–7 cm long, loosely 3- to 7(–13)-flowered, flowers opening in succession, floral bracts lanceolate, deciduous. ***Flowers*** sepals and petals pale green to yellowish-white, lip white, side-lobes, margin of claw and base of mid-lobe blackish-brown, callus cream coloured to light brown to reddish-orange to purple-brown, bordered with brownish-red, column white, base blackish-brown to brownish-red. ***Dorsal sepal*** ovate, acute, 9–13-nerved, distinct dorsal keel, margins reflexed, 3.45–4.4 × 1.8–2.2 cm. ***Lateral sepals*** ovate, acuminate, 9–12-nerved, dorsal keel rounded, constricted along lower margins, 3.4–4.4 × 1.4–1.9 cm. ***Petals*** slightly recurved, linear, acuminate, 5-nerved, 3.2–4.3 × 0.36–0.55 cm. ***Lip*** 3-lobed, 2.9–4.1 × 2.6–2.8 cm, side-lobes erect, obliquely oblong, front rounded to obtuse, extending 0.28–0.35 cm in front and slightly converging, base of front margin erose, sinus acute, mid-lobe highly convex, on short broad claw, broadly ovate, margins undulate, tip acute, callus of 3–7 keels extending from base of lip to 0.4–0.8 cm from tip, elevated, widening and fused together into a thick callus with papillae but without hairs, on mid-lobe forming a row of tightly packed irregularly rounded warts, with papillae but no hairs. ***Column*** arcuate, 1.9–2.3 cm long, hood with more or less acute apical margin, laterally notched or with 2 small cuneate projections where the wings are attached and with an additional notch above, front rounded, recurved.

DISTRIBUTION. Irian Jaya, Papau New Guinea, New Britain, Solomon Islands (Guadacanal).

HABITAT. Epiphytic, rarely terrestrial in lowland rain forest and secondary vegetation.

ALTITUDE. Sea level–1500 m.

FLOWERING. Much of the year, often more than once a year.

NOTES. This species is dedicated to Odoardo Beccari (1843–1920) collector of the type specimen in the Arfak mountains of western New Guinea and whose herbarium collections are in Florence. B. Gravendeel and E.F. de Vogel have indicated that the specimens, including variations, appeared to be the same species and so they have concluded that *C. beccarii* var. *tropidophora* and *C. beccarii* var. *micholitziana* are conspecific with *C. beccarii*. The species is recognised by the thick callus on the lip, the broad side-lobes with obtuse, rounded front margins and the linear-lanceolate leaves.

37. COELOGYNE CALOGLOSSA

Coelogyne caloglossa Schltr. in *Repert. Sp. Nov. Regni Veg. Beih.* 10: 16 (1911). Type: Sulawesi, Minahassa, Gunung Klabat, *Schlechter* 20571 (holo. B†).

DESCRIPTION. *Pseudobulbs* on a short, thread-like, smooth rhizome, close together, narrowly cylindric, 13 cm long. *Leaf* 1, erect-spreading, elliptic-lanceolate, 40–50 × 4–6 cm, narrowing into petiole. *Inflorescence* synanthous with partially to entirely developed leaf, peduncle during flowering enclosed at base by petiole and scales of young shoot, peduncle terete, smooth, slender, rachis 25 cm long, lax 6- to 12-flowered, flowers opening in succession. *Flowers* large. *Dorsal sepal* oblong, somewhat acute, base concave, smooth, thickened dorsal keel, 5 × 1.3 cm. *Lateral sepals* oblique, oblong, somewhat acute, base concave, smooth, thickened dorsal keel, 5 × 1.3 cm. *Petals* narrowly linear, obtuse, smooth. *Lip* overall 4.5 cm long, 3.2 cm across, at base broadly oval, 3-lobed, side-lobes oblique, rectangular, obtuse, mid-lobe without a claw, short, square-orbicular, obtuse, margin irregular, 1.5 cm long, 1.7 cm across, callus of 3 keels, 2 lateral keels extending from near base of lip to tip of mid-lobe, short, crested, very short median keel only near base of lip. *Column* slightly arcuate, semi-rounded, slender, 4 cm long, tip with evident bi-lobe, somewhat square, margins minutely crenulate.

DISTRIBUTION. Sulawesi (Minahassa).
HABITAT. Epiphytic in hill forest.
ALTITUDE. 1000 m.
FLOWERING. December.
NOTES. The epiphet refers to the beautiful form of the lip to the flower. This species was not included by Gravendeel & de Vogel (1999) in their revision of Section *Speciosae*, possibly because the type specimen was not found.

38. COELOGYNE CARINATA

Coelogyne carinata Rolfe in *Bull. Misc. Inf., Kew* 191 (1895). Type: unknown locality, cult. Sander s.n. (holo. K). **Plate 4E.**

Coelogyne sarrasinorum Kraenzl. in Engler, *Pflanzenr., Orch.-Mon.-Coelog.* 29 (1907). Type: New Guinea, Finisterre Mts., *Schlechter* 19110 (holo. B†; iso. Ames, E, G, K, L).
C. truncicola Schltr. in *Repert. Sp. Nov. Regni Veg. Beih.* 1: 104 (1914). Type: New Guinea, Dischore Mts., *Schlechter* 19618 (holo. B†).
C. oligantha Schltr. in *Repert. Sp. Nov. Regni Veg. Beih.* 16: 44 (1919). Type: New Guinea, Waria, *Kempf* s.n. (holo. B†).
C. alata A. Millar, *Orchids of Papau New Guinea* 75 (1978), *nom. nud.*

DESCRIPTION. *Pseudobulbs* on a tough, woody rhizome, 1–2.3 cm apart, erect, ovoid to cylindric, smooth when young, wrinkled with age and becoming 4-angled, bracts persist for first year, 4.8–11 cm long. *Leaves* 1 or 2, obovate-lanceolate, acuminate, thin, plicate, 5–7-nerved, 8.1–37 × 3.1–7.2 cm with 1–5 cm long petiole. *Inflorescence* synanthous with partially to entirely developed leaves, peduncle during flowering enclosed at base by petiole and scales of young shoot, peduncle 6.4–29 cm long, rachis (sub) erect, zig-zag, 1–9.5 cm long, 2- to 8-flowered, flowers opening in succession, floral bracts ovate, deciduous or persistent. *Flowers* sepals and petals pale green, lip pale green to green, side-lobes and mid-lobe with orange-brown markings, keels with light orange-brown sides, crest paler, column whitish-green. *Dorsal sepal* ovate, apiculate, 9-nerved, keel prominent, 1.85–2.25 × 0.55–0.95 cm. *Lateral sepals* ovate-oblong, apiculate, 9-nerved, keel prominent, 1.6–3.3 × 0.4–0.75 cm. *Petals* slightly recurved, linear, apiculate, 3-nerved, 1.5–2 × 0.1–0.25 cm. *Lip* 3-lobed, not saccate, side-lobes erect, elliptic, in front obtuse, margin at front irregularly erose, with acute sinus, mid-lobe on short erose-margined claw, orbicular to elliptic, lateral margins recurved, tip acute, callus of 3 keels extending from base of lip to near middle of mid-lobe, median keel shorter and lower elevation than lateral keels, keels widened along crest, consisting of two, often interrupted, undulating rows of irregularly rounded warts on each side of crest, with papillae but without hairs, separated by a longitudinal groove, lateral keels diverge then converge on mid-lobe. *Column* arcuate, stout, 1.15–1.3 cm long, expanding into a hood with pronounced wings, tip irregular, notched at sides near wings.

DISTRIBUTION. Sulawesi (Tomohon), Biak, Irian Jaya, Papau New Guinea, New Britain, New Ireland, Solomon Islands (New Georgia, Guadalcanal, San Cristobal).

HABITAT. Epiphytic, rarely terrestrial on moss-free smooth trunks of saplings and small trees in the lowland rain forest understory and riverine forest, logged areas, coastal vegetation on limestone rocks and flood plains.

ALTITUDE. 105–2300 m.

FLOWERING. Throughout the year.

NOTES. The epithet refers to the keels on the lip of the flower. There are specimens with flowers that are intermediate in form and size between the species *C. carinata* and *C. fragrans* which suggest possible natural hybridisation.

39. COELOGYNE CELEBENSIS

Coelogyne celebensis J.J. Sm. in *Bull. Jard. Bot. Buitenzorg* ser. 2, 25: 3 (1917). Type: Sulawesi, Kolaka, *Bünnemeijer* 10832 (holo. BO; iso. L). **Plate 4F.**

Coelogyne platyphylla Schltr. in *Notizbl. Bot. Gart. Berlin-Dahlem* 8: 118 (1922) & in *Repert. Spec. Nov. Regni Veg. Beih.* 21: 129 (1925). Type: Sulawesi, Dongala, cult. Becker s.n. (holo. B†).

DESCRIPTION. *Pseudobulbs* close together to 2.5 cm apart, cylindric, apex attenuated, 4 ridged when young, 5.5–14 cm long. *Leaves* 1 or 2, obovate-lanceolate, acuminate, coriaceous, plicate on the upper side, furrowed beneath, 33–62 × 8.1–14 cm with 1–4 cm long, grooved petiole. *Inflorescence* synanthous with partially to entirely developed leaves, peduncle during flowering enclosed at base by petiole and scales of young shoot, peduncle elongated, slender, 27–30.5 cm long, rachis (sub) erect, zig-zag, slightly arcuate, 3.5–11.5 cm long, 3- to 7-flowered, flowers opening in succession, floral bracts ovate-lanceolate, acuminate, deciduous. *Flowers* with pale green sepals and petals, mid-lobe dark brown, keels dark-brown, column green edging to orange at apex, hood margins citron yellow. *Dorsal sepal* oblong, acuminate, 11–15-nerved, 4.9–5.6 × 1.6–2.1 cm. *Lateral sepals* oblong, acuminate, 10–16-nerved, 4.6–5.3 × 1.3–1.6 cm. *Petals* oblique, linear, acuminate, 3-nerved, 4.9–5.4 × 0.28–0.35 cm. *Lip* 3-lobed, side-lobes erect, semi-orbicular, front obtuse, front margin slightly erose or entire, mid-lobe base expanding from a short broad claw, elliptic, tip acute, retuse, margins slightly erose, callus of 3 keels extending from basal quarter of lip, median keel terminates half to two thirds the way to base of mid-lobe, lateral keels terminate near tip of mid-lobe, all keels widened along crest, many slender, tapering, sometimes branched, elongate and sometimes plate-like projections, on mid-lobe lateral keels become one or two irregularly interrupted rows of the projections with papillae but without hairs. *Column* arcuate, 3.6–3.7 cm long, expanding into winged hood with truncate, rounded, irregularly dentate front margin, notched at sides.

DISTRIBUTION. Sulawesi (Kolaka, Gunung Galesong).
HABITAT. Epiphytic in forest and along roadside.
ALTITUDE. Sea level–1000 m.
FLOWERING. February–March.
NOTES. The epithet refers to the island of Sulawesi, formally the Celebes, where the type specimen was collected. The species is recognised by the dark brown lip with elongate, tapering projections on the keels and broad side-lobes with obtuse front margins.

40. COELOGYNE FORMOSA

Coelogyne formosa Schltr. in *Orchis* 6: 112 (1912). Type: Sumatra, Belawan, cult. von Fuerstenberg s.n. (holo. B†).

DESCRIPTION. *Pseudobulbs* on a strong, short rhizome, close together, ovoid, obtuse, 4-angled, 3–3.5 × 2–2.5 cm. *Leaf* 1, obtuse, erect and spreading, elliptic, acuminate, 15–17 × 5–6.5 cm, base gently narrowing into short petiole. *Inflorescence* synanthous with partially to entirely developed leaf, peduncle during flowering enclosed at base by petiole and scales of young shoot, peduncle and rachis arched and spreading, 10 cm long, few-flowered, flowers opening in succession, flower bracts deciduous. *Flowers* about 5 cm across, not wide spreading, sepals and petals flesh coloured, lip dark brown. *Dorsal sepal* oblong-ligulate, obtuse, curved inwards, dorsal keel beneath, about 4.5 cm long. *Lateral sepals* oblique, oblong-ligulate, obtuse, curved inwards, dorsal keel beneath, about 4.5 cm long. *Petals* oblique, linear, acute, about 4.5 cm long. *Lip* oval, 4.5 × 2.5 cm, base hollowed, 3-lobed, side-lobes shortened, erect, rounded, mid-lobe kidney-shaped, tip apiculate, reflexed, 1.5 × 2 cm, callus of 2 (3) keels, undulate, entire, extending from base of lip to base of mid-lobe, somewhat parallel, median keel straight and short, terminating at base of mid-lobe. *Column* semi-terete, smooth, 3 cm long, tip expanded slightly.

DISTRIBUTION. Sumatra (East Coast).
HABITAT. Epiphytic on trees in lower montane forest.
ALTITUDE. 1300 m.
FLOWERING. July.
NOTES. The epithet refers to the fine form of the flower; a quality exhibited by most sect. *Speciosae* Lindl. species. This species was not included by Gravendeel & de Vogel (1999) in their revision of Section *Speciosae*, possibly because the type specimen was not found.

41. COELOGYNE FRAGRANS

Coelogyne fragrans Schltr. in *Repert. Sp. Nov. Regni Veg. Beih.* 1: 102 (1911). Types: New Guinea, Kani Range, *Schlechter* 18083, 18838, 18216 (syn. B†; isosyn. E, K, NSW). **Plate 5A.**

DESCRIPTION. *Pseudobulbs* borne at intervals on a creeping rhizome, 1.5 cm apart, narrowly ovoid or oblong, in time 4-angled, 6.5–10 cm long. ***Leaves*** 1 or 2, erect, lanceolate, acuminate, 5–7-nerved, 21–38.5 × 3.8–6.8 cm with 3–5 cm long, grooved petiole. ***Inflorescence*** synanthous with partially to entirely developed leaves, peduncle during flowering enclosed at base by petiole and scales of young shoot, peduncle erect, bare, 18–38 cm long, rachis (sub) erect, zig-zag, 2.6–9.5 cm long, 2- to 6-flowered, flowers usually opening 2 to 4 simultaneously but sometimes in succession, floral bracts ovate-lanceolate, acute, persistent or deciduous. ***Flowers*** fragrant, with light green to golden yellow sepals and petals, lip greenish-white or cream-coloured, brownish to orange to yellow markings on mid-lobe and side-lobes, keels brown, warts on keels brown to light orange, column green shading to bright yellowish-green on hood, margin of hood orange. ***Dorsal sepal*** oblong, acuminate, 7–9-nerved, 3.4–4.9 × 0.9–1.7 cm. ***Lateral sepals*** oblique, oblong, acuminate, 6–7-nerved, 3.3–4.6 × 0.9–1.4 cm. ***Petals*** oblique, linear, acuminate, 3-nerved, 3.1–4.5 × 0.2–0.5 cm. ***Lip*** 3-lobed, side-lobes front obtuse, margin erose, mid-lobe expanding from broad, short claw, elliptic, tip acute, retuse, callus of 3 keels, curved, crenulate, fleshy, extending from base of lip to middle of mid-lobe, median keel extending to base of mid-lobe, keels consisting of two interrupted, undulating rows of warts with papillae but no hairs, median keel similar but a single row. ***Column*** lightly arcuate, glabrous, 2.4–2.7 cm long, expanding into winged hood, truncate front, rounded, notched at sides.

DISTRIBUTION. Irian Jaya, Papau New Guinea (Kani, Bismarck & Finisterre Ranges).
HABITAT. Epiphytic in lower montane forest.
ALTITUDE. 100–2000 m.
FLOWERING. April–November.
NOTES. The epithet refers to the fragrance of the flowers. See note under *C. carinata* about the possible confusion between the two species.

42. COELOGYNE FUERSTENBERGIANA

Coelogyne fuerstenbergiana Schltr. in *Orchis* 8: 131 (1914). Type: Sumatra, cult. von Fuerstenberg s.n. (holo. W). **Plate 5B.**

DESCRIPTION. *Pseudobulbs* on a stout, woody rhizome, at intervals, erect, ovoid, smooth, 4-angled, about 7 cm long, enclosed at base with bracts when young. ***Leaves*** 2, erect, lanceolate, acute, with 7–9 nerves, 30 × 4 cm with 4 cm long petiole. ***Inflorescence*** synanthous with partially to entirely developed leaves, peduncle during flowering enclosed at base by petiole and scales of young shoot, peduncle short, rachis likewise, few-flowered, flowers opening in succession. ***Flowers*** not opening widely. ***Sepals*** oblong, obtuse, dorsal keel beneath, 4.5 cm long. ***Petals*** narrowly linear, somewhat acute, slightly shorter than sepals. ***Lip*** oblong, 3-lobed, side-lobes very small, obtuse, shortened, mid-lobe ovate, somewhat acute, 4.3 cm long, 1.8 cm across, tip somewhat beaked, shortly 2-lobed, callus of 3 parallel keels extending from base of lip to tip of mid-lobe, keels increase in elevation then diminish. ***Column*** 1.7 cm long, kidney-shaped tip forming hood, rounded.

DISTRIBUTION. Sumatra.
HABITAT. Epiphytic.
ALTITUDE. Not known.
FLOWERING. November, flowered in cultivation.

NOTES. This species is named after the collector von Fuerstenberg. Gravendeel & de Vogel (1999) did not include this species in their revision of Section *Speciosae*, possibly because the type specimen was not found.

43. COELOGYNE GUAMENSIS

Coelogyne guamensis Ames in *Philipp. J. Sci.* 9: 11 (1914). Type: Mariana Islands, Guam, Experimental Stn., *J.B. Thompson* 195 (holo. AMES).

Coelogyne palawensis Tuyama in *J. Jap. Bot.* 17: 506 (1941). Type: Carolina Islands, Palau, Baobeltaob, Ngatpang (Gaspan), *Tuyama* s.n. (holo. TI).

DESCRIPTION. **Pseudobulbs** on a stout, sheathed rhizome, 1 cm apart, cylindrical-ovoid, 5.1–8 cm long. **Leaves** 2, lanceolate, acuminate-acute, 3–(5)–7-nerved, 26–36 × 3.9–9.4 cm with 2.5–5 cm long, grooved petiole. **Inflorescence** synanthous with partially to entirely developed leaves, peduncle during flowering enclosed at base by petiole and scales of young shoot, peduncle 8.5–20 cm long, rachis arcuate, zig-zag, 1.6–12 cm long, incrassate between internodes, 4- to 11-flowered, flowers opening in succession, floral bracts lanceolate to ovate-lanceolate, acuminate, deciduous. **Flowers** colour not known. **Dorsal sepal** ovate-lanceolate, acuminate, fleshy, 10–12-nerved, 3.7–4.2 × 0.8–1cm. **Lateral sepals** ovate-lanceolate, acuminate, fleshy, 9–11-nerved, 3.5–4.1 × 0.85–1.1 cm. **Petals** linear, 3.5–3.8 × 0.2–0.4 cm. **Lip** 3-lobed, side-lobes erect, obtuse, mid-lobe with short claw, broadly ovate, 1.4–1.6 × 1.3–1.7 cm, margin slightly erose, callus of 5 keels extending from base of lip, 3 median keels parallel at base of lip and become 2 keels extending onto mid-lobe, lateral keels irregular and terminate at base of mid-lobe, keels undulate, initially entire, plate-like, becoming prominent, irregularly shaped transverse ridges with papillae but no hairs. **Column** 1.2–1.9 cm long, expanding into winged hood, not notched at sides, rounded in front.
DISTRIBUTION. Palau Islands (Babel Thuap), Mariana Islands (Guam, Rota).
HABITAT. Epiphytic in damp rainforests.
ALTITUDE. 425 m.
FLOWERING. July–September.
NOTES. The epithet refers the Pacific island of Guam, the location where the type specimen was collected. The species can be recognised by the five undulating, plate-like keels and an entire margin to the lip.

44. COELOGYNE LYCASTOIDES

Coelogyne lycastoides F. Muell. & Kraenzl. in *Oesterr. Bot. Z.* 45: 179 (1895). Type: Samoa, *Betche* 24 (holo. B†). **Plate 5C.**

Coelogyne whitmeei Schltr. in *Repert. Sp. Nov. Regni Veg. Beih.* 11: 41 (1912). Type: Samao, *Whitmee* s.n. (holo. B†; iso. K).

DESCRIPTION. **Pseudobulbs** on a stout rhizome, to 1.2 cm apart, ovoid-cylindric, 4–8 cm long, obtusely 4-angled. **Leaf** 1, ovate-lanceolate, acuminate, 7–9-nerved, 26–43 × 8–11.5 cm with 1.5–3 cm long petiole. **Inflorescence** synanthous with partially to entirely developed leaves, peduncle during flowering enclosed at base by petiole and scales of young shoot, peduncle 24–27 cm long, rachis (sub) erect, zig-zag, 4–5.5 cm long, internodes incrassate, 2- to 4-flowered, flowers opening (almost) simultaneously, floral bracts ovate-oblong, acute, persistent. *Flowers* with sepals and petals pale green, lip whitish and mainly tinged orange-brown except for the margins and the apical part of the mid-lobe, keels orange-brown, column whitish-green. **Dorsal sepal** ovate-oblong, acuminate, 9-nerved, 4.2–4.5 × 1.5–1.7 cm. **Lateral Sepals** ovate-lanceolate, acuminate, 8-nerved, dorsal keel rounded, 4.2–4.6 × 1.3–1.45 cm. **Petals** slightly recurved, linear, obtuse, 4.2–4.3 × 0.35–0.7 cm. **Lip** 3-lobed, slightly saccate, side-lobes narrowly elliptic, obtuse in front, not extended in front, no sinus, margins slightly irregularly erose, mid-lobe oblong, tip acute to obtuse, callus of 5–7 keels, elevated rows of undulating, irregularly shaped, rounded warts with papillae but no hairs, the 3 median keels parallel from base of lip to base of mid-lobe then diverge and converge, centre median keel shorter and less elevated, 2–4 lateral keels extend from halfway to base of mid-lobe, median keels and 2 lateral keels continue on to mid-lobe and become a more tightly form of rounded warts and papillae with no hairs. **Column** 2.3–2.7 cm long, expanding into winged hood, notched at sides, tip rounded, recurved.

DISTRIBUTION. Vanuatu, New Caledonia, Fiji, Tonga, Samoa.
HABITAT. Epiphytic in lowland rainforest and open woodland.
ALTITUDE. 300–1550 m.
FLOWERING. January–August.
NOTES. The epithet refers to the flower having the form of a *Lycaste* species, Lycaste being the beautiful daughter of King Priam of Troy. The species may be confused with *C. macdonaldii* but it can be recognised by the unifoliate pseudobulbs and the nearly confluent rows of undulating, irregularly rounded warts on the lip. The type specimen was initially at MEL, then transferred to B.

45. COELOGYNE MACDONALDII

Coelogyne macdonaldii F. Muell. & Kraenzl. in *Oesterr. Bot. Zeit.* 44: 209 (1894). Type: Vanuata, *Macdonald* s.n. (holo. B†). **Plate 5D.**

Coelogyne lamellata Rolfe in *Bull. Misc. Inf., Kew*: 36 (1895). Type: New Hebrides [Vanuatu], cult. Sander s.n. (holo. K).

DESCRIPTION. Pseudobulbs on a thick rhizome, 1.8 cm apart, ovoid-cylindric, 5–7.5 cm long, when young 4-angled. **Leaves** 2, ovate-lanceolate, acuminate, with 5–7 main nerves, 20–29.8 × 5.1–11.5 cm with 1.6–2.5 cm long petiole. **Inflorescence** synanthous with partially to entirely developed leaves, peduncle during flowering enclosed at base by petiole and scales of young shoot, peduncle 11.5–16.5 cm long, rachis (sub) erect, zig-zag, 2.7–5.6 cm long, 3- to 5-flowered, flowers opening (mostly) simultaneously, floral bracts ovate-oblong, acuminate, persistent. **Flowers** with sepals and petals pale green to whitish, lip greenish-white, inside light orange-brown with lighter spots, keels warts cream but on mid-lobe darker brown. **Dorsal sepal** ovate-oblong, obtuse, 9-nerved, 4.2–5.2 × 1.55–2.2 cm. **Lateral sepals** ovate-oblong, obtuse, 9–10 nerves, 4–5 × 1.5–1.6 cm. **Petals** slightly recurved, linear, acute, 4.1–4.8 × 0.41–0.47 cm. **Lip** 3-lobed not saccate, side-lobes elliptic, obtuse in front, with an acute sinus, mid-lobe obovate to elliptic to ovate from a short claw, tip acute to acuminate, margin slightly erose, callus of 9–13 keels, verrucose with papillae but no hairs, extending from base of lip to middle of mid-lobe, 4–6 lateral keels not on mid-lobe. **Column** 1.9–2.3 cm long, expanding into winged hood, tip rounded, recurved.

DISTRIBUTION. Banks Islands (Vanua Lava), Vanuatu (Espiritu Santo, Ambae, Pentecost, Efate, Erromango), Fiji.

HABITAT. Epiphytic on ridge-tops in lowland rainforest and lower montane forest.

ALTITUDE. 300–1100 m.

FLOWERING. September–March.

NOTES. This species named after the collector the Rev. M. MacDonald. See note under *C. lycastoides* with regard to confusion between the species. The type specimen was initially at MEL, then transferred to B.

46. COELOGYNE RUMPHII

Coelogyne rumphii Lindl., *Fol. Orch.-Coelog.* 1: 14 (1854). Type: *Rumphius*, Herb. Amboinense 6, t. 48 (1750). **Plate 5E.**

Angraecum nervosum Rumphius in *Herb. Amboin.* 6, t. 48 (1750), *nom. pre-Linn.*
Coelogyne psittacina Rchb.f. in *Xenia Orch.* 2: 141 (1874). Type: Maluku, Ambon, *Doleschall* 90 (holo. W).
C. psittacina Rchb.f. var. *huttonii* Rchb.f. in *Gard. Chron.* 32: 1053 (1870). Type: Maluku, *Hutton* s.n. (holo. W).

DESCRIPTION. Pseudobulbs on a sheathed, stout rhizome, 1.5 cm apart, ovoid-cylindric, tapered towards tip and base, 4-angled, 5.4–13.4 cm long. **Leaf** 1, erect, lanceolate, acuminate, with 7 main nerves, plicate, 28–59 × 8–12.9 cm with a 2.5–3.8 cm long petiole. **Inflorescence** synanthous with partially to entirely developed leaf, peduncle during flowering enclosed at base by petiole and scales of young shoot, peduncle erect, 24–51 cm long, rachis (sub) erect, zig-zag, overall 1.9–7.5 cm long, 2- to 6-flowered, flowers opening in succession, floral bracts ovate-lanceolate to lanceolate, mucronate, deciduous. **Flowers** with yellowish-green to yellowish-cream sepals and petals, lip cream to whitish, at base orange-yellow, side-lobes tinged red to orange-brown, inside veined red-brown, sinus and keels red-brown, base of mid-lobe with transverse W-shaped light brown band, column cream to light green with orange margin, front with red-brown spots, hood bright yellow. **Dorsal sepal** oblong, acuminate, 9–11-nerved, dorsal keel prominent beneath, 3.9–5.07 × 1.5–1.75 cm. **Lateral sepals** oblong, acuminate, 9-nerved, median keel prominent beneath, 3.76–4.73 × 1.18–1.43 cm. **Petals** reflexed from base, linear, acute, 3.8–4.6 × 0.3–0.4 cm. **Lip** not saccate 3-lobed, divided by deep side-lobes, side-lobes erect, obtuse, short claw, mid-lobe elliptic to obrhomboid, 2 deeply divided, rounded lobes projecting forwards, tip deeply reflexed, callus of 3 keels extending from base of lip, lateral keels initially parallel then diverging and converging on mid-lobe, comprise 5 rows of slender, tapering, undulating, plate-like projections with papillae but no hairs, median keel only extends to base of mid-lobe, lateral keels terminate at middle of mid-lobe. **Column** arcuate, 2.37–2.86 cm long, expanding into winged hood, notched at sides, front rounded.

DISTRIBUTION. Maluku (Ambon, Buru, Seram).

HABITAT. Epiphytic on trees in riverine forest.

ALTITUDE. 100–1100 m.

FLOWERING. August–November.

NOTES. The species is dedicated to G.E. Rumpf (Rumphius), (1628–1702) and his association with the Island of Ambon in the Maluku group of islands. The species is recognised by the five rows of slender, tapering, undulating plate-like projections on the keels.

47. COELOGYNE SALMONICOLOR

Coelogyne salmonicolor Rchb.f. in *Gard. Chron.* ser. 2, 20: 328 (1883). Type: Sunda Islands, cult. Veitch, *Curtis* 410 (holo. W). **Plate 5F.**

Coelogyne speciosa (Blume) Lindl. var. *salmonicolor* (Rchb.f.) Schltr., *Orchideen* 146 (1915).
C. bella Schltr. in *Bot. Jahrb. Syst.* 45: 104, 5 (1911). Type: Sumatra, *Schlechter* 15921 (holo. B†).
C. salmonicolor Rchb.f. var. *virescentibus* J.J. Sm. ex Dakkus in *Orch. Ned. Ind.* 3: 89 (1935). Type: Provenance unknown (type not designated).

DESCRIPTION. ***Pseudobulbs*** on a short rhizome, to 1 cm apart, ovoid, 4 ridged, 2.5–4.2 cm long. ***Leaf*** 1, erect, lanceolate, acuminate, with 5 main nerves, 21–27 × 3.9–4.4 cm with 2.1–3.2 cm long, grooved petiole. ***Inflorescence*** synanthous with partially to entirely developed leaf, peduncle during flowering enclosed at base by petiole and scales of young shoot, peduncle slender at base, thickening towards top, 2.2–3.8 cm long, initially erect but then pendulous, rachis (sub) erect, zig-zag, 2–3.7 cm long 2- to 4-flowered, flowers opening (almost) simultaneously, floral bracts ovate-oblong, acuminate, persistent. ***Flowers*** with pale salmon-pink or creamy-white sepals and petals, lip pale salmon-pink or creamy-white, side-lobes orange-brownish veins, keels and central median streak on the lip of similar colour, column cream or creamy-yellow. ***Dorsal sepal*** ovate-oblong, obtuse to acute, 9–11-nerved, 2.8–3.3 × 1–1.4 cm. ***Lateral sepals*** oblique, ovate-oblong, obtuse, 8-nerved, 3.1–3.2 × 1–1.2 cm. ***Petals*** reflexed, linear, obtuse, 3-nerved, 3.1–3.2 × 0.23–0.28 cm. ***Lip*** 3-lobed, side-lobes erect, semi-ovate, obtuse, margin slightly erose, sinus acute, mid-lobe obovate to orbicular with broad, short claw, tip reflexed, obtuse, margins slightly erose, callus of 2–4 keels extending from base of lip, widened along crest, with two slightly elevated rows of irregularly shaped, tapering projections on each side of crest, with a longitudinal groove with hairs on rims, with papillae, lateral keels parallel near base of lip diverge then converge on mid-lobe, median keels only extend two thirds the way to base of mid-lobe and are less elevated than lateral keels, on mid-lobe lateral keels merge to form tightly packed irregularly undulating plate-like projections with papillae and hairs. ***Column*** arcuate, 1.8–2.1 cm long, expands into winged hood, irregularly dentate, notched at sides, front rounded.

DISTRIBUTION. Sumatra.

HABITAT. Epiphytic in hill forest, lower montane forest and semi-open areas on both wetter and drier side of mountains.

ALTITUDE. 900–1500 m.

FLOWERING. At intervals from February–December.

NOTES. The epiphet refers to the salmon colour of the flowers. The species is recognised by the pale pink flowers with 2–4 keels with small hairs on the lip. It differs from *C. speciosa* subsp. *incarnata* because of the longitudinally grooved keels.

48. COELOGYNE SEPTEMCOSTATA

Coelogyne septemcostata J.J. Sm. in *Icon. Bogor.* 2: 23, t. 106A (1903). Type: Sumatra, *Nieuwenhuis* s.n. (holo. BO). **Plate 6A.**

Coelogyne speciosa sensu Ridl. in *Flora* 4: 134 (1924), *non* (Blume) Lindl.
C. membranifolia Carr in *Gard. Bull. Straits Settlem.* 7: 2 (1932). Type: Peninsular Malaysia, Pahang, Tembeling, *Carr* 123 (holo. SING).

DESCRIPTION. ***Pseudobulbs*** on a creeping, stout, few-branched rhizome, 0.5–1 cm apart, ovoid, 4-angled, sides concave, pale green, to 4.3–9 cm long, initially covered with membraneous pale green bracts. ***Leaf*** 1, oblong, acuminate, plicate, with 5–7 main nerves, to 23–42 × 5.8–12.5 cm with 2.2–4.7 cm long, deeply grooved petiole. ***Inflorescence*** synanthous with partially to entirely developed leaf, peduncle during flowering enclosed at base by petiole and scales of young shoot, peduncle slender, terete, thickened upwards, 8.8–39 cm long, rachis (sub) erect to pendulous, zig-zag, gradually elongating, 2–32 cm long, 2- to 4(–22)-flowered, flowers opening in succession, floral bracts ovate-lanceolate, deciduous. ***Flowers*** with green to very pale green sepals and petals, lip whitish to medium cream, side lobes orange to dull orange-brown, keels orange to orange-brown, duller brown in front, projections on keels orange, mid-lobe creamy-white with transverse brown band at base, column cream, front tinged orange to orange-brown. ***Dorsal sepal*** ovate-oblong, mucronate, 11–13-nerved, strongly keeled, 3.4–4.8 × 1.6–2.3 cm. ***Lateral sepals*** oblong, mucronate, 9–11-nerved, strongly keeled, 3.4–4.6 × 1.5–1.9 cm. ***Petals*** slightly recurved, linear, mucronate, 3-nerved, 4.4 × 0.35 cm. ***Lip*** 3-lobed, base not saccate, side-lobes front obtuse, margin at base slightly irregularly erose to entire, sinus acute, mid-lobe convex, elliptic, 0.5–0.8 × 1.3–1.9 cm with broad, short claw, slightly raised, tip acute and with warts, margin slightly erose, recurved, callus of 5–7(–8) keels extending from base of lip, only 2 median keels continue on to mid-lobe and terminate 0.9–1.1 cm from tip, all keels more or less widened along crest, with a longitudinal grove, margins of keels drawn out in many hair-like projections 0.7–1.5 cm long, with papillae but without hairs, median keels diverge and converge on mid-lobe and become plate-like projections with papillae but without hairs. ***Column*** erect, arched from middle, 2.7–3.4 cm long, expanding to winged hood, margin irregularly dentate, front obtuse, laterally notched and with two small projections at attachment of wings.

DISTRIBUTION. Peninsular Thailand, Peninsular Malaysia (Johor, Pahang), Borneo (Kalimantan, Sarawak, Sabah, Brunei).

HABITAT. Epiphytic, rarely terrestrial in shady lowland rain forest, riverine forest, lower montane forest, usually on shrubs and trees 1–3 m above the ground.

ALTITUDE. 50–2278 m.

FLOWERING. February–July.

NOTES. The epiphet refers to the number of keels on the lip of the type specimen. However, Gravendeel noted that the number of keels varies between 5 and 8. The species is recognised by the 5–8 keels on the lip with long hairs at their margins and a strongly curved rachis.

49. COELOGYNE SPECIOSA

Coelogyne speciosa (Blume) Lindl., *Gen. Sp. Orch. Pl.* 39 (1830). Type: Java, Gunung Salak, *Blume* s.n. (holo. BO). **Plate 6B & C.**

Chelonanthera speciosa Blume, *Bijdr*, 384 (1825).
Pleione speciosa ((Blume) Lindl.) Kuntze in *Rev. Gen. Pl.* 2: 680 (1891).
Coelogyne speciosa (Blume) Lindl. var. *albicans* H.J. Veitch, *Man. Orchid Pl.* 50, pl. 6 (1890). Type: Veitch (lecto. plate cited above).
C. speciosa (Blume) Lindl. var. *alba* Hort. in *Gard. Chron.* ser. 3, 37: 205 (1905). Type: not designated.
C. speciosa (Blume) Lindl. var. *rubiginosa* Hort. in *Orch. Rev.* 30: 37 (1922). Type: not designated.

subsp. **speciosa**

DESCRIPTION. **Pseudobulbs** on a stout, sheathed rhizome, to 0.8 cm apart, ovoid, 4 flat sides and rounded angles, to 4.4–7 cm long. **Leaves** 1 or 2, lanceolate, acuminate or mucronate, with 3–5 main nerves, 22.5–35 × 3.7–8.4 cm with 2.5–4 cm long petiole. **Inflorescence** synanthous with partially to entirely developed leaves, peduncle during flowering enclosed at base by petiole and scales of young shoot, peduncle 14–20 cm long, rachis initially (sub) erect, zig-zag at which time flowers mostly hidden by the leaf (leaves) as the first flower expands at same time, but later pendulous, 4–7.5 cm long, 3- to 8-flowered, flowers open in succession, floral bracts ovate-oblong, deciduous. **Flowers** with dense light green to yellowish-green sepals and petals, transparent; lip white to cream red to orange-brown lines on the base; column light green, front of stalk tinged brownish, apex pale yellowish, wings cream coloured. **Dorsal sepal** oblong, emarginate, 9–11-nerved, 4.96–5.5 × 1.83–2.06 cm. **Lateral sepals** oblong, emarginate, 8–10-nerved, 4.75–5 × 1.46–1.7 cm. **Petals** narrowly linear, emarginate to acute, reflexed, 5.2–5.67 × 0.25–0.33 cm. **Lip** not saccate at base, 4.43–5.3 × 3.31–4.27 cm, 3-lobed, side-lobes almost covering column, front rounded to obtuse, front margin at base irregularly erose, with broadly rounded to acute sinus, mid-lobe semi-orbicular to transversally elliptic with a broad, short claw, tip retuse to emarginate, apex with warts, margin more or less erose, sides pronounced as side-lobes with radiating rows of warts, callus of 3 keels extending from base of lip to and converging on mid-lobe 0.9–1.2 cm from tip of mid-lobe, keels widened along crest and lateral keels elevated and with projections and hairs. **Column** arcuate, 2.9–3.7 cm long, expanding into hood with irregularly dentate apical margin, wings and laterally notched at base of wings, middle part rounded.

DISTRIBUTION. Java, Lesser Sunda Islands (Flores).

HABITAT. Epiphytic, rarely terrestrial on trees in hill forest, lower montane forest and in semi-open areas on both wetter and drier sides of mountains.

ALTITUDE. 760–2000 m.

FLOWERING. February, April–July, October–December.

NOTES. The epiphet refers to the showy, large nature of the flowers. There is a reference by Guillaumin (*Bull. Mus. Nat. Hist. Paris* sér. 2 (27): 233 (1955)) to *C. speciosa* Lindl. having been found in Vietnam (Da Lat). The only species found in Vietnam which could have been thought to belong to sect. *Speciosae* are *C. lawrenceana* Rolfe and *C. dichroantha* Gagnep. In my judgement both these species belong elsewhere.

subsp. **fimbriata** (J.J. Sm.) Gravendeel in *Blumea* 44: 301 (1999). Type: Sumatra, Padang Pandjang, *Storm van 's Gravesande s.n.* (holo. BO).

Coelogyne speciosa (Blume) Lindl. var. *fimbriata* J.J. Sm. in *Teysmannia* 31: 254 (1920).

DESCRIPTION. *Pseudobulbs* on a stout, sheathed rhizome, to 0.7 cm apart, ovoid, 4 flat sides and rounded angles, to 3.2–7.2 cm long. ***Leaf*** 1, obovate-lanceolate to linear-lanceolate, acuminate to mucronate, with 5 main nerves, 26–43.5 × 4.5–8 cm with 4.5–10.5 cm long petiole. ***Inflorescence*** synanthous with partially to entirely developed leaf, peduncle during flowering enclosed at base by petiole and scales of young shoot, peduncle 8–27 cm long, rachis initially (sub) erect, zig-zag at which time flowers mostly hidden by the leaf as the first flower expands at same time as leaf, but later pendulous, 2–37 cm long, 2- to 4- (20-) flowered, flowers open in succession, floral bracts ovate-oblong, deciduous. ***Flowers*** with sepals and petals ochrish-yellow, lip cream with reticulate pattern of brown lines on inside of side-lobes which is visible on outside, inside median orange, keels at base of lip orange, to the front brown; column yellowish, in front with a few brown markings. ***Dorsal sepal*** lanceolate, acuminate, 9–11-nerved, 4.7–5 × 1.5–1.9 cm. ***Lateral sepals*** ovate-lanceolate, acuminate, 8–9-nerved, 4.3–4.8 × 1.2–1.5 cm. ***Petals*** narrowly linear, acuminate, 4.4–4.9 × 0.25–0.35 cm. ***Lip*** not saccate at base, 3.3–4.4 × 2.7–3.5 cm, 3-lobed, side-lobes acute to rounded, front margin at base slightly to extremely irregularly erose, with rounded to acute sinus, mid-lobe orbicular to obrhomboid, claw (if present) broad and short, tip retuse, apex with warts, margin fringed with elongate projections, with papillae but without hairs, sides pronounced as side-lobes with a few irregularly placed warts, callus of 2–3 keels extending from base of lip to 0.9–1.2 cm from tip of mid-lobe, keels widened along crest and lateral keels elevated and with projections and hairs, median keel with papillae but without hairs. ***Column*** arcuate, 2.8–3.2 cm long, expanding into hood with more or less truncate apical margin.

DISTRIBUTION. Sumatra.
HABITAT. Epiphytic in rainforest.
ALTITUDE. 800–1100 m.
FLOWERING. January, March, May, June.
NOTES. The epithet *fimbriata* refers to the fringed margins of the mid-lobe.

subsp. **incarnata** Gravendeel in *Blumea* 44: 303 (1999). Type: Sumatra, Gunung Mamas, *de Wilde & de Wilde-Duyfjes* 15767 (holo. L; iso. K).

Coelogyne speciosa Lindl. in *Bot. Reg.* 33, t. 23 (1847), *non* (Blume) Lindl.
C. speciosa (Blume) Lindl. var. *major* C.F. Sander, F.K. Sander & L.L Sander, *Sander's Orch. Guide* 128 (1927). Type: not designated.

DESCRIPTION. *Pseudobulbs* on a stout, sheathed rhizome, to 1 cm apart, ovoid, 4 flat sides and rounded angles, to 4.7–7.3 cm long. ***Leaves*** 1 or 2, lanceolate, acuminate or mucronate, with 3–5 main nerves, 19–38 × 3.4–6 cm with 3–5.2 cm long petiole. ***Inflorescence*** synanthous with partially to entirely developed leaves, peduncle during flowering enclosed at base by petiole and scales of young shoot, peduncle 9.5–17.5 cm long, rachis initially (sub) erect, zig-zag at which time flowers mostly hidden by the leaf (leaves) as the first flower expands at same time, but later pendulous, 1.4–3.5 cm long, 2- to 3- (11-) flowered, flowers open in succession, floral bracts ovate-oblong, deciduous. ***Flowers*** with sepals and petals greenish-cream tinged salmon to salmon to brownish-salmon; lip outside salmon with a brownish tinge, hypochile deep brown to red-brown with lighter spots, at base yellowish, side-lobes brown to red-brown with creamy-white spots, lateral keels red-brown, median keel yellow, projections on keels creamy-white, mid-lobe white to creamy-white with, at base, some brown markings, margins of claw red-brown; column white to greenish-cream, front of stalk tinged brown to red-brown. ***Dorsal sepal*** oblong, emarginate, 9-nerved, dorsal nerve grooved above, 5.45–7.2 × 2–2.4 cm. *Lateral sepals* oblique, oblong, emarginate, 8–9-nerved, 5.68–6.8 × 1.75–2 cm. *Petals* linear, emarginate to acute, 1-nerved, 5.37–6.8 × 0.23–0.4 cm. *Lip* 4.59–6.1 × 4–4.5 cm, 3-lobed, side-lobes in front rounded to obtuse, front margin at base irregularly erose, with broadly rounded to acute sinus, mid-lobe semi-orbicular to transversely elliptic, 1.7–2.4 × 2.33–3.1 cm with a broad, short claw, tip retuse to emarginate, apex with warts, margin more or less erose, sides pronounced, similar to side-lobes, with radiating rows of warts, callus of 3 keels extending from base of lip to near tip of mid-lobe, widened along crest, lateral keels with projections and hairs. *Column* thick, arcuate, 3.27–4.3 cm long, expanding to form hood with wings and an irregularly dentate apical margin, notched where wings attached, middle part rounded, slightly recurved.

DISTRIBUTION. Java, Sumatra.

HABITAT. Epiphytic in hill forest, lower montane forest and semi-open areas on both wetter and drier sides of mountains.

ALTITUDE. 900–1500 m.

FLOWERING. March, October.

NOTES. The epiphet *incarnata* refers to the flesh-coloured flowers. Horticulturists often use the epiphet *salmonicolor* for *C. speciosa* subsp. *incarnata*. To avoid further confusion with the distinct species *C. salmonicolor* Rchb.f. (syn. *C. speciosa* (Blume) Lindl. var. *salmonicolor* (Rchb.f.) Schltr.), the epiphet *incarnata* has been chosen by B. Gravendeel.

50. COELOGYNE SUSANAE

Coelogyne susanae P.J. Cribb & B.A. Lewis in *Kew Bull.* 46: 317 (1991). Type: Solomon Islands, New Georgia, *Wickison* 40 (holo. K). **Plate 6D.**

DESCRIPTION. *Pseudobulbs* on a sheathed, thick rhizome, to 2 cm apart, elongate, ovate, 4–5 angled, 6.8–17 cm long. *Leaves* 2, linear-lanceolate, acuminate to mucronate, plicate, with 5–7 main nerves, 32–45 × 6.5–8.7 cm with 7 cm long petiole. *Inflorescence* synanthous with partially to entirely developed leaves, peduncle during flowering enclosed at base by petiole and scales of young shoot, peduncle erect, 7–41.5 cm long, rachis (sub) erect, zig-zag, 2- to 18-flowered, flowers opening in succession, floral bracts ovate-lanceolate, acute, deciduous. *Flowers* with sepals and petals pale yellow-green to creamy-green with orange or red-brown lip, pale yellow or white apex, column whitish with brown on ventral surface, fragrant. *Dorsal sepal* elliptic-lanceolate, obtuse, 13–15-nerved, keel at base beneath, 5–6.7 × 1.7–2.3 cm. *Lateral sepals* ovate-lanceolate, acute, 13–17-nerved, keel beneath, 5.6–6.2 × 1.6–1.95 cm. *Petals* slightly recurved, linear, acuminate, 3-nerved, 5–6.7 × 0.35–0.5 cm. *Lip* 4–5.5 × 3.1–3.6 cm, 3-lobed, not saccate, side-lobes in front obtuse, front margin at base slightly irregularly erose with obtuse sinus, mid-lobe elliptic to ovate to orbicular, 2.1–3.7 × 1.9–2.1 cm with a broad, relatively long claw, tip retuse, slightly raised, acute and with warts, margin slightly erose, recurved, callus of 11–13 keels each consisting of a row of irregularly shaped, elongate warts with papillae but without hairs on basal part of hypochile and a slightly elevated row of irregularly shaped horizontal ridges with papillae but without hairs on apical part of hypochile, not widened along crest, 5 median keels diverge at base of mid-lobe, then converge with the 3 innermost median keels continuing onto mid-lobe, 6–8 lateral keels only developed on hypochile, 7–9 keels continue onto mid-lobe with 2 longest terminating 1.2–1.5 cm from tip of mid-lobe. *Column* 2.25–2.35 cm long, expanding into winged hood with truncate apical margin, laterally notched where wings attached and above, middle part rounded, recurved.

DISTRIBUTION. New Britain, Bougainville, Shortland Islands, Solomon Islands (Guadacanal, Kolombangara, Mohira, New Georgia, San Cristobal).
HABITAT. Epiphytic, rarely terrestrial in lowland rain forest.
ALTITUDE. Sea leval–1250 m.
FLOWERING. April–January.
NOTES. This species was named after Sue Wickison, a botanical artist, who collected the type specimen in New Georgia. It is allied to *C. macdonaldii* F. Muell. & Kraenzl. but differs in having larger flowers, obtuse sinus and red- or orange-brown keels.

51. COELOGYNE TIOMANENSIS

Coelogyne tiomanensis M.R. Henderson in *Gard. Bull. Straits Settlem.* 5: 80 (1930). Type: Malaysia, Pulau Tioman, Gunong Rokan, *Henderson* 18397 (holo. SING). **Plate 6E.**

DESCRIPTION. *Pseudobulbs* on a tough, sheathed rhizome, to 3 cm apart, ovoid, 2–5 cm long. *Leaf* 1, lanceolate, acuminate to mucronate, with 2(–5) nerves, 21–32 × 3.2–6.5 cm with 3–6 cm long petiole. *Inflorescence* synanthous with partially to entirely developed leaf, peduncle during flowering enclosed at base by petiole and scales of young shoot, peduncle to 11.5–20 cm long, rachis (sub) erect, zig-zag, 1.1–10 cm long, 2- to 10-flowered, flowers opening in succession, floral bracts ovate-oblong, acuminate, deciduous. *Flowers* with sepals yellow to pale salmon, petals pale greenish to pale salmon, lip mid-lobe margin white with brown veins, centre dark brown, edges light brown, side-lobes paler with brown veins; column greenish with two faint brown streaks below, hood reddish-brown to orange. *Dorsal sepal* lanceolate, acuminate, 9-nerved, 3.1–4.7 × 0.8–1.2 cm. *Lateral sepals* ovate-lanceolate, mucronate, 8–9-nerved, 2.4–4.3 × 0.6–1.1 cm. *Petals* slightly recurved, linear, mucronate, 3-nerved, 3–4.2 × 0.15–0.19 cm. *Lip* 3-lobed, side-lobes in front rounded, front margin at base entire, with acute sinus, mid-lobe obrhomboid to obovate, 0.6–1.1 × 1.1–1.5 cm, with a broad, short claw, tip acuminate, slightly raised, without warts, margin entire, recurved, sides of mid-lobe as pronounced as side-lobes, without warts, callus of 2–3 keels extending from base of lip with lateral keels terminating 0.4–0.9 cm from tip of mid-lobe, all keels widened along crest, plate-like, undulating, with papillae but without hairs, lateral keels diverge at base of mid-lobe then converge, median keel (if present) only developed near base of lip at lower elevation than lateral keels. *Column* 1.8–2.7 cm long, expanding into winged hood with truncate apical margin, laterally notched, where wings attached, additional notch above, middle part rounded, slightly recurved.
DISTRIBUTION. Peninsular Malaysia (Pulau Tioman).
HABITAT. Epiphytic and lithophytic in montane dwarf forest.
ALTITUDE. 600–1040 m.
FLOWERING. May, August.
NOTES. The epiphet refers to the island of Tioman, which is off the east coast of Peninsular Malaysia where the type specimen was collected. The species is recognised by the dark brown lip and undulating, plate-like keels.

52. COELOGYNE TOMMII

Coelogyne tommii Gravendeel & O'Byrne in *Blumea* 44(2): 310–313 (1999). Type: cult. Singapore s.n. (holo. K; iso. L). **Plate 6F.**

Coelogyne tomiensis O'Byrne in *Malayan Orchid Rev.* 29: 33–35 (1995), *nom. invalid.*

DESCRIPTION. Pseudobulbs on a stout, sheathed, creeping rhizome, to 1.2 cm apart, cylindric to ovoid, obtusely 4-angled, 4.1–7.3 cm long. **Leaf** 1, lanceolate, acuminate, with 5 main nerves, 20.5–25 × 4.3–5.2 cm with 2.5–3.5 cm long petiole. **Inflorescence** proteranthous, peduncle during flowering enclosed at base by petiole and scales of young shoot, peduncle to 6 cm long, rachis (sub) erect, zig-zag, c. 3.8 cm long, 4- to 5-flowered, flowers opening (almost) simultaneously, floral bracts ovate-oblong, acute, deciduous. **Flowers** fragrant, sepals and petals light green to creamy-yellow, transparent, lip light brown to rusty brown, side-lobes darker red-brown veins on outside, forward margins almost black, mid-lobe light brown with creamy-white tip, keels brown tipped, column creamy-yellow to light green, front of stalk with 3 brown lines, tip white, wings cream coloured. **Dorsal sepal** lanceolate, retuse to acute, 9-nerved, 4–4.1 × 1.3–1.35 cm. **Lateral sepals** ovate-lanceolate, acute, 8–9-nerved, 3.7–3.9 × 1–1.1 cm. **Petals** slightly recurved, linear, acute, 3–5-nerved, 3.6–3.8 × 0.3–0.35 cm. **Lip** 3.3–3.5 × 2.4–2.6 cm, 3-lobed, side-lobes in front rounded, slightly diverging, front margin at base irregularly erose, with acute sinus, mid-lobe obrhomboid to orbicular, 1.3–1.4 × 1.8–1.85 cm, with broad, short claw, tip acute, with warts, slightly raised, margin slightly erose, recurved, sides of mid-lobe as pronounced as side-lobes, with warts, callus of 3 keels extending from base of lip, median keel terminates at base of mid-lobe, lateral keels terminate 1 cm from tip of mid-lobe, keels widened along crest, heavily fringed margin, with papillae, but without hairs, halfway along hypochile keels become slender, tapering, either or not branched, irregularly shaped projections with undulating irregular margin, with papillae, but without hairs, slightly divergent at base of mid-lobe then convergent on mid-lobe, median keel less elevated than lateral keels, a further 2 keels (if present) develop on and near base the mid-lobe only. **Column** arcuate, 2.2–2.7 cm long, expanding into winged hood with more or less truncate apical margin, irregularly dentate, laterally notched where wings attached, middle part rounded, slightly recurved.

DISTRIBUTION. Unknown, possibly Sumatra.
HABITAT. Not known.
ALTITUDE. Not known.
FLOWERING. April, October–December.
NOTES. The epiphet was changed from *tomiensis* to *tommii*, the correct orthographic form. It is named after Mr Tommy Sng who cultivated the type specimen in Singapore.

53. COELOGYNE USITANA

Coelogyne usitana Röth & Gruss in *Die Orchidee* 52(2): 79-80 (2001). Type: Philippines, Central-East Mindanao, coll. *Vilmoor Usita*, cult. Jürgen Röth 083742 (holo. HAL). **Plate 7A.**

DESCRIPTION. Pseudobulbs close together on a thick rhizome, narrowly ovoid, four-sharp-edges, to 10.5 cm long, to 1.5 cm diameter, 3–4 membraneous bracts from the base of the pseudobulb, sometimes to 10 cm long, bracts narrowly ovoid, converge to a narrow tip, central nerve is evident at tip. **Leaf** 1, broadly elliptical, acute, with 5–7 main nerves, to 37 × 10.5 cm with the base converging to a short petiole 1–3 cm long. **Inflorescence** synanthous, peduncle initially erect but as it extends, pendulous, to 30 cm long, 0.2 cm across, rachis zig-zag, slightly flattened, internodes 1.5–2 cm apart, to 25 cm long so that overall length to 55 cm, to 20-flowered, flowers opening in succession, but maybe two open together, floral bracts membranous, narrowly ovoid, c. 3.2 × 2.2 cm, deciduous. *Flowers* widely open, to 7 cm across, sepals and petals creamy-greenish, distinctly greenish keels, tips with some diffused orange, lip dark-brown overall, column dark brown at base becoming brownish towards tip, anther cream coloured. ***Dorsal sepal*** elliptic, acuminate, concave, strongly kneeled, with many nerves, 4.5 × 1.9 cm. ***Lateral sepals*** elliptic, acuminate, oblique, concave, strongly kneeled, with many nerves, to 4.2 × 1.6 cm. ***Petals*** widespread, linear, obtuse, with a few nerves, to 4 × 0.3 cm. ***Lip*** 3-lobed, with a knee-like curve at the base, to 4.8 cm long and across side-lobes, 4 cm broad, side-lobes front rounded, 3.1 cm long, margins entire, mid-lobe without claw, tip strongly reflexed, margins erose, tip mucronate, callus of 3 keels, lateral keels extends from c. 0.6 cm from the base to the mid-lobe to well onto the mid-lobe, keels somewhat elevated, fleshy in the form of upright cylindrical teeth in two series, median keel shorter than lateral keels and less pronounced with a single row of teeth. [In my judgement the keels take the form of a series of cylindrical teeth or pustules arranged as rosettes along the length of the keels]. ***Column*** arcuate, 3.8 cm long, at base 0.3 cm across and expands into hood with wings, 1 cm across, front slightly erose in the form of short teeth.

DISTRIBUTION. Philippines (Central-east Mindanao).
HABITAT. Epiphytic and lithophytic.
ALTITUDE. 800 m.
FLOWERING. Probably over an extended period at different times of the year.
NOTES. The species is named after the collector, Vilmoor Usita. This seems to be the first record of a sect. *Speciosae* species from the Philippines. The type has an inflorescence 55 cm long and with 20 flowers. There are a number of species in sect. *Speciosae* which produce an extended inflorescence and develop a number of flowers. However, the same species can be found with just a few flowers and this could be the case with this new species, both in its natural habitat or in cultivation.

54. COELOGYNE XYREKES

Coelogyne xyrekes Ridl. in *J. Fed. Malay States Mus.* 6, 3: 181 (1915). Type: Peninsular Malaysia, Pahang, Gunong Tahan, *Ridley* s.n. (holo. SING.; iso. K). **Plate 7B**.

Coelogyne xanthoglossa Ridl. in *J. Fed. Malay States Mus.* 6, 3: 180 (1915). Type: Peninsular
    Malaysia, Pahang, Mt. Tahan, *Ridley* s.n. (holo. SING; iso. K).

DESCRIPTION. ***Pseudobulbs*** on a stout, sheathed rhizome, to 1.5 cm apart, cylindric, 4-angled, about 4–5.5 cm long. ***Leaf*** 1, obovate-oblanceolate, acuminate, with 5 nerves, 20.5–36.5 × 4.1–9.2 cm with 3.5–9 cm long petiole. ***Inflorescence*** synanthous with partially to entirely developed leaf, peduncle during flowering enclosed at base by petiole and scales of young shoot, peduncle 7–15 cm long, rachis (sub) erect, zig-zag, 1.8–5.8 cm long, 2- to 4-flowered, flowers opening in succession, floral bracts ovate-oblong, acute, deciduous. ***Flowers*** with sepals and petals brownish-salmon tinged olive on the outside, transparent; lip pale salmon, heavily reticulated with dark brown inside, keels blackish-brown, outside greyish-salmon; column cream coloured tinged salmon. ***Dorsal sepal*** ovate-oblong, acute, 7–13-nerved, keel prominent beneath, 4.3–5.6 × 1.6–2.35 cm. ***Lateral sepals*** obovate-lanceolate, acute, 6–10-nerved, keel prominent beneath, 4.1–5.9 × 1.4–1.6 cm. ***Petals*** slightly recurved, narrowly linear, 3-nerved, 4.2–5.5 × 0.23–0.35 cm. ***Lip*** 3–3.6 × 2.9–4 cm, 3-lobed, side-lobes in front rounded, slightly divergent, front margin at base irregularly erose, with acute sinus, mid-lobe broadly ovate to orbicular, 1.2–1.45 × 1.85–2.1 cm, with broad, short claw, tip obtuse, with warts, margin slightly erose, recurved, sides of mid-lobe pronounced as side-lobes, with rows of warts, callus of 2 keels extending from base of lip to 0.6–0.75 cm from tip of mid-lobe, widened along crest, plate-like, undulating, covered with elongate papillae, keels initially parallel, diverging at base of mid-lobe, then converging on mid-lobe. **Column** arcuate, 2.5–3.9 cm long, expanding into winged hood with more or less truncate apical margin, laterally notched where wings attached and additional notch above, middle part rounded, recurved.

DISTRIBUTION. Thailand, Peninsular Malaysia, Sumatra.
HABITAT. Epiphytic in lower montane forest.
ALTITUDE. 700–1900 m.
FLOWERING. February–December.
NOTES. The Greek epiphet means smooth or clean-shaven, which presumably was chosen to differentiate this species from some of the other taxa in sect. *Speciosae* which have relatively long hairs on the keels. The species can be recognised by the purple colour of the young leaves and undulating, plate-like keels on the lip with long papillae.

SECTION *BICELLAE*

Section 7. **Bicellae** J.J. Sm. in *Bull. Jard. Bot. Buitenzorg.* ser. 3 (1928). Type: *Coelogyne bicamerata* J.J. Sm.

Pseudobulbs close together, towards oblong. Inflorescence synanthous, peduncle bare, rachis zig-zag. Lip adnate to the base of the column, hypochile extending saccate at the back of the lip where median keel(s) divide the base into compartments. Distribution map 8.

Map 8. Distribution of *Coelogyne* Sect. *Bicellae*; (1) *C. bicamerata*, (2) *C. calcarata*.

THE GENUS COELOGYNE

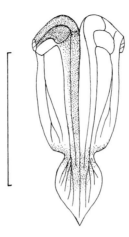

Fig. 11. **Section 7. Bicellae** J.J. Sm.
1. **Coelogyne bicamerata** J.J. Sm. [*Lack & Grimes* 1704]. Drawn by Linda Gurr. Scale bar = 1 cm.

KEY TO SPECIES IN SECTION BICELLAE

Lip with 3 undulate keels terminating at the middle of the mid-lobe. Flowers translucent flesh-coloured to pale brown, column with narrow wings .... ***C. bicamerata***
Lip with 4 scarcely prominent keels terminating at the middle of the mid-lobe. Flowers white with lip black, column with fleshy wings .... ***C. calcarata***

55. COELOGYNE BICAMERATA

Coelogyne bicamerata J.J. Sm. in *Bull. Jard. Bot. Buitenzorg* ser. 3, 10: 7 (1928). Type: Sulawesi, Gunung Lompobatang, *Bünnemeijer* 11622 (holo. BO; iso. L). **Plate 7C.**

DESCRIPTION. *Pseudobulbs* close together, oblong-ovoid to elongated, 3–7.5 cm long. ***Leaves*** 2, erect, lanceolate, acute-acuminate, with 5 major nerves, dorsal prominent, coriaceous, 9.3–31 × 1.85–4 cm, narrows into 1–5 cm long petiole. ***Inflorescence*** synanthous, peduncle bare, flattened, elliptical section, 4–10 cm long, rachis straight to slightly zig-zag, 2.4–8.5 cm long, several-flowered, flowers opening in succession, floral bracts deciduous. ***Flowers*** 2 cm across, uniformly translucent flesh-salmon to pale brown with slightly darker dorsal keel on sepals. ***Dorsal sepal*** oblong-ovate, acuminate, obtuse, 5–7 nerved, 1.75–1.95 × 0.8–0.9 cm. ***Lateral sepals*** lanceolate-ovate, acute-acuminate, 5–7 nerved, 1.8–2–15 × 0.8–0.95 cm. ***Petals*** elliptic-lanceolate, acuminate, 3–5 nerved, 1.5–1.7 × 0.6–0.8 cm. ***Lip*** base saccate, adnate to the column at the base, 3-lobed, side-lobes erect and equal length of column, front rounded, mid-lobe reflexed with column rising above, oblong to 5-angled, tip acute, reflexed, callus of 3 parallel undulate keels from base of lip to middle of mid-lobe, lateral keels stretch into side-lobes and thicken. ***Column*** slightly curved, 0.95–1 cm long, narrow wings, apex 3-lobed, mid lobe semi-orbicular, side lobes auriculate.
    DISTRIBUTION. Sulawesi (Gunung Lompobatang).
    HABITAT. Epiphytic on mossy trees 1–2 m from ground, in shade, in lower and upper montane forest.

ALTITUDE. 1100–2000 m.

FLOWERING. May–June.

NOTES. The epithet refers to the two chambers created by the adnate base of the lip and the median keel. Other authors consider there to be a single species in the subgenus Bicellae. I have grouped it with *C. calcarata* J.J. Sm., see Note below.

56. COELOGYNE CALCARATA

Coelogyne calcarata J.J. Sm. in *Bull. Jard. Bot. Buitenzorg* ser. 3, 10: 104 (1928). Type: Maluku, Seram, Gunung Murkele, *L. Rutten* 1475 (holo. BO; iso. L).

DESCRIPTION. **Pseudobulbs** overlapping, well sheathed, obtuse, sometimes ending in a sharp point, to 4 cm long, in a dry state with dense prominent nerves. **Leaves** 2, elliptic to lanceolate-elliptic, obtuse, with 9 prominent nerves, 6.6–7.75 × 2–2.3 cm. **Inflorescence** synanthous, erect, peduncle bare, 8.5 cm long, rachis zig-zag, 5.5 cm long, about 6-flowered, flowers opening in succession. **Flowers** thin-textured, meagre looking, 2.2 cm long, white with black lip. **Dorsal sepal** ovate-oblong, obtuse, minutely apiculate, 5-nerved with dorsal keel prominent, about 2 × 0.7 cm. **Lateral sepals** obliquely oblong, obtuse, apiculate, falcate, partly channelled and lightly convex, 5-nerved with prominent dorsal keel, 2.1 × 0.65 cm. **Petals** oblique, lanceolate, obtuse, 5-nerved with dorsal keel grooved, 2 × 0.55 cm. **Lip** around column and slightly reflexed above, base of lip saccate, adnate near the base, 3-lobed, side-lobes longer than column, front nearly rounded, mid-lobe oblong, tip rounded, margins narrowly curving irregularly inwards, callus of 4 scarcely prominent keels extending from base of lip to tip of mid-lobe. **Column** straight, 0.65 cm long, thickened apex wedge-shaped and widening into fleshy wings, apex 3-lobed, mid-lobe small and rounded, side-lobes spreading and nearly rounded with two notches.

DISTRIBUTION. Maluku (Seram, Gunung Murkele).

HABITAT. Epiphytic or terrestrial in montane forest and on limestone.

ALTITUDE. 2000–2500 m.

FLOWERING. July.

NOTES. The epithet refers to the spur-like form at the base of the lip to the flower. I have found no reference associating this species with a section. J.J. Smith, in his type description of *C. bicamerata*, indicated that he thought *C. bicamerata* and *C. calcarata* were similar except for the adnate form of the bicamerate, saccate portion of the base, the different keels and much longer column of *C. bicamerata*. I have placed the two species in the same section.

SECTION *MONILIFORMES*

Section 8. *Moniliformes* Carr in *Gard. Bull. Straits Settlem.* 6: 206 (1935). Type: *Coelogyne monilirachis* Carr.

Pseudobulbs close together on the rhizome, elongated and slender, one leaf at the apex which is tender. Inflorescence hysteranthous or synanthous, peduncle bare and slender, rachis thickened with distinctly swollen, very short internodes. Generally many-flowered with the flowers long-lasting and opening in succession. Keels of the lip entire but may be irregular cristate. Distribution map 9.

The Genus Coelogyne

Map 9. Distribution of *Coelogyne* Sect. *Moniliformes*.

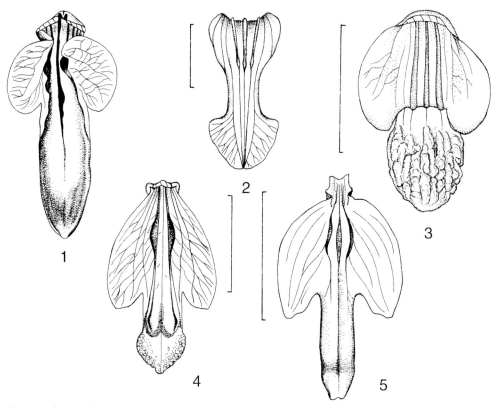

Fig. 12a. **Section 8. Moniliformes** Carr
1. Coelogyne crassiloba J.J. Sm. [after J.J. Sm.]; **2. C. gibbifera** J.J. Sm. [after J.J. Sm.]; **3. C. harana** J.J. Sm. [after de Vogel]; **4. C. incrassata** (Blume) Lindl. var. **incrassata** [Cult. D.A. Clayton Spirit DAC 7]; **5. C. incrassata** (Blume) Lindl. var. **valida** J.J. Sm. [after Eleanor Catherine]. Drawn by Linda Gurr. Scale bar = 1 cm.

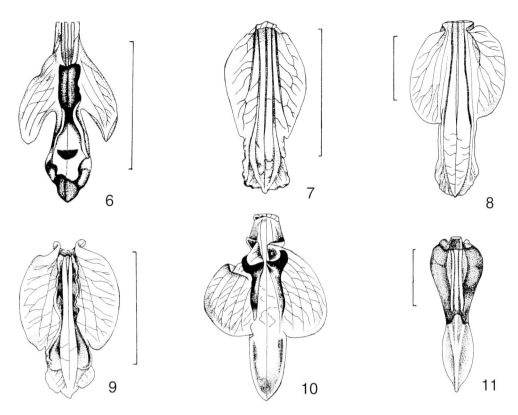

Fig. 12b. **Section 8. Moniliformes** Carr **contd.**
6. C. kelamensis J.J. Sm. [after J.J. Sm]; **7. C. longpasiaensis** J.J. Wood & C.L. Chan [after Eleanor Catherine]; **8. C. monilirachis** Carr [after C.L. Chan]. **9. C. naja** J.J. Sm. [after Cult. Leiden 30238]; **10. C. tenuis** Rolfe [after J.J. Sm.]; **11. C. vermicularis** J.J. Sm. [after J.J. Sm.]. Drawn by Linda Gurr. Scale bar = 1 cm.

KEY TO SPECIES IN SECTION MONILIFORMES

1. Inflorescence synanthous 2
   Inflorescence hysteranthous 4

2. Flowers with sepals 2.5–3.5 cm long, not widely open. [Lip with 3 keels.] 3
   Flowers with sepals about 1.5 cm long, [Lip mid-lobe obovate-oblong or subrhomboid, 2 keels diverge and converge into papillate mass at the tip of the mid-lobe.] ....
   .... *C. kelamensis*

3. Lip mid-lobe clawed, suboblong, margins rolled back, 3 keels, and median keel terminating two thirds onto mid-lobe .... *C. gibbifera*
   Lip mid-lobe clawed, quadrate, 3 parallel keels extending from a half to a third the way from the base of lip to tip of mid-lobe, becoming thickened and flattened ...... *C. harana*

4. Lip S-shaped 5
   Lip not obviously S-shaped 6

5. Lip mid-lobe ovate, 3 keels, erect, dentate, terminating at the tip of mid-lobe. [Flowers pale salmon.] .... *C. tenuis*
   Lip mid-lobe oblong. [3 keels, elevated with erose margin, terminating a short way from the base of lip, merging and flattening.] .... *C. crassiloba*

6. Lip mid-lobe ovate, oblong-ovate or oblong, 3 keels 7
   Lip mid-lobe quadrate, 2 keels. [Flowers greenish, tinged rose, lip with light brown markings.] .... *C. naja*

7. Lip mid-lobe oblong-ovate or oblong. Flowers translucent 8
   Lip mid-lobe with long claw. [Lip with 3 keels, fleshy and entire, terminating at the base of the mid-lobe.] .... *C. monilirachis*

8. Lip oblong 9
   Lip oblong-ovate. Flowers generally translucent pale brown: Lip whitish with transverse red band at centre of mid-lobe, 3 straight keels, terminating at the base of mid-lobe .... *C. incrassata* var. *incrassata*

9. Lip whitish with transverse brown bar, side-lobe yellowish, 2 conspicuous keels merging near base of mid-lobe, becoming smaller and terminating at the tip of mid-lobe .... *C. incrassata* var. *valida*
   Lip pale whitish with transverse zig-zag orange-cinnamen band, 3 parallel, straight, fleshy keels terminating just before base of mid-lobe .... *C. incrassata* var. *sumatrana*

10. Lip mid-lobe thickened with sides folded inwards. Flowers semi-translucent pale grey, lip yellow, lip with 3 smooth keels terminating at the middle of mid-lobe ... *C. vermicularis*
    Lip mid-lobe truncate. Flowers small, pale green or yellowish-green, lip cream or pale yellowish, 3 keels, lateral keels extending from base of lip to middle of mid-lobe, median keel near base of lip only .... *C. longpasiaensis*

57. COELOGYNE CRASSILOBA

Coelogyne crassiloba J.J. Sm. in *Mitt. Inst. Allg. Bot. Hamburg* 7: 28, t. 4, f. 18 (1927). Type: Borneo, Kalimantan, Bukit Mulu, *Winkler* 481 (holo. HBG, drawing at BO).

DESCRIPTION. **Pseudobulbs** on a short stout rhizome, 1–1.3 cm apart, elongated, lower end thickened and quadrangular, 9.5–11 cm long. **Leaf** 1, oblong-elliptic, shortly acuminate-acute, 7-nerved prominent beneath, papery when dry, 25.5–27 × 7–7.75 cm, base acute, tapering below into 1–1.5 cm long, grooved petiole. **Inflorescence** hysteranthous, peduncle bare, slender, thickened towards top, strongly angled, 11–14.5 cm long, rachis thickened, zig-zag, 15 cm long, many-flowered, long lasting, flowers opening in succession. **Flowers** colour not known. **Dorsal sepal** ovate-oblong, apiculate, 7-nerved, minutely spotted, 2.3 × 1.1 cm. **Lateral sepals** oblique, somewhat ovate-oblong, obtuse, scarcely apiculate, angular grooves, 5-nerved, minutely spotted on underside, 2.3 × 0.8 cm. **Petals** oblique, linear scarcely falcate, narrowing towards tip, truncate, minute apiculate, 3-nerved, partly dotted, 2.1 × 0.175 cm. **Lip** 3-lobed, base saccate-concave, then recurved to form S-shape, slightly fleshy, side-lobes erect, rising above column, rounded, tip slightly rounded-elongated, slightly fleshy, mid-lobe straight, oblong, very fleshy, callus of 3 keels flat at base of lip then elevated with irregular margins, especially on the lateral keels, fleshy to about one third the distance from base of lip, then merging and becoming flattened. **Column** slender, arcuate, 0.85 cm long, tip narrowly winged and fleshy.

DISTRIBUTION. Borneo (Kalimantan, Sarawak).
HABITAT. Epiphytic in lowland rain forest.
ALTITUDE. 600–700 m.
FLOWERING. December.
NOTES. The epithet refers to the very fleshy form of the lip.

58. COELOGYNE GIBBIFERA

Coelogyne gibbifera J.J. Sm. in *Bull. Jard. Bot. Buitenzorg* ser. 2, 3: 51–53 (1912). Type: Borneo, Sarawak, Batu Lawai, Ulu Limbang, *Moulton* 12 (holo. BO). **Plate 7D.**

C. macroloba J.J. Sm. in *Mitt. Inst. Allg. Bot. Hamburg* 7: 30 (1927). Type: Borneo, Kalimantan, Bukit Raya, *Winkler* 860 (holo. HBG; drawing at BO).

DESCRIPTION. *Pseudobulbs* about 0.9 cm apart on rhizome, erect, elongated, ellipsoid, narrowing at base and apex, 12.7–13.7 × 0.6–0.65 cm. *Leaf* 1, lanceolate, acuminate, 6–7-nerved, finely coriaceous, about 18.5–21 cm with petiole. *Inflorescence* synanthous, erect, about 8–10 cm long, rachis zig-zag, about 12-flowered, flowers opening in succession. *Flowers* large, about 4.8 cm across. *Dorsal sepal* erect, lanceolate, narrowly obtuse, dorsal nerve prominent beneath, 3.5 × 1 cm. *Lateral sepals* oblique-lanceolate, obtuse, apiculate, reflexed, dorsal nerve prominent, 2.75 × 0.73 cm. *Petals* linear, reflexed, 3-nerved, (basal part only present). *Lip* 3-lobed, side-lobes erect, suborbicular in the front, not at all extended, mid-lobe large, porrect, rolled back at the edges, clawed, somewhat oblong, 0.475 × 0.25 cm, callus of 3 keels, median keel extending from the base of lip to two-thirds to base of mid-lobe. *Column* slender, arcuate, 1 cm long.
DISTRIBUTION. Borneo (Kalimantan, Sarawak, Sabah, Brunei).
HABITAT. Epiphytic in hill forest and lower montane forest.
ALTITUDE. 600–1600 m.
FLOWERING. May, October–December.
NOTES. The epithet refers to the very prominent median nerves on the back of the sepals and petals.

59. COELOGYNE HARANA

Coelogyne harana J.J. Sm. in *Mitt. Inst. Allg. Bot. Hamburg* 7: 27, t. 4, f. 17 (1927). Type: Borneo, Kalimantan, Lebang Hara, *Winkler* 246 (holo. HBG, drawing at BO). **Plate 7E.**

DESCRIPTION. *Pseudobulbs* on a robust rhizome, close together, elongated, 12–15.5 cm long. *Leaf* 1, erect, elliptic-lanceolate, triangular to acuminate-acute, with 7 nerves prominent beneath, papery when dry, 33.5 × 8–8.25 cm with 1–1.2 cm long, grooved petiole. *Inflorescence* synanthous, peduncle bare, thickened towards top, 22–23 cm long, rachis thickened, zig-zag, 5 cm long, many-flowered, long lasting, flowers opening in succession. *Flowers* large, not widely open, golden-brown lip. *Dorsal sepal* oblong, obtuse-angled, tip reflexed, on both sides minutely spotted, 7-nerved, 2.45 × 1.13 cm. *Lateral sepals* oblique, oblong, tip narrowing gradually, obtuse, grooved, 7-nerved, 2.3 × 0.8 cm. *Petals* oblique, linear, tip shortly oblique, somewhat acute, base partly covered with minute spots, 3-nerved, 2.2 × 0.17 cm. *Lip* porrect below column, slightly reflexed, 3-lobed, slightly fleshy, base saccate-concave, side-lobes erect, porrect, column at same level, rounded-quadrangular, mid-lobe also porrect and at base clawed, quadrangular, tip shortly obtuse, callus of 3 parallel keels extending from about a quarter the distance to about one third the distance from base of lip, elevated and becoming thickened, then flattening and extending to tip of mid-lobe. **Column** slender, arcuate, 1.6 cm long, narrowly winged, tip inconspicuously 3-lobed and slightly reflexed.

DISTRIBUTION. Borneo (Kalimantan, Sarawak).
HABITAT. Epiphytic in lowland rain forest.
ALTITUDE. 100–200 m.
FLOWERING. November.
NOTES. The epithet refers to the locality, Limbang Hara, Kalimantan, Borneo where the type was collected.

60. COELOGYNE INCRASSATA

Coelogyne incrassata (Blume) Lindl. in *Gen. Sp. Orch. Pl.* 40 (1830). Type: Java, Pantjar, *Blume* s.n. (holo. L). **Plate 8A & B.**

Chelonanthera incrassata Blume, *Bijdr*. 384 (1825).
Pleione incrassata (Blume) Kuntze in *Rev. Gen. Pl.*. 2: 680 (1891).

var. **incrassata**

DESCRIPTION. **Pseudobulbs** 8.5 cm long, with a long drawn-out tip, smooth. **Leaf** 1, copper-coloured when young, maturing to a dark green, sharply acuminate, 17 × 4 cm with a 1.25 cm long petiole. **Inflorescence** hysteranthous, peduncle thin, semi-pendulous, 6 cm long, the rachis much thicker, expanding to 10 cm or more long. **Flowers** not opening very widely, about 3 cm across, sepals translucent pale brown, column greenish-yellow, margin golden-yellow edge. **Sepals** lanceolate, acute, 2 cm long. **Petals** linear, straight. **Lip** 3-lobed, side-lobes rounded, mid-lobe oblong, tip convex, white with transverse red band in the centre, callus of 3 short, straight keels extending from base of lip to base of mid-lobe. **Column** slender, arcuate, 1 cm long, expanding very slightly into hood, tip entire, rounded.

DISTRIBUTION. Borneo (Kalimantan, Sarawak, Brunei), Java (West), Sumatra.
HABITAT. Epiphytic in lowland tropical rainforest.
ALTITUDE. 600–1000 m.
FLOWERING. April.
NOTES. The epithet refers to the thickened nodes on the rachis of the inflorescence.

var. **sumatrana** J.J. Sm. in *Bull. Jard. Bot. Buitenzorg* ser. 2, 25: 1 (1914). Type: Sumatra, Bengkulu, Pegunungan Barisan, cult. Bogor, *E. Jacobsen* 775, 863 (holo. BO; iso. L).

DESCRIPTION. **Pseudobulbs** about 0.4–0.8 cm apart with a long drawn-out tip, obtuse angled, green with dark tinges, 5.5–11.5 × 1.2–1.8 cm. **Leaf** 1, oblong-elliptic, acute-acuminate, dark grey-green with dull texture, 18–25 × 5–8.75 cm, at base obtuse or abruptly rounded into 0.5–0.9 cm long petiole. **Inflorescence** hysteranthous, semi-pendulous, the peduncle thin and rachis much thicker, expanding eventually to 8–13.5 cm, many-flowered, inconspicuous, flowers opening in succession. **Flowers** not opening very widely, about 2.5 cm across, pale brown. **Dorsal sepal** somewhat elliptic, obtuse, 5-nerved, 1.65 × 0.8 cm. **Lateral** *sepals* spreading, somewhat ovate-oblong, subacute, 1.5 × 0.6 cm with longitudinal groove. *Petals* very widely spread, linear, somewhat obtuse or retuse, 1.4 × 0.075–0.17 cm. **Lip** 3-lobed, side-lobes oblique, oblong, mid-lobe oblong, white with transverse zig-zag orange-cinnamen band, callus of 3 parallel, straight, fleshy keels on the mid-lobe extending from the base of lip but the lateral keels terminate before end of side-lobes. **Column** arcuate (no other detail).

DISTRIBUTION. Sumatra (West Coast).
HABITAT. Epiphytic in lowland tropical rainforest.
ALTITUDE. 1000–1200 m.
FLOWERING. Not known.
NOTES. The epithet refers to Sumatra where this variation of the type was collected.

var. **valida** J.J. Sm. in *Bull. Jard. Bot. Buitenzorg* ser. 3, 11: 92 (1930). Type: Borneo, Kalimantan, Liang Gagang, *Hallier* 3054 (holo. L; iso. BO).

DESCRIPTION. Pseudobulbs about 1 cm apart, slender, elongated, 8.5–11 cm long. **Leaves** 1 or ?2, oblong. acute-acuminate, undulate, with 7 prominent nerves beneath, short brownish-green, 24–26 × 6.5–7.75 cm with 0.5–0.7cm long, grooved petiole. **Inflorescence** hysteranthous, nodding, peduncle thread-like, 13–15 cm long, rachis thickened, zig-zag, to 22 cm long. **Flowers** lip nearly white with transverse brown bar, lateral segments yellow, column salmon colour. **Dorsal sepal** oblong, strongly concave, 1.95 × 0.74 cm. **Lateral sepals** oblique, linear-lanceolate, fleshy, 2 × 0.45 cm. **Petals** linear, 1.95 × 0.1 cm. **Lip** base hollowed out, 3-lobed, side-lobes porrect, obtuse, front free, fleshy, callus of 2 conspicuous keels merging with thickened margins of mid-lobe at base then becoming smaller all the way to tip of mid-lobe. **Column** slender, arcuate, narrowly winged, stretched rounded apex, 1 cm long.

DISTRIBUTION. Borneo (Kalimantan, Sarawak, Sabah).
HABITAT. Epiphytic in riverine and tropical rainforest.
ALTITUDE. 400 m.
FLOWERING. September.
NOTES. The epithet refers to the robust growth of this variation when compared with var. *incrassata*.

61. COELOGYNE KELAMENSIS

Coelogyne kelamensis J.J. Sm. in *Icon. Bogor.* 4: 5, t. 302 (1910). Type: Borneo, Kalimantan, Bukit Kelam, *Hallier* 2489 (holo. BO). **Plate 8C.**

DESCRIPTION. Pseudobulbs on a stout rhizome, 3–3.5 cm apart, elongated, slender, 5–7 cm long. **Leaf** 1, erect, oblong-elliptic, obovate, acuminate-acute, undulate, 5-nerved, prominent beneath, coriaceous, 10–13 × 3.6–4.3 cm with a 2.5–3 cm long, grooved petiole. **Inflorescence** synanthous, peduncle slender, bare, 5.5–6.5 cm long, rachis zig-zag, to 5.5 cm long, 11-flowered, flowers opening in succession. **Flowers** large, yellow. **Dorsal sepal** oblong, obtuse, apiculate, 5-nerved, 1.65 × 0.675 cm. **Lateral sepals** oblique, oblong, obtuse, apiculate, fleshy, 1.6 × 0.6cm. **Petals** linear, acute, 3-nerved, 1.5 × 0.15 cm. **Lip** 3-lobed, base hollowed and thickened, concave, also folded, side-lobes erect, wrapped around column, falcate-triangular, obtuse, mid-lobe obovate-oblong, subrhomboid, obtuse, callus of 2 keels diverging and then converging and thickened into a warty mass at tip of mid-lobe. **Column** slender, arcuate, 1.3 cm long, hood winged, front 3-lobed, side-lobes broadly rounded, apex truncated and crenulate.

DISTRIBUTION. Borneo (Kalimantan, Sarawak).
HABITAT. Terrestrial in heath (Kerangas) forest.
ALTITUDE. Sea level.
FLOWERING. Not known.
NOTES. The epithet refers to the type locality, Kelam, Kalimantan, Borneo.

62. COELOGYNE LONGPASIAENSIS

Coelogyne longpasiaensis J.J. Wood & C.L. Chan in *Lindleyana* 5, 2: 87, f. 4 (1990). Type: Borneo, Sabah, Sipitang District, 4 km S. of Long Pa Sia, *J.J. Vermeulen & Duistermaat* 946 (holo. K; iso. L). **Plate 8D.**

DESCRIPTION. *Pseudobulbs* on a slender rhizome, close together, narrowly cylindrical, (1.5) 2.5–(8.5) × 0.5–0.8 cm. ***Leaf*** 1, narrowly elliptic, acute to acuminate, somewhat plicate, margin undulate, 8–15 × 1.5–3.8 cm with 0.2–0.5 cm long petiole. ***Inflorescence*** hysteranthous, peduncle slender with wings, naked, pendulous, 4–10 cm long, rachis thickened, fleshy, with prominent abscission scars, 1–6 cm long, many-flowered, 1- to 2- flowers opened together, flowers opening in succession, floral bracts deciduous. ***Flowers*** slightly fragrant, sepals and petals pale green or yellowish-green, lip cream or pale yellowish, column cream edged brown distally, just with orange-tan basal blotch. ***Dorsal sepal*** oblong-elliptic, acute, slightly concave, 1.5 × 0.6 cm. ***Lateral sepals*** narrowly oblong-elliptic, subacute or acute, 1.4 × 0.4 cm. ***Petals*** linear, obtuse, 1.3 × 0.08 cm. ***Lip*** shallowly 3-lobed, 1.3 × 0.7 cm, side-lobes erect, rounded, 0.8–0.9 cm long, mid-lobe oblong, truncate, 0.4 × 0.4 cm, 3 keels, callus of 2 lateral keels, slightly raised, extending from base of lip to middle of mid-lobe, median keel extending only a short distance from base of lip. ***Column*** gently curved, 1 cm long.

DISTRIBUTION. Borneo (Kalimantan, Sarawak, Sabah).
HABITAT. Epiphytic in low, very open and dry heath forest on steep sandstone ridges, with open places thickly overgrown with ferns; also in lower montane forest.
ALTITUDE. 1200–1300 m.
FLOWERING. October.
NOTES. The epithet refers to the locality Long Pa Sia, Sabah, Borneo where the type specimen was collected.

63. COELOGYNE MONILIRACHIS

Coelogyne monilirachis Carr in *Gard. Bull. Straits Settlem.* 6: 206 (1935). Type: Borneo, Sabah, Mt. Kinabalu, Tenompok, *C.E. Carr* 3366 in SFN 27230 (holo. SING; iso. AMES, K). **Plate 8E.**

DESCRIPTION. *Pseudobulbs* on a creeping 0.6 cm dia. rhizome, 1.3 cm apart, elongate-ovoid, to 13.5 cm long. ***Leaf*** 1, elliptic, shortly acuminate, acute, membraneous, with 7 strong nerves, green often suffused red, to 27 × 10 cm with a 1.5 cm long, grooved petiole. ***Inflorescence*** hysteranthous, elongate, arched, peduncle bare, very slender, slightly dilate towards apex, to 17.5 cm long, rachis dilate, smooth, contracted at nodes and resembling a necklace when fresh, sinuous and wrinkled when dry, at first imbricating bracts at apex of rachis, then caducous, 40-flowered, flowers opening in succession. ***Flowers*** with semi-transparent pale salmon pink sepals and petals, lip pale pink with paler tip, column pink. ***Dorsal sepal*** oblong, triangular-acuminate, very shortly apiculate or tri-dentate and narrowly truncate, base cymbiform, keeled beneath, 7-nerved, 2.9–3.6 × 1.55–1.75 cm. ***Lateral sepals*** oblong-lanceolate, obtuse, minutely apiculate, somewhat S-shaped, keeled beneath, 7-nerved, margin near front shortly deflexed, 2.7–3.3 × 1–1.3 cm. ***Petals*** narrowly linear-lanceolate, subacute, grooved on top, 3-nerved, 2.85–3.15 × 0.23–0.37 cm. ***Lip*** 3-lobed, shortly saccate at base, 2.5–2.75 × 1.55–1.68 cm, side-lobes erect, short, broad, rounded, mid-lobe with long oblong claw, 1.28–1.45 × 0.85 cm, margins minutely erose, sometimes undulate, oblique groove on each side, callus of 3 keels, median keel extending from base of lip to base of mid-lobe then becoming raised nerve to the tip of mid-lobe, 2 lateral keels terminating abruptly at claw of mid-lobe, all keels fleshy, entire. ***Column*** arcuate, 0.215 cm long, wings slightly roundly dilate from middle and of column and with a very small recurved tooth, hood rounded, minutely crenulate.

DISTRIBUTION. Borneo (Sarawak, Sabah).
HABITAT. Epiphytic in shade in lower montane forest.
ALTITUDE. 1200–1700 m.
FLOWERING. February, May, August, November.
NOTES. The epithet refers to the characteristic contraction of the rachis at the nodes to give it a necklace appearance.

64. COELOGYNE NAJA

Coelogyne naja J.J. Sm. in *Bull. Jard. Bot. Buitenzorg* ser. 3, 11: 93 (1931). Type: Borneo, Kalimantan, Gunung Kemoel, *Endert* 3708 (holo. L; iso. BO). **Plate 8F.**

DESCRIPTION. ***Pseudobulbs*** on a short rooted, creeping rhizome, close together, slender, elongated, 10–11 × 0.45 cm (in dried state). ***Leaf*** 1, oblong-elliptic, acute-acuminate, with 7 very prominent nerves beneath, 18–19.5 × 4.5 cm, base acute and tapering into 0.6–10.8cm long, grooved petiole. ***Inflorescence*** hysteranthous, peduncle thread-like, top thickening, 9 cm long, rachis thickened, zig-zag to 8 cm long, flowers develop slowly in succession. ***Flowers*** greenish, somewhat tinged rose, inside of lip light brown markings, outside of column brownish-orange. ***Dorsal sepal*** oblong, obtuse, apiculate, 7-nerved, 1.6 × 0.5 cm. ***Lateral sepals*** oblique, lanceolate, obtuse, apiculate, 1.6 × 0.4 cm. ***Petals*** oblique, linear, 3-nerved, 1.4 cm long, at base 0.125 cm wide. ***Lip*** 3-lobed, base strongly hollowed out to concave, side-lobes widen from the base, not prominent in front, nearly rectangular, erect, obtuse, mid-lobe quadrangular, expanding into rounded-obtuse apex, callus of 2 longitudinal keels stand well up, abruptly terminated at base of lip and open out on mid-lobe until they terminate at the middle of mid-lobe. ***Column*** slender, arcuate, 1.25 cm long, divides into upper wings, in outline wings rhombic, apex obtuse.

DISTRIBUTION. Borneo (Kalimantan, Sarawak).
HABITAT. Epiphytic on ridges in lower montane forest.
ALTITUDE. 600–1500 m.
FLOWERING. September.
NOTES. The epithet presumably refers to *Naja* (cobra), a member of which species was possibly present when the type was collected. The cobra species found in Borneo is the *Naja sumatrana* (Equatorial spitting cobra), possibly known in Borneo by the synonym *N. n. miolepis*.

65. COELOGYNE TENUIS

Coelogyne tenuis Rolfe in *Bull. Misc. Inf., Kew* 171 (1893). Type: Borneo, *Linden* s.n. (holo. K).

Coelogyne bihamata J.J. Sm. in *Mitt. Inst. Allg. Bot. Hamburg* 7: 29, t. 4, f. 19 (1927).

DESCRIPTION. *Pseudobulbs* on a short rhizome, close together, conical-elongated, 5–7.75 cm long. *Leaf* 1, oblong to elliptic-lanceolate, acuminate-acute, with 5–7 major nerves, prominent beneath, papery when dry, 14–17.5 × 3.5–5.2 cm with 0.7–0.8cm long, grooved petiole. *Inflorescence* hysteranthous, peduncle bare, thread-like, top thickened, 10–14.5 cm long, rachis thickened, pendulous, zig-zag, 12 cm long, few to many-flowered, long lasting, flowers opening in succession. *Flowers* large, pale salmon. *Dorsal sepal* reflexed, oblong-ovate, tip narrowing, strongly concave, 7-nerved, 1.3 × 0.55 cm. *Lateral sepals* oblique, oblong-ovate, subacute, concave, 5-nerved, 1.3 × 0.55 cm. *Petals* oblique, linear, subacute, 3-nerved, 1.2 × 0.175 cm. *Lip* porrect in an S-shape with column well above, 3-lobed, side-lobes erect, semi-orbicular, lightly concave margin in front, on inside convex, thickened, rounded, mid-lobe ovate with broad apex, callus of 3 keels extending from base of lip to tip of mid-lobe, dentate, erect. *Column* slender, rectangular, lightly thickening from base, 0.875 cm long, narrowly winged, tip lightly contracted and rounded-truncate.

DISTRIBUTION. Borneo (Kalimantan).
HABITAT. Epiphytic in hill forest, lower and upper montane forest.
ALTITUDE. 500–2150 m.
FLOWERING. November.
NOTES. The epiphet refers to the thread-like nature of the inflorescence.

66. COELOGYNE VERMICULARIS

Coelogyne vermicularis J.J. Sm. in *Icon. Bogor.* 3: 9, t. 204 (1906). Type: Borneo, *Nieuwenhuis* s.n. (holo. BO). **Plate 9A**.

Chelonistele vermicularis (J.J. Sm.) Kraenzl. in Engler, *Pflanzenr. Orch.-Mon.-Coelog.* 163 (1907).

DESCRIPTION. *Pseudobulbs* on a short rhizome, erect, elongated with an attenuated tip, obtuse angled to rhombic in cross-section, 17 cm long, dull green colour. *Leaf* 1, oblong, acute-acuminate, margin undulate, with 7 nerves, strongly sulcate on top surface, prominent beneath, dull green but polished appearance, 26 × 9.5 cm with 1 cm long, folded-grooved petiole. *Inflorescence* hysteranthous, peduncle dull brown colour, thin and tapering towards apex, 17.5 cm long, rachis slightly thickened, zig-zag, dull brown colour, 10 cm long, many-flowered, flowers opening in succession. *Flowers* 5 cm across, sepals and petals shinning to semi-translucent, pale grey colour with veins yellow, petals brownish at base, lip yellowish with 3 deep yellow-brown keels, column pale brown and fleshy turning white towards base, apex yellow-brown. *Dorsal sepal* obtuse, oblong-lanceolate, apiculate, fleshy, 7-nerved, 3.1 × 1.2 cm. *Lateral sepals* oblong-lanceolate, falcate, acute, fleshy, 5-nerved, 2.8 × 1.8 cm. *Petals* widely spread, linear, somewhat obtuse, 1-nerved, sulcate. *Lip* 3-lobed, side-lobes erect, short and rounded, mid-lobe at base concave, thickened with both sides folded inwards, oblong-ovate, front of mid-lobe extending beyond side-lobes, rounded with tip attenuated, callus of 3 keels glabrous, extend from base to middle of mid-lobe, flattened. **Column** slender, slightly curved, 1.2 cm long, narrowly winged, wings entire.

DISTRIBUTION. Sumatra (West Coast), Borneo (Kalimantan, Sabah).
HABITAT. Epiphytic in lower montane forest.
ALTITUDE. 1400 m.
FLOWERING. Not known.
NOTES. The epiphet refers to the thickened worm-like nature of the rachis of the inflorescence.

SECTION *LONGIFOLIAE*

Section 9. ***Longifoliae*** Pfitzer & Kraenzlin in Engler, *Pflanzenr. Orch.-Mon.-Coelog.*: 21 (1907). Type: *Coelogyne longifolia* (Blume) Lindl.

Pseudobulbs on a stout, sheathed rhizome either close together or quite far apart, ovoid, cylindric, fusiform or pyriform with 2 coriaceous leaves at the apex. The inflorescence is synanthous or hysteranthous and the peduncle is bare, mainly stiff and long, flattened, sometimes with wings. The rachis is also long, generally arcuate, zig-zag. The internodes on the rachis are slender and long. The flowers open in succession over a considerable period of time. The flowers are small, keels entire or slightly crenate. Distribution map 10.

Map 10. Distribution of *Coelogyne* Sect. *Longifoliae*.

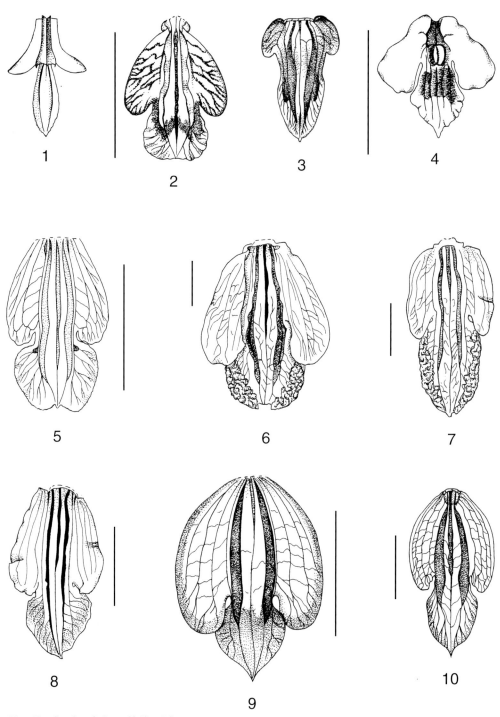

Fig. 13a. **Section 9. Longifoliae** Pfitzer
1. Coelogyne bilamellata Lindl. [Type]; **2. C. borneensis** Rolfe [Kew Spirit Coll. 1892]; **3. C. brachygyne** J.J. Sm. [after J.J. Sm.]; **4. C. candoonensis** Ames [after Ames (1920)]; **5. C. compressicaulis** Ames & C. Schweinf. [Kew Spirit Coll. 24866]; **6. C. cuprea** H. Wendl. & Kraenzl. var. **cuprea** [*Bailes & Cribb* 618]; **7. C. cuprea** H. Wendl. & Kraenzl. var. **planiscapa** J. J. Wood & C.L. Chan [Kew Spirit Coll. 48607]; **8. C. dulitensis** Carr [*Synge* S342]; **9. C. endertii** J.J. Sm. [Kew Spirit Coll. S502]; **10. C. kinabaluensis** Ames & C. Schweinf. [*A. Lamb* AL1412/92]. Drawn by Linda Gurr. Scale bar = 1 cm.

Section Longifoliae

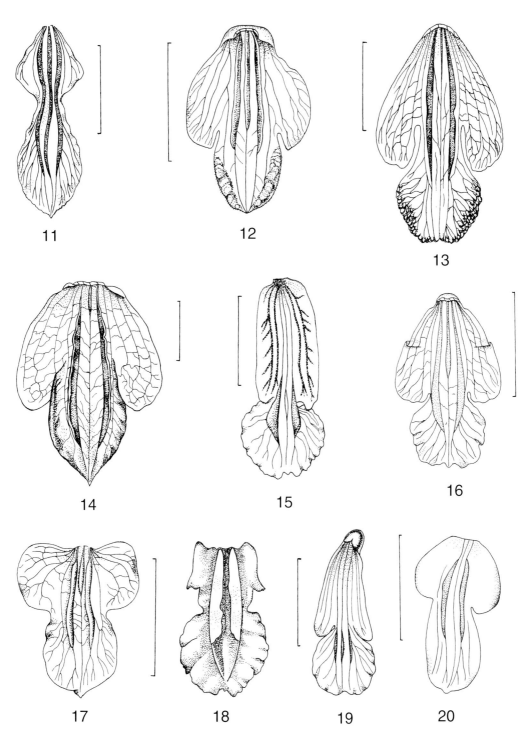

Fig. 13b. **Section 9. Longifoliae** Pfitzer **contd.**
11. C. longifolia (Blume) Lindl. [Hooker 1867]; **12. C. motleyi** Rolfe ex J.J. Wood, D.A. Clayton & C.L. Chan [*Franklin & Ross* 20858]; **13. C. planiscapa** Carr var. **planiscapa** [*Synge* S.556]; **14. C. planiscapa** Carr var. **grandis** Carr [after C.L. Chan]; **15. C. prasina** Ridl. [after C.L. Chan]; **16. C. tenompokensis** Carr [*Carr* 3270]; **17. C. radicosa** Ridl. [after Seidenfaden & Wood (1992)]; **18. C. remediosae** Ames & Quisumb. [after Ames (1932)]; **19. C. stenochila** Hook.f. [after Seidenfaden & Wood (1992)]. **20. C. tumida** J.J. Sm. [after de Vogel]. Drawn by Linda Gurr. Scale bar = 1 cm.

Key to species in section Longifoliae

1. Lip not 3-lobed or indistinctly 3-lobed 2
   Lip distinctly 3-lobed 6

2. Pseudobulbs close together or not more than 2 cm apart on the rhizome 3
   Pseudobulbs a distance apart on the rhizome, cylindric. [Pseudobulbs 6–13 cm long. Inflorescence peduncle and rachis erect, lip entire, oblong-lingulate, 3 keels, median keel thickened, lateral keels terminating at the tip of the lip. Sepals and petals salmon-pink, lip brown.] .... *C. stenobulbon*

3. Pseudobulbs cylindric to fusiform or pyriform 4
   Pseudobulbs ovoid. [Lip ovate, entire, no prominent keels. Sepals and petals white, lip with gold-yellow veins.] .... *C. integra*

4. Pseudobulbs cylindric to fusiform and close together or no more than 1.5 cm apart 5
   Pseudobulbs narrowly pyriform and close together. [Lip simple, undivided, transverse elliptic, abbreviated keels near the base of the lip only. Sepals and petals orange, lip orange and white.] .... *C. quinquelamellata*

5. Pseudobulb to 1.5 cm apart, flattened cylindric, 6.5 cm long. [Lip indistinctly 3-lobed, mid-lobe clawed, orbicular, margins undulate, keels not very evident, median nerve thickens and merges with thickened lateral keels on the mid-lobe. Sepals and petals olive yellow, lip salmon pink.] .... *C. compressicaulis*
   Pseudobulbs close together, very slender, to 12 cm long. [Lip indistinctly 3-lobed, obovate, keels merge as a papillose mass. Flowers white and brown.] .... *C. candoonensis*

6. Pseudobulbs less than 4 cm long 7
   Pseudobulbs more than 4.5 cm long 9

7. Pseudobulbs ovoid, less than 3 cm long 8
   Pseudobulbs pyriform, 3.5 cm long. [Lip mid-lobe clawed, orbicular, margins undulate, 2 simple, straight, entire keels terminating at the middle of the mid-lobe. Flowers light chocolate brown, white veins.] .... *C. borneensis*

8. Pseudobulbs ovoid, 1.25–3 cm long. [Lip indistinctly 3-lobed, orbicular, 3 keels, and lateral keels terminating at the tip of the lip. Flowers white, 2 yellow-brown spots on the lip.] .... *C. stenochila*
   Pseudobulbs ovoid, 2–3 cm long. [Lip mid-lobe 6-angled, tip tri-lobed, 5 simple, smooth keels terminating a quarter way onto the mid-lobe. Flowers light brown, lip translucent red-brown.] .... *C. trilobulata*

9. Pseudobulbs close together or very slightly apart on the rhizome 10
   Pseudobulbs 3 cm or more apart on the rhizome 18

10. Pseudobulbs ovoid to oblong-ovoid or elongated ovoid 11
    Pseudobulbs cylindric to fusiform or pyriform 12

11. Pseudobulbs elongated ovoid, more than 8 cm long. [Lip mid-lobe clawed, ovate, margins fleshy, wrinkled, 3 entire keels, median keel terminating half way to the base of the mid-lobe, lateral keels terminating at the base of the mid-lobe. Flowers pale pink.] .... *C. dulitensis*
    Pseudobulbs ovoid to oblong-ovoid, 9.5 cm long. [Lip mid-lobe oblong-elliptic, apiculate, papillae on the mid-lobe, 3 simple keels, median keel terminates at the base of the mid-lobe, lateral keels continue to the middle of the mid-lobe. Flowers salmon pink, tips of keels yellow-brown.] .... *C. tenompokensis*

12. Pseudobulbs cylindric to fusiform 13
    Pseudobulbs pyriform 16

13. Inflorescence synanthous 14
    Inflorescence hysteranthous or initially synanthous but becoming hysteranthous as the rachis continues to extend and produces more flowers 15

14. Inflorescence with slender, erect peduncle, to 8 cm long, rachis arcuate, 8-flowered. [Lip mid-lobe clawed, orbicular, 3 keels, median shorter. Flowers greenish.] .... *C. elmeri*
    Inflorescence peduncle terete, to 17 cm long, rachis slender, few-flowered. [Lip mid-lobe triangular, margins fleshy, recurved, 3 keels extending from the base of the lip, median keel terminates at the base of the mid-lobe, lateral keels extend towards middle of the mid-lobe. Flowers olive-ochre, keel crests brown.] .... *C. motleyi*

15. Inflorescence with erect peduncle, rachis zig-zag, many-flowered. [Lip mid-lobe clawed, 3 keels, median keel slender, becomes elevated towards the tip of the mid-lobe, lateral keels conspicuous near the base of the lip, and disappear three quarters the way onto the mid-lobe. Flowers white, suffused pale salmon pink.] .... *C. brachygyne*
    Inflorescence with erect peduncle, rachis distinctly zig-zag, many-flowered. [Lip side-lobes extending to more than half the length of the lip, front rounded, 2 lateral keels terminating at the middle of the mid-lobe, median keel present only near the tip of the mid-lobe. Flowers pale dark salmon pink or white, lip deep salmon pink.] ... *C. radicosa*

16. Leaves oblong-lanceolate or linear-oblong 17
    Leaves oblong-elliptic. [Lip rounded, reflexed, fleshy tip, 2 keels, wing-shaped, terminating at the middle of the mid-lobe. Sepals and petals light brown to white, lip white with brown veins.] .... *C. endertii*

17. Leaves oblong-lanceolate. [Inflorescence stiffened, flattened, over 60-flowered. Lip mid-lobe cuneate-obovate, 2 keels terminating at the tip of the mid-lobe. Flowers yellowish-brown and white.] .... *C. bilamellata*
    Leaves elliptic-lanceolate. [Inflorescence stiff, erect, over 60-flowered. Lip mid-lobe rounded at the base, triangular apex, tri-lobed, keels grooved. Flower colour not known.] .... *C. longirachis*

18. Inflorescence hysteranthous or initially synanthous but becoming hysteranthous as the rachis continues to extend and produces more flowers 19
    Inflorescence synanthous. [4-flowered, lip mid-lobe suborbicular, 5 low keels at the base of the lip, 2 increase in elevation and undulate.] .... *C. remediosae*

19. Leaves oblong-lanceolate or linear-oblong 20
    Leaves obovate, ovate, ovate-elliptic or elliptic 22

20. Sepals less than 1.5 cm long 21
    Sepals more than 2.25 cm long. [Lip mid-lobe ovate-orbicular, 3 keels, median keel near base of the lip only, lateral keels slender, terminating at the middle of the mid-lobe.] ...... *C. steenisii*

21. Sepals 1.4 cm long. [Lip mid-lobe straight, oblong, 3 parallel, simple keels terminating at the tip of the mid-lobe, lateral keels thickened. Flowers light brown.] .... *C. contractipetala*
    Sepals 1 cm long. [Lip mid-lobe circular, 4 keels, 2 short, 2 very short. Flowers pinkish-brown.] .... *C. tumida*

22. Pseudobulbs ovoid to elongated ovoid 23
    Pseudobulbs cylindric to fusiform or pyriform 24

23. Pseudobulbs ovoid, to 7 cm long. Inflorescence peduncle strongly flattened, winged, 17 cm long, rachis zig-zag, 24 cm long, many-flowered. Lip mid-lobe clawed, ovate, margins thickened, wrinkled, 3 entire keels extending from the base of the lip, median keel terminates one third the way on the hypochile, lateral keels terminate at the middle of the mid-lobe. Flowers shell pink, keels pink .... *C. planiscapa* var. *planiscapa*
    Pseudobulbs elongated ovoid, 13 cm long. Inflorescence peduncle strongly flattened, narrowly winged, 30 cm long, rachis zig-zag, 24 cm long, many-flowered. Lip mid-lobe clawed, ovate, margin thickened, wrinkled, 3 entire keels extending from base of lip, median keel terminates one third the way on the hypochile, lateral keels terminate at the middle of the mid-lobe, 2 extra keels near margin at the base of the mid-lobe. Flowers salmon pink with orange, lip deep salmon pink .... *C. planiscapa* var. *grandis*

24. Pseudobulbs pyriform 25
    Pseudobulbs fusiform to cylindric 26

25. Pseudobulbs to 10 cm long. Lip mid-lobe clawed, 2 straight parallel keels terminating two-thirds way onto the mid-lobe, further 4 weak keels disposed laterally. [Flowers pale salmon.] .... *C. longifolia*
    Pseudobulbs 5–10 cm long. Lip mid-lobe clawed, orbicular, 3 keels, 2 simple lateral keels terminating at the middle of the mid-lobe, median keel near base of the lip only. [Flowers pale or olive green, may be greenish-brown or salmon-brown.] .... *C. prasina*

26. Pseudobulbs 20–28 cm long. Inflorescence peduncle erect, 8–11 cm long, rachis zig-zag 8–14 cm long, 36-flowered. Lip mid-lobe ovate-lanceolate, 3 keels, median keel very short, lateral keels thin and flat. [Sepals pale flesh coloured with yellowish-green keel, petals pale yellowish, lip pale rose, keels light ochre.] .... *C. kinabaluensis*
    Pseudobulbs 5–20 cm long. Not this combination 27

27. Inflorescence peduncle terete, 7.5–13 cm long, rachis thickened, stiff, not zig-zag, prominent internodes, 17 to 24-flowered. Lip mid-lobe orbicular, margins crenulate, 3 plate-like keels, median keel less so, terminating on the mid-lobe. [Dorsal sepal 2.5–3 cm long. Flowers copper to salmon-brown, lip with darker veins and keels.] .... *C. cuprea* var. *cuprea*

Inflorescence peduncle flattened, rachis zig-zag, overall to 50 cm long, many-flowered. Lip mid-lobe probably orbicular, margins crenulate. [Dorsal sepal 3.5–4 cm long. Sepals and petals pale translucent salmon pink, lip darker salmon pink.] ....
.... *C. cuprea* var. *planiscapa*

67. COELOGYNE BILAMELLATA

Coelogyne bilamellata Lindl., *Fol. Orch.-Coelog.* 14 (1854). Type: Philippines, *Cuming* s.n. (holo. K-LINDL). **Plate 9B.**

Panisea bilamellata (Lindl.) Rchb.f. in *Walp. Ann.* 6: 240 (1861).
Pleione bilamellata (Lindl.) Kuntze, *Rev. Gen. Pl.* 2: 680 (1891).

DESCRIPTION. ***Pseudobulbs*** on a stout, sheathed rhizome, close together, gradually tapering upwards from a broad base, 12–19 × 1.4 cm, bare at base. ***Leaves*** 2, oblong-lanceolate, acute, with 5 nerves, 10–20 × 2.5–5 cm with a 1.6 cm long, grooved petiole. ***Inflorescence*** hysteranthous, peduncle erect, stiff, flattened, to 28 cm long, rachis slender, pendulous, zig-zag, up to 65-flowered, flowers open in succession, floral bracts deciduous. ***Flowers*** yellowish-brown and white, the horseshoe callus and sac of the 3-lobed lip are defining characteristics. ***Dorsal sepal*** oblong, acute, fleshy, keeled, 1.4 cm long. ***Lateral sepals*** oblong, acute, fleshy, keeled, 1.4 cm long. ***Petals*** linear, 1.4 × 0.1cm. ***Lip*** with an S-shaped curve, 3-lobed, side-lobes oblong, somewhat acute, 0.6 cm long, falcate, mid-lobe cuneate-obovate, 0.3cm across, callus of 2 keels erect, extending from base of lip to tip of mid-lobe, truncate at tip, tip rounded and curled. ***Column*** 1.1 cm long.

DISTRIBUTION. Philippines (Leyte, Luzon (Laguna, Quezon), Mindanao (Agusan, Bukidnon, Surigao, Zamboanga)). Reported from Borneo.
HABITAT. Epiphytic, lithophytic on trees and rocks in forests.
ALTITUDE. 60–1000 m.
FLOWERING. April.
NOTES. The epithet refers to the two keels on the lip of the flower. *Nabaluia* de Vogel is said to be a genus allied to *Coelogyne* Lindl. through this species.

68. COELOGYNE BORNEENSIS

Coelogyne borneensis Rolfe in *Bull. Misc. Inf., Kew* 62 (1893). Type: Borneo, cult. Kew, ex Linden 22 (holo. K).

DESCRIPTION. ***Pseudobulbs*** on a short rhizome, roots moderately thick, close together, flask-shaped, at base thickened, when dry ellipsoid, 3.5 × 1.6 cm (at neck). ***Leaves*** 2, oblong-lanceolate, shortly triangular-acuminate, acute, base narrowing to long cuneate, with 7 prominent nerves beneath, 11–12 × 4.5–4.8 cm with 1.7–2 cm long, grooved petiole. ***Inflorescence*** hysteranthous, peduncle bare, strong, somewhat compressed, 16 cm long, rachis zig-zag, compressed, 35 cm long, internodes 1 cm long, many-flowered, in two rows, opening in succession, floral bracts deciduous, coriaceous, 2.6 cm long. ***Flowers*** about 5 cm across, fragrant, light chocolate-brown with dull-white nerves. ***Dorsal sepal*** oblong, narrowed towards tip, acute, strongly grooved, mainly at base, 5-nerved, 1.7 × 0.6 cm. ***Lateral sepals*** oblique, oblong-lanceolate, narrowing towards tip, acute, scarcely falcate, base concave, 5-nerved, 1.5 × 0.5 cm. **Petals** oblique, linear, somewhat falcate, shortly obliquely subacute, 1-nerved, 1.45 × 0.1 cm. **Lip** porrect, column well above, 3-lobed, concave at base, base shortly saccate, clawed with transverse fold from side to side, side-lobes erect, somewhat semi-orbicular, in front scarcely extended, obtuse, somewhat retuse, veined, 0.85 cm long, 0.925 cm across, mid-lobe porrect, broadly clawed, orbicular, ovate, tip shortly obtuse and extended, margin undulate, 0.45 cm long, 0.45 cm across, callus of 2 simple, entire, straight keels extending from base of lip and prominent but diminishing and terminating in middle of mid-lobe, 3 strongly branched nerves develop on the exterior part of mid-lobe and extend to the tip, mid-lobe is broadly oval. **Column** slender, halfway from base rounded-somewhat rectangular, becoming winged, concave below.

DISTRIBUTION. Borneo (Kalimantan (Bukit Tilung)).
HABITAT. Epiphytic on trees in lowland rain forest.
ALTITUDE. 800 m.
FLOWERING. July–August.
NOTES. The epithet refers to Borneo where the type was collected.

69. COELOGYNE BRACHYGYNE

Coelogyne brachygyne J.J. Sm. in *Bull. Jard. Bot. Buitenzorg* ser. 3, 2: 25 (1920). Type: Sumatra, Ophir District, Gunung Ophir, *Bünnemeijer* 715 (holo. BO; iso. L). **Plate 9C.**

DESCRIPTION. Pseudobulbs on a stout, sheathed rhizome, close together, elongated, 4–11 × 1.2 cm, with elongated bracts at base. **Leaves** 2, erect, oblong-lanceolate to obovate-oblong, acute-acuminate, with 7 prominent nerves beneath, 15–25 × 2.8–6 cm, narrowing gently into 4.5–5.5 cm long, channelled petiole. **Inflorescence** initially synanthous but probably becoming hysteranthous, peduncle erect, 18–33 cm long, rachis zig-zag, 50 cm or more long, many-flowered, flowers open in succession. **Flowers** colour not known. **Dorsal sepal** erect, ovate-lanceolate, shortly acute, 1.5 × 0.5 cm. **Lateral sepals** parallel, porrect, ovate-lanceolate, 1.5 × 0.4 cm. **Petals** strongly recurved, linear, acute, 1.4 × 0.15 cm. **Lip** 3-lobed, side-lobes parallel sided and lying flat, mid-lobe saccate near base, callus of 3 keels, median keel slender, glabrous, gradually increasing in height toward tip to mid-lobe, lateral keels conspicuous at base of lip but gradually disappearing about a quarter of way from tip of lip. **Column** short and scarcely curved, about 0.4 cm long, winged.

DISTRIBUTION. Sumatra (Gunung Ophir).
HABITAT. Epiphytic in lower montane forest.
ALTITUDE. 1250–1850 m.
FLOWERING. May.
NOTES. The epiphet refers to the short column of the flowers. Similar to *C. integra* Schltr. and *C. stenobulbon* Schltr.

70. COELOGYNE CANDOONENSIS

Coelogyne candoonensis Ames, *Orch.* 6: 18 (1923). Type: Philippines, Mindanao, Mt. Candoon, *M. Ramos & G. Edano* 38894 (holo. PNH†; syn. AMES).

DESCRIPTION. *Pseudobulbs* on a stout, sheathed rhizome, close together, very slender, tapering gradually to the apex, 7.5–12 cm long when mature, near base up to 1.8 cm diam., base loosely enclosed with a pair of bracts. *Leaves* 2, narrowly elliptic-lanceolate, acute, with 5 nerves prominent beneath, further 4 nerves slightly less prominent, 24 × 3.8 cm, contracted into a short, grooved petiole. *Inflorescence* initially synanthous but probably becoming hysteranthous, peduncle initially enclosed in developing leaves and bracts at base, 17.5–22.5 cm long, rachis arched, zig-zag, about 20 cm long, flowers opening in succession, floral bracts deciduous. *Flowers* white and brown. *Dorsal sepal* oblong, acute, arching forward over column, 1.5 × 0.4 cm. *Lateral sepals* oblong, acute, 1.4 × 0.4 cm. *Petals* linear-triangular, acute, tapering gradually to the tip, 3-nerved, 1.3 × 0.2 cm. *Lip* conspicuously 3-lobed, 1.2 cm long, side-lobes 0.5 cm long from tip to sinus formed with mid-lobe, upper surface minutely and densely glandulose, mid-lobe 0.9 × 0.6 cm, obovate, contracted at the base of lip in a triangular apiculate, upper half minutely and densely glandulose, hypochile with erect more-or-less sulcate callus which merges with a longitudinal keel. *Column* arcuate, 0.9 cm long, conspicuously winged.

DISTRIBUTION. Philippines (Mindanao, Bukidnon).
HABITAT. Epiphytic in lower montane forest, on trees on mossy forest slopes.
ALTITUDE. 1675–1830 m.
FLOWERING. June–July.
NOTES. The epithet refers to the locality, Mount Candoon, Mindanao, Philippines where the type specimen was collected.

71. COELOGYNE COMPRESSICAULIS

Coelogyne compressicaulis Ames & C. Schweinf., *Orch.* 6: 25 (1920). Type: Borneo, Sabah, Mt. Kinabalu, *J. Clemens* s.n. (holo. AMES). **Plate 9D.**

DESCRIPTION. *Pseudobulbs* on a thick and woody rhizome, 1.5 cm apart, fusiform or narrowly cylindric, about 6.5 × 1.1 cm. *Leaves* 2, linear-oblong, gently narrowed to each end, acute, mucronate from the extended mid vein, shining above, dull below, with 5 prominent nerves beneath, variable in size, about 20 × 2.5 cm, narrowed into 3–4 cm long, grooved petiole. *Inflorescence* hysteranthous, peduncle 20.5 cm, strongly bilaterally flattened and narrowly winged, rigid, shinning yellow, rachis strongly zig-zag, 6.5 cm long, many flowered, flowers opening in succession, floral bracts deciduous. *Flowers* with olive-yellow sepals and petals suffused salmon towards the base, lip pale salmon pink, side-lobes bright olive-yellow, mid-lobe olive-yellow suffused pale salmon towards the base, keels bright salmon, column bright salmon pink. *Dorsal sepal* narrowly lanceolate, acuminate, mucronate, cymbiform, gibbous-saccate at the base, 5-nerved, distinct dorsal keel, especially at apex, 1.85 × 0.65 cm. *Lateral sepals* narrowly lanceolate, acuminate, mucronate, cymbiform, gibbous-saccate at the base, 5-nerved, dorsal keel evanescent, apex backward pointing with sharp tip, nearly 2 × 0.55 cm. *Petals* narrowly linear, unequally and sharply bi-dentate at apex, 1-nerved. *Lip* about 1.5 cm long, from short broad claw abruptly dilated into orbicular mid-lobe, slightly retuse and undulate at margins, somewhat thickened on lateral nerves, callus of 3 keels, median keel evident at base of lip then thickened ridge along mid-lobe till it meets continuations of lateral keels. *Column* arcuate, 1.15 cm, prominently winged, shallowly toothed.

DISTRIBUTION. Borneo (Sabah (Mount Kinabalu), Sarawak).

HABITAT. Epiphytic in low mossy and dry scrub on ultramafic substrate in upper montane forest.
ALTITUDE. 800–2400 m.
FLOWERING. June, October.
NOTES. The epithet refers to the flattened nature of the peduncle. This is not an uncommon characteristic in sect. *Longifoliae* species and taken by itself it can lead to confusion.

72. COELOGYNE CONTRACTIPETALA

Coelogyne contractipetala J.J. Sm. in *Bull. Jard. Bot. Buitenzorg* ser. 3, 12: 107 (1932). Type: Sumatra, Jambi, Gunung Kerinci, *Bünnemeijer* 8608 (holo. BO; iso. L).

DESCRIPTION. *Pseudobulbs* on a strong, creeping, sheathed rhizome, forming acute angle with rhizome, 2.5–4 cm apart, elongated fusiform, 6.5–7.5 cm long, enclosed at base with 5–6.25 cm long imbricate bracts marked with black spots. *Leaves* 2, narrowly lanceolate, acute, with 5 nerves, 12.5–22.5 × 1.75–2.75 cm with a 1.25–3.25 cm long, grooved petiole. *Inflorescence* initially synanthous but probably becoming hysteranthous, peduncle strong, erect, 12.75–26 cm long, rachis zig-zag, 33 cm or longer, a long time to flowering, lax many-flowered, flowers opening in succession. *Flowers* very small, when expanded 1.5 cm across, light brown. *Dorsal sepal* oblong-ovate to lanceolate, shortly acute, 7-nerved, 1.35 × 0.6 cm. *Lateral sepals* spreading, oblique, ovate to lanceolate, acute, obtusely-angled inwards, 6-nerved, 1.4 × 0.5 cm. *Petals* at base lanceolate swelling and becoming linear, gently curved outwards, 3-nerved, 1.1 × 0.45 cm, widest at swelling near base, becoming 0.07 cm wide. *Lip* above base S-shaped and sharply curved backwards, 3-lobed, side-lobes erect, rounded, curved outwards, mid-lobe straight, oblong, becoming curved inwards about halfway, second half somewhat narrower with straight parallel sides, tip acute, margin in front obtuse angled, curved inwards, callus of 3 keels extending from base of lip where they are inconspicuous and the lateral keels are parallel, simple and continue along line of the side-lobe to tip of mid-lobe and are flattened. *Column* slender, gradually thickening, 0.55 cm long, middle obtuse-angled, curved inwards, tip narrowing, fleshy, winged, wings obtuse-angled with tip making rounded, retuse, uneven margin.
DISTRIBUTION. Sumatra (West Coast (Gunung Kerinci)).
HABITAT. Epiphytic in lower montane forest.
ALTITUDE. 1200 m.
FLOWERING. March.
NOTE. The epithet refers to the contracted base of the flower petals.

73. COELOGYNE CUPREA

Coelogyne cuprea H. Wendl. & Kraenzl. in *Gard. Chron.* ser. 3, 11: 619 (1892). Type: Provenance unknown, cult. Sander & Co. (holo. B†). **Plate 9E & F.**

var. **cuprea**

DESCRIPTION. *Pseudobulbs* on a stout, sheathed, creeping rhizome, 3–4 cm apart, ovoid to almost cylindric, 5–11 × 1–1.5 cm, loosely enclosed with bracts at base. *Leaves* 2, ovate to elliptic, obtuse to acute, with 7–9 nerves, 8–30 × 3.5–5 cm with 2.5–7 cm long, grooved petiole. *Inflorescence* hysteranthous, peduncle erect, terete, bare, 7.5–13 cm long, rachis thickened, stiff, not zig-zag but pronounced nodes, 17- to 24-flowered, flowers opening in succession, floral bracts deciduous. *Flowers* about 3 cm across, coppery to salmon-brown, lip with darker veins and keels. *Dorsal sepal* oblanceolate, obtuse, concave at base, 11-nerved, dorsal lightly keeled, 2.5–3 × 1 cm. *Lateral sepals* oblique, oblanceolate, obtuse, 9-nerved, dorsal prominently keeled, 2.5–3 × 1 cm. *Petals* linear, obtuse, 3-nerved, 2.5–3 cm long. *Lip* 3-lobed, side-lobes elliptic, rounded in front, margins entire, mid-lobe orbicular, margins crenulate, callus of 3 keels extending from base of lip to just beyond middle of mid-lobe, median keel only extends to near base of mid-lobe, median keel less pronounced than lateral keels, lateral keels plate-like, elevated, undulate, slightly crenulate on most elevated section. *Column* thickened at base, slightly arcuate, 2.75 cm long, expanding into winged hood.

DISTRIBUTION. Sumatra (West Coast), Borneo (Sarawak, Sabah).
HABITAT. Epiphytic in low montane forest on ultramafic substrate.
ALTITUDE. 800–1700 m.
FLOWERING. September–October.
NOTES. The epithet refers to the colour of the flowers.

var. **planiscapa** J. J. Wood & C. L. Chan in *Lindleyana* 5, 2: 84, f. 3 (1990). Type: Borneo, Sabah, Gunung Alab, *J.J. Wood* 784 (holo. K).

DESCRIPTION. *Pseudobulbs* on a sheathed rhizome, 5 cm apart, conical-fusiform, attenuate above, sulcate when mature, becoming curved and stained purple-red, 5–20 × 1.5–2.5 cm. *Leaves* 2, ovate, obovate or oblong-ovate, obtuse, coriaceous, margin undulate, with 9 nerves, dull dark green above, often flushed purple-red below, (6–)12–15 × 6–8 cm with (1–)3–5 cm, petiole often stained purple-red. *Inflorescence* hysteranthous, peduncle distinctly flattened, rachis zig-zag, overall (8–)18–50 cm long, many flowered, flowers opening in succession. *Flowers* with pale translucent salmon-pink sepals and petals, lip darker salmon-pink to salmon-buff, column pale salmon-pink. *Dorsal sepal* 3.5–4 × 1.3–1.4 cm. *Lateral sepals* 3.5–4 × 0.9–1 cm. *Petals* 3.1–3.5 × 0.2 cm. *Lip* 3–3.4 cm, 3-lobed, callus of 3-keels, lateral keels prominent, elevated, extending from base of mid-lobe to about three quarters along mid-lobe, median keel usually less prominent and only slightly elevated, extending a short way from the base of the lip. *Column* straight, then arcuate, 2.3–2.4 cm long, expanding halfway along into hood, incisions at either end of rounded tip.

DISTRIBUTION. Borneo (Sarawak, Sabah).
HABITAT. Epiphytic in lower montane ridge forest and oak-laurel forest.
ALTITUDE. 1200–2000 m.
FLOWERING. May.
NOTES. The typical variety has a terete peduncle whereas in this variety the peduncle is flattened.

74. COELOGYNE DULITENSIS

Coelogyne dulitensis Carr in *Gard. Bull. Straits Settlem.* 8: 73 (1935). Type: Borneo, Sarawak, Mt. Dulit, *Synge* S. 342 (holo. SING; iso. K).

DESCRIPTION. *Pseudobulbs* on a sheathed rhizome, 1 cm apart, narrowly elongate-ovoid, slightly laterally flattened, at first covered with imbricating bracts, 9.25 × 0.8 cm at base, 0.3 cm near tip (dried state). *Leaves* 2, oblong-lanceolate, acute, leathery, with 5 strong nerves, 10 × 3.35 cm with 1.75 cm long, grooved and 3-ribbed petiole. *Inflorescence* hysteranthous, peduncle bare, slender, elliptic in cross-section, 5–12 cm long, rachis zig-zag, about 5–30 cm long, 10- to 42-flowered, flowers opening in succession, floral bracts deciduous. *Flowers* with a pale pink lip, column dirty white with brown tip, hood pale pink. *Dorsal sepal* oblong, triangular-acuminate, acute, 5–7-nerved, 2.3 × 0.7 cm. *Lateral sepals* narrowly oblong-ovate, acute, somewhat S-shaped, keel beneath, 5-nerved with outer nerves branched above base, 2.3 × 0.6 cm. *Petals* narrowly linear, acuminate to acute, 1-nerved, 1.9 × 0.13 cm (wide at base). *Lip* 3-lobed, 2.2 cm long, 1.25 cm across side-lobes, side-lobes broad, roundly triangular, very obtuse, mid-lobe 1 cm long, 0.8 cm across, shortly clawed with claw transversally oblong and dilate slightly upwards, broadly ovate, obtuse, minutely apiculate, fleshy, and transversally wrinkled towards the margins, callus of 3 keels extending from base of lip, median keel terminates halfway to mid-lobe, lateral keels terminate at base of mid-lobe, keels entire. *Column* arcuate, 1.5 cm long, wings slightly roundly dilate from about the middle, hood rounded, entire.

DISTRIBUTION. Borneo (Sarawak, Sabah).
HABITAT. Epiphytic on small trees in hill forest.
ALTITUDE. 610 m.
FLOWERING. August.
NOTES. The epithet refers to Mount Dulit, Sarawak, Borneo, the locality where the type specimen was collected.

75. COELOGYNE ELMERI

Coelogyne elmeri Ames in *Leafl. Philipp. Bot.* 5: 1556 (1912). Type: Philippines, Mindanao, Todaya, Mt. Apo, *Elmer* 10694w (holo. PNH†; iso. AMES, K).

DESCRIPTION. *Pseudobulbs* on a stout, creeping rhizome which has nodes 2.1 cm. apart and only a short sheath, 4–10 cm apart, cylindric or somewhat fusiform, 6.5–7.5 × 1 cm, wrinkled when dry, bare of bracts at base except when developing. *Leaves* 2, elliptic-lanceolate, acute, with 9 main nerves, dorsal prominent beneath, somewhat leathery, 9–16 × 2.8–4.5 cm with 2.4–2.7 cm long, grooved petiole. *Inflorescence* Synanthous, peduncle bare, slender, erect, 8 cm long, rachis arching, zig-zag, 7.2 cm long, 8-flowered, flowers opening in succession, floral bracts deciduous. *Flowers* greenish. *Dorsal sepal* lanceolate, hooded, 1.2 × 0.35 cm. *Lateral sepals* lanceolate, acuminate-acute, 1 × 0.35 cm. *Petals* narrowly linear, 1.1 × 0.1 cm (near base), 0.05 cm (near tip). *Lip* 1.1 cm long, obovate with a deep recess on either side, 3-lobed, pandurate, side-lobes rounded, mid-lobe separated by a short claw, somewhat orbicular, callus of 3 keels, 2 lateral keels and a short median keel near base of lip. *Column* arcuate, 0.5 cm long.

DISTRIBUTION. Philippines (Mindanao).
HABITAT. Epiphyte in clumps on moss-covered tree trunks in hill forest.
ALTITUDE. 1100–1300 m.
FLOWERING. May–July.
NOTES. This species is named after A.D.E. Elmer, the collector of the type specimen.

76. COELOGYNE ENDERTII

Coelogyne endertii J.J. Sm. in *Bull. Jard. Bot. Buitenzorg* ser. 3, 11: 94 (1931). Type: Borneo, Kalimantan, Long Petak, *Endert* 3200 (holo. L, iso. BO). **Plate 10A**.

DESCRIPTION. ***Pseudobulbs*** on a stout rhizome close together, erect, pyriform, neck thickened, elongated base, 7.5–16 × 0.8 cm (at base when dried). ***Leaves*** 2, erect, oblong-elliptic, acuminate-acute, with 9 nerves, 13.5–19 × 4.4–4.7 cm, when dry papery, base gently narrows into 1.2–2.2 cm long, grooved petiole. ***Inflorescence*** initially synanthous but probably becoming hysteranthous, peduncle elliptic cross-section, 13–21 cm long, rachis zig-zag, 10 cm long, loosely many flowered, flowers opening in succession, floral bracts deciduous. ***Flowers*** with light brown to nearly white sepals and petals, lip white with dark-brown longitudinal markings, column light salmon coloured. ***Dorsal sepal*** narrowly oblong, tip reflexed, groove along dorsal keel, convex, 5-nerved, 2.5 × 0.75 cm. ***Lateral sepals*** oblique, lanceolate, acuminate, obtuse-angled, concave, 5-nerved, 2.3 × 0.63 cm. ***Petals*** reflexed, oblique, linear, acute, falcate, 1-nerved, 2.1 × 0.2 cm. ***Lip*** parallel to column, gently reflexed, 3-lobed, side-lobes erect, 1.4 cm long, broad base, porrect, front rounded, 4-angled, margins gently retuse, mid-lobe porrect to strongly reflexed, rounded, fleshy tip, callus of 2 undivided keels narrow wing-shape which thicken and terminate in middle of mid-lobe. ***Column*** slender, arcuate, 1.4 cm long, narrowly winged.

DISTRIBUTION. Borneo (Kalimantan, Sarawak).

HABITAT. Epiphytic on small trees in hill forest.

ALTITUDE. 800–900 m.

FLOWERING. September.

NOTES. This species is named after Frederik Endert, the collector of the type specimen.

77. COELOGYNE INTEGRA

Coelogyne integra Schltr. in *Bot. Jahrb. Syst.* 45, (104): 3 (1911). Type: Sumatra, Padang Panjang, *Schlechter* 15908, 16014 (syn. B†; isosyn. BO, L).

DESCRIPTION. ***Pseudobulbs*** on a thick, sheathed, creeping, branching rhizome, 1 cm apart, ovoid at base, 4-angled, glossy, 3–7 × 1.2–1.5 cm. ***Leaves*** 2, erect-spreading, elliptic to ovate-elliptic, acuminate, leathery, 7–25 × 2.5–6 cm, narrowing into petiole. ***Inflorescence*** initially synanthous but probably becoming hysteranthous, peduncle slender, rachis arching, slightly zig-zag, elongating, many-flowered, flowers opening in succession, floral bracts deciduous. ***Flowers*** about 2.5 cm across, fragrant, rather waxy, sepals and petals white, lip veined with golden-yellow. ***Dorsal sepal*** oblong, somewhat acute, smooth, 1.2 × 0.4 cm. ***Lateral sepals*** oblique, oblong, somewhat acute, dorsal keel prominent, 1.2 × 0.4 cm. ***Petals*** spreading, narrowly linear, obtuse, 1 × 0.1 cm. ***Lip*** indistinctly 3-lobed, mid-lobe ovate, tip acute, margins at base reflexed, entire, keels not prominent. ***Column*** short, 0.4 cm long, semi-terete, expanding to form winged hood, tip tri-lobed.

DISTRIBUTION. Sumatra (West Coast), ?Philippines.

HABITAT. Not known.

ALTITUDE. 1200–2200 m.

FLOWERING. June–September.

NOTES. The epithet refers to the lack of distinct side-lobes on the lip of the flowers.

78. COELOGYNE KINABALUENSIS

Coelogyne kinabaluensis Ames & C. Schweinf., *Orch.* 6: 30 (1920). Type: Borneo, Sabah, Mt. Kinabalu, Marai Parai Spur, *J. Clemens* 229 (holo. AMES; iso. BO). **Plate 10B**.

DESCRIPTION. *Pseudobulbs* on a thick, woody, creeping rhizome, at intervals, very nearly cylindric and stem-like, 20–28 × 0.5–0.85 cm, enclosed in large bracts at base. *Leaves* 2, narrowly elliptic, sharply acuminate, coriaceous, many nerved with 5–9 prominent beneath, 18 × 5.5 cm with 2.2 cm long, indistinct petiole. *Inflorescence* hysteranthous, peduncle rigid, erect, 8.5–11 cm long, rachis rigid, erect, zig-zag, 8–13.5 cm long, 36-flowered, flowers opening in succession, floral bracts deciduous. *Flowers* large, sepals pale flesh coloured with yellow-green keel, petals pale yellowish, lip pale rose coloured with keels bright ochre-yellow along summit, column pale pinkish salmon, darker near the whitish tip, base whitish. *Dorsal sepal* lanceolate, nearly acute, carinate, 4.3 × 1.45 cm. *Lateral sepals* oblong-lanceolate, broadest at the base, sharply acute, slightly falcate, carinate, 4 × 1.25 cm. *Petals* very narrowly linear, abruptly acute, lower portion with 3 nerves, upper part with 1 nerve, 3.8 × 0.175 cm. *Lip* deeply 3-lobed, side-lobes ovate, erect, slightly exceeding half length of mid-lobe, mid-lobe ovate-lanceolate, rounded at apex with a broadly obtuse sharp tip, lateral margins rolled back throughout, mid-lobe covered with parallel, irregular ridges and 2 thin, flat keels about 0.2 cm in height and extending about three fifths the length of mid-lobe, additional short median keel. *Column* slightly shorter than side-lobes, arcuate, winged and rounded apex.

DISTRIBUTION. Borneo (Sarawak, Sabah).

HABITAT. Epiphytic in lower montane oak-laurel forest, sometimes on ultramafic substrate.

ALTITUDE. 900–1800 m.

FLOWERING. April.

NOTES. The epiphet refers to the origin of this species Mount Kinabalu, Sabah, Borneo.

79. COELOGYNE LONGIFOLIA

Coelogyne longifolia (Blume) Lindl., *Gen. Sp. Orch. Pl.* 42 (1830). Type: Java, Gunung Arjuna, *Lobb* s.n. (holo. K). **Plate 10C**.

Chelonanthera longifolia Blume in *Bijdr.* 385 (1825).
Cymbidium stenopetalum Reinw. ex Lindl., *Fol.-Orch. Coelog.* 14 (1854). Type: provenance unknown (holo. K-LINDL).
Pleione longifolia (Blume) Kuntze, *Rev. Gen. Pl.* 2: 680 (1891).

DESCRIPTION. *Pseudobulbs* on a thick, creeping rhizome, 3 cm apart, oval or flask-shaped, slightly compressed, somewhat 4-angled, about 10 × 3 cm. *Leaves* 2, narrowly obovate-lanceolate, acute to acuminate, undulate, with 3 nerves prominent below, papery, 50 × 5.5 cm with 10 cm long petiole. *Inflorescence* initially synanthous but becoming hysteranthous when flowers finally open, peduncle and rachis of equal length, rachis zig-zag, overall extending to about 65 cm, many-flowered, flowers opening in succession, floral bracts ovate, coriaceous, 4 cm long, deciduous. *Flowers* about 3.5 cm across, pale salmon-coloured. *Dorsal sepal* oblong-lanceolate, somewhat acute, concave, about 2 cm long. *Lateral sepals* narrower, distinctly keeled, 2 cm long. *Petals* spreading, linear and straight, 2 cm long. *Lip* shortly clawed, obscurely 3-lobed, side-lobes erect, small and located at the lip base, front rounded, mid-lobe obovate, gradually opening to an obtuse apex, margin reflexed, callus of 2 straight parallel keels extending from the base of lip for two thirds length of lip, a further 4 weak keels disposed laterally. *Column* short, slender, arcuate, tip narrowly winged.

DISTRIBUTION. Burma, Sumatra (Aceh (Gunung Leuser)), Java (East (Gunung Arjuna)).

HABITAT. Epiphytic, sometime terrestrial in lower and upper montane forest and open areas; often forms huge dense clumps.

ALTITUDE. 1400–2100 m.

FLOWERING. July–November.

NOTES. The epithet refers to the long leaves.

80. COELOGYNE LONGIRACHIS

Coelogyne longirachis Ames, *Orch.* 7: 88 (1922). Type: Philippines, Mindanao, Bukidnon, Mahilucal River, *M. Ramos & G. Edano* 38671 (holo. PNH†; syn. AMES; iso. K).

DESCRIPTION. *Pseudobulbs* in series on a stout, sheathed rhizome, 1.5 cm apart, narrowly pyriform, 10–17 × 1.2–2 cm, thicker in lower half and attenuated towards tip, bare of bracts at base except when developing. *Leaves* 2, elliptic-lanceolate, gently acute, with 5 nerves prominent beneath, many additional nerves evident, 15.5–30 × 3.5–5 cm with 2–3.7 cm long, compressed petiole. *Inflorescence* initially synanthous but probably becoming hysteranthous, peduncle erect, stiff, 20–30 cm long, rachis arching, zig-zag, 30 cm long, to 63-flowered, flowers opening in succession, floral bracts deciduous. *Flowers* colour not known. *Dorsal sepal* narrowly elliptic, abruptly acute, slightly fleshy, 0.7 × 0.6 cm. *Lateral sepals* similar. *Petals* linear, acute, slightly curved and gently attenuated from near base towards tip, 1.1 × 0.1 cm. *Lip* 1.2 cm long, slightly fleshy, prominent 3-lobed, side-lobes nearly square, 0.35 × 0.35 cm, margins with unequal triangular teeth, front rounded, mid-lobe 0.65 × 0.3 cm, base slightly rounded, apex triangular-acute, tip with 3 teeth, keels sulcate. *Column* 0.8 cm long, fleshy, apex with wings, wings with small teeth.

DISTRIBUTION. Philippines (Mindanao).

HABITAT. Epiphytic in riverine forest.

ALTITUDE. 1200–1300 m.

FLOWERING. July.

NOTES. The epithet refers to the long rachis of the inflorescence.

81. COELOGYNE MOTLEYI

Coelogyne motleyi Rolfe ex J.J. Wood, D.A. Clayton & C.L. Chan in *Sandakania* 11: 35–42 (1998). Type: Indonesia, Kalimantan Timur, near Balikapapan, alt. 50–200 m, 1981, *Franken & Roos* s.n., cult. Leiden. no. 20858 (holo. L, alcohol material only). **Plate 10D.**

DESCRIPTION. *Pseudobulbs* on a creeping, sheathed rhizome, 0.8– 2.5 cm apart, cylindrical or fusiform, (3–)5–7 × 1–1.5 cm, cataphylls subtending pseudobulbs and leaves (3–)4, ovate to triangular-ovate, acute to acuminate, 1–7 cm long. *Leaves* 2, mature blade narrowly elliptic, acute to shortly acuminate, thin-textured, coriaceous, with 5 main nerves, 3 remaining prominent in dried material, (12–)20–24.5 × (1.4–)2.5–3 cm with conduplicate, (1–)2–3.5(–5) cm long petiole. *Inflorescence* synanthous, erect to gently curving, peduncle terete, naked, 8–17 cm long; rachis slender, zig-zag, 3–6.5 cm long, 2- to 3-flowered, flowers opening simultaneously, floral bracts narrowly elliptic or ovate-elliptic, subobtuse to acute, deciduous, 2.5–4 cm long. *Flowers* appearing not to open widely, pale olive-ochre, keels on the lip with a brown crest, anther-cap yellow with brown along margins. *Dorsal sepal* ovate-elliptic, obtuse to subacute, concave, 8–10-nerved, 2.6–2.7 × (0.8)1 cm. *Lateral sepals* narrowly oblong-elliptic, slightly oblique, acute, concave distally, sigmoid, 7-nerved, 2.5–2.6 × 0.8 cm. *Petals* linear, obtuse, 3-nerved, 2.5–2.6 × 0.12–0.19 cm. *Lip* 3-lobed, gently curved, concave at base, 2.2–2.3 cm long, 1.3 cm wide across side lobes, side lobes erect, semi-orbicular, rounded, margin entire, thin-textured, nervose, 1.2–1.3 cm long, c. 0.5 cm long, mid-lobe narrowly ovate, obtuse, apiculate, margins thickened, fleshy and uneven, 1.1–1.2 × 0.8–0.85 cm, callus of 3 elevated keels, running along the 3 central nerves, extending from base of lip, laterals terminating a quarter way onto mid-lobe, median terminating at base of mid-lobe, with 8 main nerves. *Column* slightly curved, truncate, wings entire, 1.7–1.8 cm long, tip retuse.

DISTRIBUTION. Borneo (Kalimantan (Timur, Selatan), Sarawak (Bau, Lawas Districts), Sabah (Keningau to Sepulot Rd).

HABITAT. Terrestrial, ?epiphytic in lowland mixed dipterocarp forest; podzolic dipterocarp/*Dacrydium* forest on very wet sandy soil; rocky forest.

ALTITUDE. 50–400 m.

FLOWERING. August.

NOTES. The specific epithet honours James Motley, a British civil engineer who was appointed superintendent of the coal mining operations of a private company in the territory of the Sultan of 'Bandjermasin' (now Banjarmasin). He was an amateur naturalist who collected and cultivated orchids in Borneo. Natives at Bangkal near Banjarmasin murdered him in 1859 only a short time after making the earliest known collections of this *Coelogyne*.

82. COELOGYNE PLANISCAPA

Coelogyne planiscapa Carr in *Gard. Bull. Straits Settlem.* 8: 74 (1935). Type: Borneo, Sarawak, Dulit ridge, *Synge* S.419 (holo. SING; iso. K). **Plate 10E.**

var. **planiscapa**

DESCRIPTION. *Pseudobulbs* on a sheathed rhizome, 3 cm apart, elongate-ovoid, slightly laterally flattened, ribbed, covered at first with fleshy tubular sheaths, 7 × 1.5 cm dia., at base, 0.6cm dia., just below apex. *Leaves* 2, linear-oblanceolate, acute, rigid, leathery, with 5 strong nerves, 21.5 × 2.7 cm with 4–5 cm long, grooved petiole. *Inflorescence* hysteranthous, elongate, stout, peduncle strongly laterally flattened and narrowly winged, narrowly elliptic in cross-section, 17 cm long, rachis zig-zag, 24 cm long, many flowered, flowers opening in succession. *Flowers* widely spreading, shell-pink, keels pink, tip of column brown. *Dorsal sepal* oblong, acuminate, obtuse, minutely apiculate, 5-nerved, outer nerves branched, 3.1 × 1.28 cm. *Lateral sepals* ovate, acuminate, narrowly obtuse, minutely apiculate, keeled beneath, 5-nerved, outer nerves branched, slightly longer and broader than dorsal sepal. *Petals* linear, subacute, minutely apiculate, 3-nerved, 2.83 × 0.27 cm. *Lip* 3-lobed, 2.75 × 1.3 cm, side-lobes erect, S-shaped, roundly oblong-triangular, very obtuse, mid-lobe clawed, with claw subquadrate and slightly cuneate, dilated upwards, ovate, 0.6 cm long, obtusely apiculate, thickened and transversely wrinkled towards the margins with the median nerve elevated near the tip, callus of 3 entire keels extending from base of lip, a low median keel extending to one third the distance to mid-lobe, 2 elevated, rather thin becoming fleshy lateral keels extending to and ending abruptly at middle of mid-lobe. *Column* arcuate, 1.85 cm long, wings gently dilate from middle, shortly roundly triangular at tip, hood slightly rounded subquadrate, nearly truncate, entire.

DISTRIBUTION. Borneo (Sarawak, Sabah).

HABITAT. Epiphytic, terrestrial, lithophytic in mossy lower montane forest; low and rather open, wet, somewhat podzolic heath forest with dense undergrowth of pandanus and rattan palms on sandstone and shale outcrops.

ALTITUDE. 800–1800 m.

FLOWERING. September–October.

NOTES. The epithet refers to the flattened form of the peduncle of the inflorescence.

var. **grandis** Carr in *Gard. Bull. Straits Settlem.* 8: 202 (1935). Type: Borneo, Sabah, Mt. Kinabalu, Penibukan ridge, *Carr* 3120 in SFN 27464 (holo. SING, iso. K).

DESCRIPTION. *Pseudobulbs* on a sheathed rhizome, 8 cm apart, elongate-ovoid, slightly laterally flattened, ribbed, covered at first with fleshy tubular bracts, 13 × 1.5 cm dia. at base, 0.6 cm dia. just below tip. *Leaves* 2, oblong-oblanceolate, shortly acuminate, rigid, coriaceous, with 5 strong nerves, 33–46 × 4–7 cm with 7.5 cm long, grooved petiole. *Inflorescence* hysteranthous, elongate, stout, peduncle strongly laterally flattened and narrowly winged, narrowly elliptic in cross-section, 30 cm long, rachis zig-zag, 24 cm long, many flowered, flowers opening in succession. *Flowers* wide spreading, sepals and petals salmon-pink tinged with orange, lip deep salmon-pink, column white, tinged pale-salmon. *Dorsal sepal* oblong, acuminate, obtuse, minutely apiculate, 5-nerved, outer nerves branched, 4–4.5 × 1.2–1.75 cm. *Lateral sepals* ovate, acuminate, narrowly obtuse, minutely apiculate, keeled beneath, 5-nerved, outer nerves branched, 4–4.3 × 1.05–1.4 cm. *Petals* linear, subacute, minutely apiculate, 3-nerved, 3.7–4 × 0.25–0.28 cm. *Lip* 3-lobed, 2.75 × 1.3 cm, side-lobes erect, S-shaped, roundly oblong-triangular, very obtuse, mid-lobe clawed, with claw subquadrate and slightly cuneate, dilated upwards, ovate, 0.6 cm long, obtusely apiculate, thickened and transversely wrinkled towards the margins with the median nerve elevated near the tip, callus of 3 entire keels extending from base of lip, a low median keel extending to one third the distance to mid-lobe, 2 very elevated, rather thin becoming fleshy lateral keels extending to and ending abruptly at middle of mid-lobe, sometimes extra 2 keels near margins at base. *Column* arcuate, 2.2–2.3 cm long, wings gently dilate from middle, shortly roundly triangular at tip, hood slightly rounded subquadrate, nearly truncate, entire.

DISTRIBUTION. Borneo (Sabah).

HABITAT. Epiphytic in lower and upper montane forest, oak-laurel forest, sometimes on ultramafic substrate.

ALTITUDE. 1300–2000 m.

FLOWERING. March, November.

NOTES. The epithet for this variety refers to the larger form of the plant when compared with the typical variety.

83. COELOGYNE PRASINA

Coelogyne prasina Ridl. in *Linn. Soc., Bot.* 32: 326 (1896). Type: Peninsular Malaysia, Kedah Peak, *Ridley* 5131 (holo. SING). **Plate 10F.**

Coelogyne rhizomotosa J.J. Sm. in *Recueil Trav. Bot. Néerl.* 1: 146 (1904). Type: Sulawesi, Tomohan, *Förstermann* s.n. (holo. L).
C. modesta J.J. Sm., *Orchids Java* 141 (1905). Type: Java, Gunung Guntur, *Bünnemeijer* 908 (holo. B†; iso. L).
C. vagans Schltr. in *Bot. Jahrb. Syst.* 45(104): 5 (1911). Type: Sumatra, Fort de Cock, Bukittinggi, *Schlechter* 15927 (holo. B†; iso. AMES, BO).
C. rhizomatosa J.J. Sm. var. *quinquelobata* J.J. Sm. in *Bull. Jard. Bot. Buitenzorg* ser. 3, 9: 140 (1927). Type: Maluku, Ternate, *Beguin* 1463 (holo. BO; iso. L).

DESCRIPTION. *Pseudobulbs* on a narrow, elongate, bare rhizome, 6–15 cm apart, ovoid-pyriform, tapering to narrow tip, to 10 cm long, thickest one third from base, tapering to narrow apex. *Leaves* 2, elliptic, acute, plicate, 7-nerved, 12–28 × 4.5–7.5 cm with 2–4 cm long, grooved petiole. *Inflorescence* initially synanthous and sometimes becoming hysteranthous, peduncle slender, bare, 9 cm long, rachis very irregular, zig-zag, several- to many- flowered, internodes about 1 cm apart, flowers opening in succession, floral bracts deciduous. *Flowers* about 2.5 cm across, uniformly pale green or olive-green. *Sepals* 1.2–1.4 cm long. *Petals* 1.2–1.4 × 0.1–0.2 cm, widest at base tapering to very narrow tip. *Lip* 3-lobed, side-lobes erect, rounded, mid-lobe orbicular, 0.5–0.6 cm across, slightly longer than wide, slightly widening from base, tip broadly rounded, notched, callus of 3 keels, lateral keels extending from near base of lip to middle of mid-lobe, curved, smooth, median keel near base of lip only. *Column* about 0.6 cm long.

DISTRIBUTION. Peninsular Malaysia (Kedah), Maluku (Ternate), Sumatra (East Coast, West of Lake Toba), Borneo (Sabah), Java (West (Gunung Guntur)), Sulawesi.

HABITAT. Epiphytic, terrestrial on trees in hill forest and terrestrial on mossy clumps in heath (Kerangas) forest.

ALTITUDE. 1000–1500 m.

FLOWERING. June, September–November, January.

NOTES. The epithet refers to the leek-green colour of the rhizomes and pseudobulbs. Following conversations and correspondence with E.F. de Vogel, *C. prasina* Ridl., *C. rhizomotosa* J.J. Sm., *C. modesta* J.J. Sm. and *C. vagans* Schltr. are considered to be conspecific by virtue of the similar vegetative parts and similar form of the lip.

84. COELOGYNE QUINQUELAMELLATA

Coelogyne quinquelamellata Ames, *Orch.* 6: 280 (1920). Type: Philippines, Mindanao, Surigao Province, *Wenzel* 1206 (holo. AMES). **Plate 11A & B.**

DESCRIPTION. *Pseudobulbs* on a stout rhizome, close together, narrowly pyriform to flask-shaped, somewhat 4-angled, 4–9 × 1.5 cm, thickest near base, enclosing bracts persistent, fibrous. *Leaves* 2, oblong-elliptic, attenuated at base and tip, acute, strongly coriaceous, 10–30

× 1.5–4.5 cm with 3.5–4.8 cm long petiole. *Inflorescence* initially synanthous but becoming hysteranthous, peduncle bare, erect, flattened, 35 cm long, rachis slender, arching zig-zag, 17.5 cm long, 28-flowered, flowers opening in succession, floral bracts deciduous. *Flowers* with reddish or orange sepals and petals, lip orange and white. *Dorsal sepal* ovate-lanceolate, acute, strongly concave, suberect, 1.2 × 0.7 cm. *Lateral sepals* lanceolate, acute, 1.1 × 0.4 cm. *Petals* linear, 0.9 × 0.1 cm. *Lip* simple and undivided, 0.7 × 0.1 cm, transversely elliptical, small tip, 5 abbreviated keels at base. *Column* arcuate, apex broadly winged.

DISTRIBUTION. Philippines (Mindanao).
HABITAT. Epiphytic in lowland rain forest.
ALTITUDE. 150 m.
FLOWERING. August–September.
NOTES. The epithet refers to the five keels to be found on the lip of the flower.

85. COELOGYNE RADICOSA

Coelogyne radicosa Ridl. in *J. Fed. Malay States Mus.* 6: 57 (1915). Type: Peninsular Malaysia, Perak, *Scortechini* 1342 (holo. K). **Plate 11C.**

Coelogyne carnea Hook.f., *Fl. Brit. India* 5: 838 (1890), *non* (Blume) Rchb.f. (1861).
C. stipitibulbum Holttum in *Gard. Bull. Straits Settlem.* 11: 278 (1947). Type: Peninsular Malaysia, Cameron Highlands, collector unknown (holo. SING).

DESCRIPTION. *Pseudobulbs* on a stiff, creeping, sheathed rhizome, 1–3 cm apart, base slender, swelling and somewhat flattened in the apical half and with irregular groves when old, 8 × 0.7 cm, enclosed by loosely sheathing bracts at base. *Leaves* 2, elliptic, acute, 9-nerved, dorsal prominent beneath, 9–20 × 2.5–4 cm with 3.5–5 cm long, grooved petiole. *Inflorescence* hysteranthous, peduncle erect, flat, bare, to 12 cm long, rachis distinctly zig-zag, elongating to 30 cm long, 10- to 30-flowered, flowers opening in succession, floral bracts deciduous. *Flowers* pale dark salmon-pink or white, lip side lobes brown-veined. *Dorsal sepal* 1.7–2.5 cm long. *Lateral sepals* 1.7–2.5 cm long. *Petals* 1.7–2.5 × 0.1–0.2 cm. *Lip* 3-lobed, nearly 2.5 cm long, side-lobes extending to rather more than half the length of lip, free ends short and broadly rounded, mid-lobe nearly orbicular, widening slightly from base, callus of 3 keels, 2 lateral keels extending from base of lip to just beyond the middle of mid-lobe, elevated, smooth, median keel, very short, near tip only. *Column* 1.8 cm long.

DISTRIBUTION. Thailand (Gunung Ina, Betong), Peninsular Malaysia, Borneo (Sabah).
HABITAT. Epiphytic on mountain ridges in podsol forest, lower montane dipterocarp forest.
ALTITUDE. 1000–1300 m.
FLOWERING. August.
NOTES. The epithet refers to the large number of roots. *C. carnea* Hook.f. should take precedence over *C. radicosa* Ridl. However, the name had been previously used as *C. carnea* (Blume) Rchb.f. in *Walpers Ann.* 6: 237 (1861) which in turn was incorrect as the specimen was one of the variations of *Pholidota carnea* (Blume) Lindl. (*Gen. Spec. Orch.* 37 (1830)). I have retained *C. radicosa* to avoid confusion.

86. COELOGYNE REMEDIOSAE

Coelogyne remediosae Ames & Quisumb. in *Philipp. J. Sci.* 49: 484 (1932). Type: Philippines, without exact location, cult. Quisumbing 82232 (holo. AMES).

DESCRIPTION. *Pseudobulbs* on a stout, sheathed, creeping rhizome, 2.6–3.4 cm apart, narrowly cylindric, 11–13 × 2–2.7 cm, tapered at both ends, bare of bracts at base except when developing. *Leaves* 2, elliptic-oblong or elliptic, acute, with 5 nerves prominent beneath, 18–26 × 5.5–8.5 cm with 4.5–5 cm long petiole. *Inflorescence* synanthous, peduncle bare, erect, flattened, 15.5 cm long, rachis slender, erect or curved at top, 6.5 cm long, 4-flowered, flowers opening in succession. *Flowers* sepals sea-foam yellow and a deep sea-foam green in the middle, petals sea-foam green, column glass green, lip marguerite yellow except for side-lobes and centre of mid-lobe which are brown. *Dorsal sepal* oblong-lanceolate, acute, 2–2.2 × 0.8–1 cm. *Lateral sepals* oblong, acute, 2.5–2.9 × 1–1.3 cm. *Petals* linear, obtuse, minutely apiculate at the oblique tip, 2.5–2.6 × 0.25 cm. *Lip* somewhat fleshy, 3-lobed, about 2.5 cm long, side-lobes rather short, front broadly obtuse or rounded, mid-lobe suborbicular, 1.5 × 1.5 cm, margins undulate, callus of 5 keels at base of lip, low, 2 keels increasingly elevated, undulate. *Column* arcuate, 1.8 cm long, fleshy, clearly winged.

DISTRIBUTION. Philippines (?Luzon (Sierra Madre).
HABITAT. Not known.
ALTITUDE. Not known.
FLOWERING. Not known.
NOTES. This species dedicated to Mrs Remedios C. Gonzales who cultivated the type in her garden. The plant was obtained from a peddler and it was thought to have come from the Sierra Madre, Luzon.

87. COELOGYNE STEENISII

Coelogyne steenisii J.J. Sm. in *Bull. Jard. Bot. Buitenzorg* ser. 3, 12: 108 (1932). Type: Sumatra, Palembang, Ranau, summit of Gunung Pakiwang, *Steenis* 3886 (holo. BO; iso. L).

DESCRIPTION. *Pseudobulbs* on a strong, sheathed rhizome, close together, elongated fusiform from thickened base, 5–11 cm long. *Leaves* 2, narrowly lanceolate, acute to acuminate, with 5–7 nerves, prominent slender dorsal nerve, 19–25 × 2.8–3 cm with 2–5.5 cm long, grooved petiole. *Inflorescence* initially synanthous but probably becoming hysteranthous, peduncle compressed, 12.5–25 cm long, rachis zig-zag, 12.5 cm or longer, a long time coming into flower, many flowered, flowers opening in succession. *Flowers* large, pale salmon-coloured. *Dorsal sepal* oblong-ovate, somewhat obtuse, acuminate, 7 nerves, dorsal nerve prominent, 2.35 × 0.8 cm. *Lateral sepals* oblique, ovate-lanceolate, somewhat acute, 7 nerves, dorsal nerve fleshy, prominent, 2.3 × 0.65 cm. *Petals* oblong swelling at base becoming linear, shortly acuminate, 3 nerves, dorsal grooved, 2 × 0.1 cm. *Lip* arcuate, somewhat S-shaped, 3-lobed, base of lip transversally grooved and folded, when expanded oblong-ovate, 2.2 cm long, side-lobes 1.2 × 1.2 cm, rise above column, erect, tip recurved, mid-lobe conspicuous, margins strongly recurved, nerves form furrowed and wrinkled appearance on mid-lobe, ovate-orbicular, obtuse, callus of 2 conspicuous keels slender at base of lip and gradually becoming elevated and ending abruptly at middle of mid-lobe, keels erect, irregular, rounded, swelling, inconspicuous median keel vanished at base of mid-lobe. ***Column*** scarcely arcuate, 1.2 cm long, winged, tip narrowing, reflexed, obtuse, irregular margin.

DISTRIBUTION. Sumatra (East Central (Palembang).
HABITAT. Epiphytic on moss-covered trees in upper montane forest.
ALTITUDE. 1600 m.
FLOWERING. November.
NOTES. The species is named after C.G.G.S. van Steenis, the collector of the type specimen.

88. COELOGYNE STENOBULBON

Coelogyne stenobulbon Schltr. in *Bot. Jahrb. Syst.* 45(104): 4 (1911). Type: Sumatra, Padang Panjang, *Schlechter* 15997 (holo. B†; iso. BO).

DESCRIPTION. ***Pseudobulbs*** loosely arranged on a rising rhizome, cylindrical, 6–13 × 3.5–4 cm. ***Leaves*** 2, erect, spreading, broadly elliptical, acute or shortly acuminate, 7–12 × 3.4–5 cm, base tapering but petiole not very evident. ***Inflorescence*** synanthous, peduncle and rachis erect, slender, flowers open in succession. ***Flowers*** with salmon-pink sepals and petals, lip brown. ***Dorsal sepal*** ovate-lanceolate, subacute, fleshy, glabrous 2 cm long. ***Lateral sepals*** oblique, fleshy, glabrous, 2 cm long. ***Petals*** narrowly linear, acute. ***Lip*** circular oblong-lingulate, entire, curved, tip subacute, callus of 3 keels, 2 extending from base of mid-lobe to tip of mid-lobe, median keel thickened. ***Column*** shortly curved, 0.7 cm long, tri-lobed.

DISTRIBUTION. Sumatra (West coast).
HABITAT. Not known.
ALTITUDE. 1200 m.
FLOWERING. January.
NOTES. The epiphet refers to the narrowly cylindrical form of the pseudobulbs.

89. COELOGYNE STENOCHILA

Coelogyne stenochila Hook.f., *Fl. Brit. India* 5: 837 (1890). Type: Peninsular Malaysia, Perak, summit of Gunung Batu Puteh, *L. Wray* 884 (holo. K). **Plate 11D.**

DESCRIPTION. ***Pseudobulbs*** on a creeping, sheathed rhizome, 0.5 cm apart, small, ovoid, sulcate with age, 1.25–3 × 1 cm. ***Leaves*** 2, small, elliptic, subacute, coriaceous, 5 × 1.5 cm with 1.5–2 cm long petiole. ***Inflorescence*** hysteranthous, peduncle bare, 4 cm long, rachis pendulous, 50 cm long, 25-flowered, flowers opening in succession, floral bracts deciduous. ***Flowers*** 2.5 cm across, white lip, 2 yellow-brown spots on mid-lobe, orange-yellow column. ***Sepals*** ovate-lanceolate, acuminate, 1.5 cm long. ***Petals*** narrowly linear, 0.2 cm wide. ***Lip*** entire or slightly 3-lobed, side-lobes elliptic, front rounded, margins minutely dentate, mid-lobe orbicular, margins erose, minutely dentate, callus of 3 keels extending from base of lip to tip of mid-lobe, lateral keels diverging then converging on mid-lobe. ***Column*** slender, straight, 1–1.2 cm long, tip slightly winged.

DISTRIBUTION. Peninsular Malaysia (Perak).
HABITAT. Epiphytic on exposed ridges in upper montane forest.
ALTITUDE. 1430–2000 m.
FLOWERING. August.
NOTES. The epiphet refers to the narrow lip of the flower.

90. COELOGYNE TENOMPOKENSIS

Coelogyne tenompokensis Carr in *Gard. Bull. Straits Settlem.* 8: 203 (1935). Type: Borneo, Sabah, Mt. Kinabalu, main spur above Tenompok, *Carr* 3270 in SFN 27501 (holo. SING; iso. AMES, BM, K). **Plate 11E.**

DESCRIPTION. *Pseudobulbs* on a stout, creeping rhizome, 2 cm apart, ovoid or oblong-ovoid, rather strongly laterally flattened, to 9.5 × 1.7 cm. *Leaves* 2, lanceolate, linear-lanceolate or linear-oblanceolate, acuminate, acute, rather thin in texture, 9.75–29 × 1.3 cm with 4 cm long, petiole. *Inflorescence* initially synanthous but probably becoming hysteranthous, elongate, peduncle bare, strongly laterally flattened, to 20 cm long, rachis sinuous, slender, 35 cm long, at first with imbricating bracts at apex, then caducous, up to more than 50-flowered, flowers opening singly or in pairs. *Flowers* large, salmon-pink with tip of keels on lip yellow-brown. *Dorsal sepal* oblong-lanceolate, acute, apiculate, inconspicuously keeled, 7-nerved, 2.3 × 0.55 cm. *Lateral sepals* oblong-lanceolate, acute, apiculate, S-shaped inconspicuously keeled, 7-nerved, 2.3 × 0.45 cm. *Petals* narrowly subulate, long acuminate, very acute, 1-nerved, 2.1 × 0.07 cm. *Lip* 3-lobed, 2.05 × 0.93 cm, side-lobes short and broad, truncate or subretuse tip, mid-lobe oblong-elliptic, minutely apiculate, minutely papillose inside with undulate margins, 1 × 0.6 cm, callus of 3 simple keels, dilate upwards, median keel reaches base of mid-lobe, lateral keels terminating abruptly in middle of mid-lobe. *Column* 1.25 cm long, wings conspicuously rounded, dilate, hood with irregular margins and truncate tip.
DISTRIBUTION. Borneo (Sarawak, Sabah).
HABITAT. Epiphytic in mossy lower and upper montane forest; oak-laurel forest.
ALTITUDE. 1000–2100 m.
FLOWERING. April.
NOTES. The epiphet refers to the region Tenompok, Sabah, Borneo where the type specimen was collected.

91. COELOGYNE TRILOBULATA

Coelogyne trilobulata J.J. Sm. in *Bull. Jard. Bot. Buitenzorg* ser. 3, 10: 28 (1928). Type: Sumatra, Gunung Kerinci, *Bünnemeijer* 10090 (holo. BO; iso. L).

DESCRIPTION. *Pseudobulbs* close together, ovoid with unequal thickened angles, 2.2–3.2 × 1.2 cm. *Leaves* 2, lanceolate, base and apex acute, lightly undulate, mid rib on the upper side and a prominent channel on underside, coriaceous, 7.5–9.5 × 1.3–1.5 cm, changes abruptly into 1.3 cm long, folded petiole. *Inflorescence* hysteranthous, peduncle slender, bare, 7.5–9 cm long, rachis zig-zag and almost flattened, 12–21 cm long, 12- to 30-flowered, flowers opening in succession. *Flowers* about 3.4 cm across, light brown, lip less glossy and transparent, reddish-brown. *Dorsal sepal* ovate-lanceolate, acute-acuminate, obtuse, concave, 7-nerved, 2.6 × 0.9 cm. *Lateral sepals* oblique, ovate-lanceolate, acute, 7-nerved, 2.55 × 0.6 cm. *Petals* oblique, linear, gently narrows to minute bi-dentate tip, 3-nerved, 2.15 × 0.23 cm. *Lip* 3-lobed, base of lip rounded, saccate, side-lobes large, erect, apex rounded and elongated with column rising above, mid-lobe six-angled, slightly irregular margin, tip truncated to form 3 lobes, outer lobes obtuse, mid-lobe narrow triangular-acute, callus of 5 simple glabrous keels extend from base of lip which increase in height for about one quarter of mid-lobe, then becoming more prostrate towards rising tip of lip. *Column* slightly arcuate, club-shaped apex, 1 cm long.

DISTRIBUTION. Sumatra (West coast (Gunung Kerinci)).
HABITAT. Epiphytic in upper montane forest.
ALTITUDE. 2200 m.
FLOWERING. May.
NOTES. The epiphet refers to the three-lobed form of the lip mid-lobe.

92. COELOGYNE TUMIDA

Coelogyne tumida J.J. Sm., *Orchids of Java* 141 (1905). Type: Java, West, Gunung Guntur, *M. Raciborski* s.n. (holo. BO; iso. L).

DESCRIPTION. *Pseudobulbs* with 4-flat sides, about 10 cm long, narrowed towards tip on a rhizome, 3 cm apart. *Leaves* 2, narrowly lanceolate, acute, 34 cm long with 9 cm long petiole. *Inflorescence* hysteranthous, peduncle and rachis up to 50 cm long, to 36-flowered, flowers opening in succession, floral bracts deciduous. *Flowers* pinkish-brown, 2 cm across. *Sepals* lanceolate, 1 cm long. *Petals* linear, 1 cm long. *Lip* 3-lobed, side-lobes small and rounded, mid-lobe almost circular and shortly apiculate, callus of 4 keels, 2 main keels along the lip which are abruptly elevated for a short distance, further 2 short keels at the base of the lip. *Column* (no detail).

DISTRIBUTION. Java (West).
HABITAT. Epiphytic in lower montane forest.
ALTITUDE. 1000–1500 m.
FLOWERING. Not known.
NOTES. The epiphet refers to the swollen base of the pseudobulbs.

SECTION *CYATHOGYNE*

Section 10. *Cyathogyne* (Schltr.) D.A. Clayton **stat. nov.** Type: *Coelogyne multiflora* Schltr. *Coelogyne* subgenus *Cyathogyne* Schltr. in *Feddes Repert. Sp. Nov. Regni Veg.* 10: 18 (1911).

Pseudobulbs on a thick, short rhizome, cylindric with semi-ovoid base and tip. Inflorescence synanthous, 25 cm long, 5 cm dia. Flowers densely packed, spirally arranged in cylindrical form, white. Lip mid-lobe large, quadrate, 3 keels with papillae. Distribution map 11.

A monotypic section.

THE GENUS COELOGYNE

Map 11. Distribution of *Coelogyne* Sect. *Cyathagyne*.

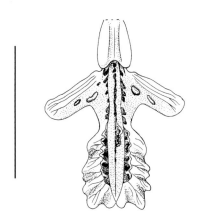

Fig. 14. **Section 10. Cyathogyne** (Schltr.) D.A. Clayton
1. Coelogyne multiflora Schltr. [*van Balgooy* 3128]. Drawn by Linda Gurr. Scale bar = 1 cm.

93. COELOGYNE MULTIFLORA

Coelogyne multiflora Schltr. in *Repert. Sp. Nov. Regni Veg. Beih.* 10: 18 (1911). Type: Sulawesi, Palu, Sopu Valley, *van Balgooy* 3128 (lecto. L; isolecto. K). **Plate 11F.**

DESCRIPTION. ***Pseudobulbs*** on a thick, short rhizome, roots moderately thick, close together, cylindric with tip and base semi-ovoid, compressed, longitudinally sulcate, 15 × 3–3.5 cm. ***Leaves*** 2, erect-spreading, lanceolate-elliptic, acute, coriaceous, many nerved, with 7 prominent beneath, 40–55 × 7–8 cm, base narrowing into 7–9 cm long petiole. ***Inflorescence*** synanthous, peduncle bare, terete, 15 cm, rachis very many flowered, flowers tightly packed in cylindrical form to 25 cm long, 5 cm dia. *Flowers* white. *Dorsal sepal* ovate, obtuse, 1.3 cm long. *Lateral sepals* oblique, oblong, somewhat acute, 1.3 cm long. *Petals* oblique, narrowly linear, obtuse, base slightly narrower. *Lip* 1.2 cm long, 1.3 cm across, shortly S-shape to broadly cuneate, 3-lobed, side-lobes diverging, erect, oblique, oblong, obtuse, mid-lobe very large, somewhat quadrate, in front cut out, margin very undulate, 0.7 cm long, 0.6 cm across, lip totally smooth, callus of 3 papillose, parallel keels from near base of lip to middle of mid-lobe. *Column* semi-circular, smooth, 0.65 cm long, thickening and spreading from base to hood with tip 3-lobed, mid-lobe largely cut-out, winged, margins somewhat tubular or cup-shaped.

DISTRIBUTION. Sulawesi (Minahassa).
HABITAT. Epiphytic on trunks of fallen trees in lower montane forest.
ALTITUDE. 1200 m.
FLOWERING. December.
NOTES. The epithet refers to the many-flowered inflorescence.

SECTION *VERRUCOSAE*

Section 11. ***Verrucosae*** Pfitzer & Kraenzlin in Engler, *Pflanzenr. Orch.-Mon.-Coelog.*: 73 (1907). S.E.C. Sierra, B. Gravendeel & E.F. de Vogel in *Blumea* 45: 275–318 (2000). Type: *Coelogyne asperata* Lindl.

Pseudobulbs medium-sized to large on a stout, sheathed rhizome, generally elliptic, conical or pyriform and flattened to varying degrees. Leaves 2, elliptic to lanceolate, large and plicate. Inflorescence proteranthous to synanthous, peduncle generally enclosed with bracts and emerging leaves. The flowers open simultaneously and the ovaries are slightly hairy. Flowers spaced on the rachis in two opposite rows with the nodes pronounced to produce a zig-zag rachis. The mid-lobe of the lip is equal to or greater than half the length of the lip and there are verrucae or papillae on the mid-lobe. Distribution map 12.

Map 12. Distribution of *Coelogyne* Sect. *Verrucosae*.

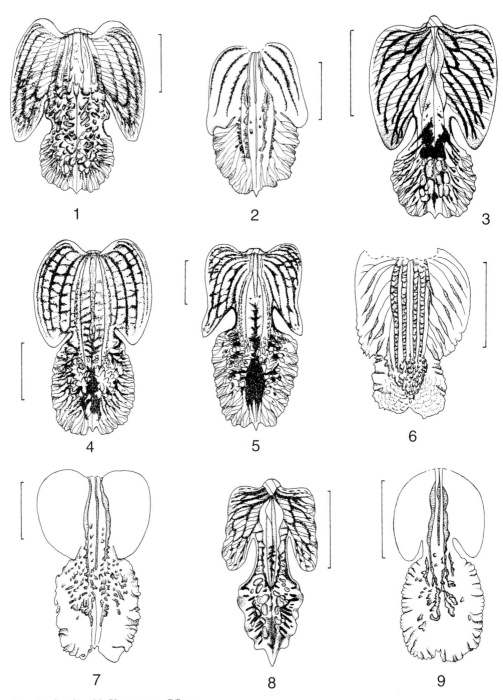

Fig. 15. **Section 11. Verrucosae** Pfitzer
1. Coelogyne asperata Lindl. [Cult. D.A. Clayton DAC 47]; **2. C. marthae** S.E.C. Sierra [Cult Leiden 27497]; **3. C. imbricans** J.J. Sm.; **4. C. mayeriana** Rchb.f. [Cult. H. Lowe DAC 43]; **5. C. pandurata** Lindl. [Cult. D.A. Clayton DAC 51]; **6. C. papillosa** Ridl. ex Stapf [*Wood* 607, after Eleanor Catherine]; **7. C. peltastes** Rchb.f. [after Sierra (2000)]; **8. C. verrucosa** S.E.C. Sierra [*Vermeulen & Lamb* 322]; **9. C. zurowetzii** Carr [after C.L. Chan]. Drawn by Linda Gurr. Scale bar = 1 cm.

KEY TO SPECIES IN SECTION VERRUCOSAE
(after Sierra *et al.*)

1. Pseudobulbs circular in cross-section 2
   Pseudobulbs slightly to strongly flattened 4

2. Pseudobulbs broadly pyriform 3
   Pseudobulbs cylindrical. [Inflorescence synanthous, peduncle bare and stiff, rachis slender and arcuate, floral bract deciduous. Flowers white, lip with yellow-brown markings.] .... *C. papillosa*

3. Flowers cream-coloured with orange-brown on lip. [Inflorescence proteranthous, peduncle slender, enclosed in large bracts, rachis straight, floral bracts persistent. Lip mid-lobe ovate, acute, margins undulate, 2 verrucose keels terminating just onto the mid-lobe. Sepals and petal pale green, side-lobes of the lip white and marked inside with cinnamen, mid-lobe yellow and spotted cinnamen, keels pale green, becoming cinnamen towards the mid-lobe.] .... *C. asperata*
   Flowers green with black on lip. [Inflorescence proteranthous, peduncle slender and enclosed in bracts, rachis slender, floral bracts deciduous. Lip mid-lobe semi-elliptic, margins undulate and crenulate, 3 keels terminating at the middle of the mid-lobe, median keel slender, scarcely elevated and terminates at the base of the mid-lobe, lateral keels distinctly elevated and verrucose, 2 verrucose patches extend from the termination of the keels.] .... *C. mayeriana*

4. Claw on mid-lobe of lip distinct 5
   Claw on mid-lobe of lip absent or indistinct 6

5. Pseudobulbs oblong to suborbicular, 7–13 cm long. [Inflorescence proteranthous, peduncle very stout, rachis stout, slightly arcuate, zig-zag, floral bracts persistent. Lip mid-lobe broad, pandurate, bi-lobed; claw one third to a quarter the length of the lip, 2 verrucose, elevated keels in the form of plates terminating at the base of the mid-lobe, on the mid-lobe a verrucose patch. Flowers bright green, lip marked with black markings.] .... *C. pandurata*
   Pseudobulbs ovate, 5–10 cm long. [Inflorescence proteranthous or synanthous. Lip mid-lobe irregularly rectangular, claw rectangular with margins straight, 3 keels initially elevated and rounded, lateral keels becoming widened and verrucose. Flowers light greenish, lip with brown patches.] .... *C. verrucosa*

6. Claw on mid-lobe of lip present; 2, 4 or 6 swollen nerves on the claw and base of mid-lobe to lip 7
   Claw on mid-lobe of lip absent; swollen nerves absent at base of mid-lobe to lip 8

7. Margin of mid-lobe very finely undulating; keels on mid-lobe become short rows or patches of scattered, single or connected, molar-like warts. [Inflorescence synanthous, peduncle arcuate, rachis more or less hairy. Flowers greenish-yellow, hypochile whitish, mid-lobe suffused pale chestnut near the base.] .... *C. zurowetzii*
   Margin of mid-lobe broadly undulating; keels on mid-lobe becoming two irregular flattened callus, which together have a more or less ovate shape. [Inflorescence synanthous, peduncle thickened, rachis arcuate. Sepals and petal pale green, lip side-lobes whitish, inside marked cinnamen, mid-lobe yellow and spotted cinnamen.] ....
   .... *C. imbricans*

8. Median keel of the lip continuing on the base of mid-lobe; keels on mid-lobe becoming patch of tooth-like, more or less flattened warts, often arranged in radiating rows. [Lip side-lobes elliptic, rounded, undulate; mid-lobe orbicular, margins erose and undulate. Flowers whitish or greenish-yellow, lip veined yellow-brown, 4 dark red veins become 5 on the side-lobes, similar veins radiate from the tuberculae.] .... ***C. peltastes***

Median keel on the lip not reaching base of the mid-lobe, keels on mid-lobe becoming patch or 4–6 single crested, parallel keels which are broken up into flat irregular teeth or warts. [Lip side-lobes triangular-ligulate, mid-lobe quadrangular, margins undulate. Flower sepals and petals light green, lip white, tinged green with brown crests to keels.] .... ***C. marthae***

94. COELOGYNE ASPERATA

Coelogyne asperata Lindl. in *J. Hort. Soc. London* 4: 221 (1849). Type: Malaya, cult. Twisden Hodges s.n. (holo. K-LINDL). **Plate 12A.**

Coelogyne lowii Paxt. in *Paxton's Mag. Bot.* 16: 225 (1849). Type: Borneo, Sarawak, icon. S. Holden (holo. illustration cited above).
?*C. macrophylla* Teijsm. & Binn. in *Tijdschr. Ned.-Indie* 29: 241 (1867), *nom. nud.*
C. pustulosa Ridl. in *J. Bot.* 24: 353 (1886). Type: New Guinea, Sogeri region, *Forbes* s.n. (holo. BM).
C. edelfeldtii F. Muell. & Kraenzl. in *Oesterr. Bot. Z.* 44: 421 (1894). Type: New Guinea, Moresby, *Lauterbach* 1598 (holo. B†).
Pleione asperata (Lindl.) Kuntze in *Rev. Gen. Pl.* 2: 680 (1891).

DESCRIPTION. ***Pseudobulbs*** on a 1–1.8 cm thick, sheathed rhizome, close together to 6.5 cm apart, somewhat compressed, ribbed, broadly conical, to 15 × 2.5–3 cm, enclosed in persistent bracts at base. ***Leaves*** 2, oblanceolate, acute, mucronate tip, to 100 × 12 cm with 21 cm long, grooved and prominently ridged petiole. ***Inflorescence*** proteranthous, peduncle slender, arcuate, bracts large and persistent, to 15 cm long, rachis straight, to 15 cm long, 8-flowered, flowers opening simultaneously, floral bracts persistent. ***Flowers*** fragrant, sepals and petals creamy-white to yellowish, side-lobes white, veined and splashed light-brown, mid-lobe splashed and marked with red-orange or red-brown, 2 creamy-yellow spots about the middle, bright yellow spot at base of lip, verrucose keels rich brown, column pale yellow. ***Dorsal sepal*** lanceolate, acute, 5-nerved, 3.1–4 × 1 cm. ***Lateral sepals*** oblique, lanceolate, acute, 5-nerved, 3.1–4 × 1 cm. ***Petals*** linear-lanceolate, obtuse, 3–3.5 × 0.5 cm. ***Lip*** 3-lobed, side-lobes erect, narrow, front triangular, margin entire, mid-lobe reflexed, ovate, margins undulate, tip acute, callus of 2 verrucose keels extending from near base of lip to just beyond middle of mid-lobe where they become tuberculate. ***Column*** slender, slightly arcuate, 1.25 cm long, remains slender then expands to form hood then narrows to obtuse tip.

DISTRIBUTION. Peninsular Malaysia (Perak), Sumatra (East & West coasts), Borneo (Kalimantan, Sarawak, Sabah, Brunei), Philippines (Mindanao), Sulawesi (West & South), Maluku (Central Seram), New Guinea (Irian Jaya, Papau New Guinea), Solomon Islands, Santa Cruz Islands.

HABITAT. Epiphytic, sometimes terrestrial or lithophytic, in riverine or hill forest on open limestone rock faces or near waterfalls.

ALTITUDE. Sea level–2042 m.

FLOWERING. January, April–July. Also at other times of the year depending on geographical location and altitude.

NOTES. The epithet refers to the tuberculate patch on the lip mid-lobe.

95. COELOGYNE IMBRICANS

Coelogyne imbricans J.J. Sm. in *Bull. Jard. Bot. Buitenzorg.* ser. 3, 2: 26 (1920). Type: Borneo, Kalimantan, Landak, Sungai Menjeeke, cult. Bogor, *Gravenhorst* 1917 (holo. L; iso. BO).

DESCRIPTION. *Pseudobulbs* on a rising rhizome, laterally flattened, in two rows, 4.5–5 cm apart, thickened in the middle, rounded base, light green, 8.5–11.5 × 6.5–8 cm. *Leaves* 2, elliptic to obovate-elliptic, shortly acute to acuminate, with 5 nerves, the 3 middle nerves sulcate above, prominent below, glossy light green, 11.75–15.5 × 4.5–6.5 cm with 6.5 cm long petiole. *Inflorescence* synanthous, peduncle thickened, 6.5 cm long, rachis arcuate, 6.5 cm long, laxly many-flowered. *Flowers* about 5.5–6 cm across, sepals and petals pale green, lip side-lobes white with inside finely marked with cinnamon spots except margin, mid-lobe yellow spotted cinnamon, keels pale green near base becoming cinnamon coloured. *Dorsal sepal* erect, oblong-elliptic to lanceolate, somewhat acute, 3.1–3.5 × 1.1 cm. *Lateral sepals* widely spreading, oblong or oblanceolate, acute, 3–3.2 × 0.9–1 cm. *Petals* widely spreading, oblanceolate, acute, 2.9–3.3 × 0.2–0.7 cm. *Lip* erect, 3-lobed, base concave, side-lobes erect, shorter than column, concave, trapezium, tip free, broadly triangular, obtuse, margin reflexed, mid-lobe more or less reflexed, undulate, concave at base, shortly clawed, trapezium shaped, tip shortly rounded, bi-lobed, callus of 3 inconspicuous keels, extending from base of lip, pubescent, slowly growing to become erect, simple, smooth. *Column* arcuate, 1.75 cm long, apex winged, tip rounded.

DISTRIBUTION. Borneo (Kalimantan, Sarawak).
HABITAT. Epiphytic on tree trunks. Habitat unknown.
ALTITUDE. Not known.
FLOWERING. Not known.
NOTES. The epithet refers to the pronounced imbricate sheathing on the rhizomes.

96. COELOGYNE MARTHAE

Coelogyne marthae S.E.C. Sierra in *Blumea* 45: 295–297 (2000). Type: Borneo, Sarawak. *J.J. Vermeulen* 1156 (holo. L). **Plate 12B.**

DESCRIPTION. *Pseudobulbs* on a climbing, stout, sheathed rhizome, 0.6–1.3 cm apart, in cross-section flattened, in outline ovate-oblong, 1.5–5 × 0.7–2.3 cm bracts covering pseudobulb. *Leaves* 2, lanceolate, with 3–5 main nerves, 6–28 × 1.2–4.5 cm with a 1–4 cm long petiole. *Inflorescence* proteranthous or synanthous, peduncle enclosed in developing young shoot, 3.7–8 cm long, rachis 7–16.2 cm long, 3- to 5-flowered, flowers opening simultaneously, floral bracts persistent. *Flowers* with light green sepals and petals, lip white tinged with green, at base of lip orange. Side-lobes with 4–5 brown veins, mid-lobe at margins green in middle brown rows of papillae, keels initially light green, becoming yellow with brown crests. *Dorsal sepal* ovate-lanceolate, 11-nerved, dorsal prominent, 3.4–3.8 × 0.9–1.3

cm. *Lateral sepals* slightly falcate, ovate-lanceolate, 7-nerved, 3.1–3.5 × 0.9–1 cm. *Petals* lanceolate, 5-nerved, dorsal prominent, 2.9–3.2 × 0.4–0.6 cm. *Lip* 3-lobed, side-lobes triangular-ligulate, in front broadly rounded, mid-lobe with claw absent, quadrangular, tip acute, notched at either side, margin broadly undulating, callus of 3 keels, low and rounded at base of lip, median keel becomes elevated and then plate-like, extending two thirds the way to base of mid-lobe, lateral keels become more elevated than median keel, widen into plate-like sections, double crested and then become 4–6 single crested parallel keels to about half way onto mid-lobe and then become broken into irregular teeth or warts. *Column* 1.4–1.7 cm long expanding into rectangular to broadly rounded hood, tip broadly rounded, margin irregular.

DISTRIBUTION. Borneo (Sarawak (Bahagian Kuching)).

HABITAT. Epiphytic on lower part of trunks of undergrowth trees in heath forest on level terrain with deep sandy soil overlain by a layer of raw humus, locally with pools of stagnant brown water.

ALTITUDE. 50–300 m.

FLOWERING. March, December.

NOTES. The species is named after Dr. Martha Tilaar, the benefactor of the newly established Martha Tilaar Chair of Ethnobotanical Knowledge Systems with special reference to Medicinal Plants in Developing Countries at Leiden University, The Netherlands.

97. COELOGYNE MAYERIANA

Coelogyne mayeriana Rchb.f. in *Gard. Chron.* ser. 2, 8: 134 (1877). Type: Java, Nusa Kambangan, cult. Mayer s.n. (holo. W). **Plate 12C.**

DESCRIPTION. *Pseudobulbs* on a stout, sheathed rhizome, 5–10 cm apart, circular in cross-section to very slightly flattened, pyriform, 6 × 2 cm, enclosed with bracts at base. *Leaves* 2, lanceolate or elliptic, acute, with 7 nerves, about 20–30 × 4–9 cm with 3 cm long, grooved petiole. *Inflorescence* proteranthous, peduncle slender enclosed in undeveloped leaves and bracts, 7–8 cm long, rachis slender, 10–20 cm long, flowers open just as the leaves are expanding, 4- to 10-flowered, flowers opening simultaneously, floral bracts deciduous. *Flowers* 6.5 cm across, sepals and petals green, lip pale green with brown-black markings mostly radiating outwards with two patches of whitish tuberculae in the centre of mid-lobe and 4 longitudinal and parallel dark brown stripes on side-lobes, keels green. *Dorsal sepal* lanceolate, acuminate, 5-nerved, 3.3 × 0.9 cm. *Lateral sepals* oblique, lanceolate, acuminate, 3.3 × 0.9 cm. *Petals* lanceolate, acuminate, 3-nerved, 3 × 0.5 cm. *Lip* 3-lobed, the hypochile longer (c.f. *C. pandurata*), scarcely any claw, side-lobes erect, obtuse, mid-lobe semi-elliptic, margins undulate and coarsely crenulate, callus of 3 keels extending from base of lip to middle of mid-lobe, scarcely elevated on hypochile, median keel terminates at base of mid-lobe, lateral keels distinctly elevated and verrucose, median keel slender, glabrous, 2 verrucose patches extend from termination of keels. *Column* slender, arcuate, 1.3 cm long, broadens to form hooded tip.

DISTRIBUTION. Peninsular Malaysia (South), Sumatra (West coast, Pulau Bangka), Java (West & Central).

HABITAT. Epiphytic on trees and terrestrial in humus on ground in lowland rain forest.

ALTITUDE. Sea level–100 m.

FLOWERING. July.

NOTES. This species dedicated to Mayer, the grower of the type specimen.

98. COELOGYNE PANDURATA

Coelogyne pandurata Lindl. in *Gard. Chron.* 1853: 791 (1853). Type: Borneo, *Low* s.n. (holo. K-LINDL). **Plate 12D.**

Pleione pandurata (Lindl.) Kuntze, *Rev. Gen. Pl.* 2: 680 (1891).
Coelogyne peltastes Rchb.f. var. *unguiculata* J.J. Sm. in *Mitt. Allg. Bot. Hamburg* 7: 33, t. 33, f. 23 (1927). Type: Borneo. Kalimantan, Lebang Hara, *Winkler* 347 (holo. HBG; iso. BO).

DESCRIPTION. **Pseudobulbs** on a 0.9–1.3 cm, sheathed rhizome, 3–10 cm apart, strongly compressed, oblong or suborbicular, slight tapering towards tip, sulcate, 6.5–12.5 × 2–3 cm, sheathed with persistent bracts at base. **Leaves** 2, elliptic-lanceolate, acuminate, rigid, with 5–7 nerves, 20–45 × 4–7 cm with a 4 cm long, stiff, grooved petiole. **Inflorescence** proteranthous or synanthous, peduncle very stout, enclosed in developing leaves and bracts, to 17 cm long, rachis stout, slightly arcuate, zig-zag, to 25 cm long, 6-flowered, flowers opening simultaneously, floral bracts persistent. **Flowers** fragrant, 7–10 cm across, clear bright green, lip appearing black-mottled. **Dorsal sepal** linear-lanceolate, acuminate, with 5 prominent nerves, 3.5–5.8 × 1.3 cm. **Lateral sepals** oblique, linear-lanceolate, acuminate, with 5 prominent nerves, 3.5–5 × 1.3 cm. **Petals** subspathulate, clawed, acute, 4.5 × 1.2 cm. **Lip** 3-lobed, pandurate, cordate at base, 4 × 1.5 cm, claw nearly one quarter to one third length of lip, side-lobes small, curved upwards, front acute, mid-lobe broad, margins crisped-undulate, callus of 3 keels at base of lip, lateral keels becoming verrucose and extending from base of lip to base of mid-lobe, on mid-lobe they are elevated to form plates and become verrucose patch. **Column** slender, arcuate, 2.4 cm long, widening towards tip to form hood.

DISTRIBUTION. Peninsular Malaysia, Sumatra (West coast), Borneo (Kalimantan, Sarawak, Sabah, Brunei), Java.

HABITAT. Epiphytic, terrestrial, lithophytic, usually on large trees in lowland rain forest and hill forest. Also in lower and upper montane forest, among rocks, in open or in scrub, on granite or ultramafic substrate.

ALTITUDE. Sea level–1200 m (1800–3700 m).

FLOWERING. January–March.

NOTES. The epithet refers to the fiddle-shape of the lip.

99. COELOGYNE PAPILLOSA

Coelogyne papillosa Ridl. ex Stapf in *Trans. Linn. Soc. London, Bot.* ser. 2, 4: 238 (1894). Type: Borneo, Sabah, Mt. Kinabalu, *G.D. Haviland* 1098 (holo. SING; iso. K). **Plate 12E.**

DESCRIPTION. **Pseudobulbs** on a woody, moderately thick rhizome, close together, elongate cylindric, 7.5 cm long, enclosed with bracts at base. **Leaves** 2, lanceolate, acute, with 7 main nerves, 8–27 × 2.3–4.3 cm with a 7.5–8 cm long, grooved petiole. **Inflorescence** heteranthous or synanthous, peduncle bare, stiff, erect, to 38 cm long, rachis slender, arching, zig-zag, to 10 cm long, 6- to 18-flowered, flowers opening simultaneously, floral bracts deciduous. **Flowers** with snow-white sepals and petals, hypochile yellow-brown, side-lobes white with yellow-brown nerves, mid-lobe white with large median brown spot and white papillae, keels yellow-brown tipped white, column white suffused pale salmon along middle of back and pale yellow-brown inside, anther salmon-pink. **Dorsal sepal** lanceolate, acute, 2 cm long. *Lateral sepals* oblique, lanceolate, acute, 2 cm long. *Petals* oblique, lanceolate, obtuse, 2 cm long. *Lip* oblong, 2 cm long, 3-lobed, side-lobes small, obtuse, scarcely separated, mid-lobe ovate, densely papillose, callus of 9 keels, strongly undulate, lying between side-lobes. *Column* initially arched upwards, narrow at base, thickening into hood with large wings, triangular, somewhat acute.

DISTRIBUTION. Borneo (Sabah).

HABITAT. Terrestrial, lithophytic in upper montane scrub or tropical subalpine scrub in granite crevices or ultramafic substrate.

ALTITUDE. 1800–3700 m.

FLOWERING. February (June).

NOTES. The epithet refers to the lip being covered with papillae.

100. COELOGYNE PELTASTES

Coelogyne peltastes Rchb.f. in *Gard. Chron.* ser. 2, 14: 296 (1880). Type: Borneo, *Veitch* s.n. (holo. W). **Plate 12F.**

DESCRIPTION. *Pseudobulbs* on a stout, creeping, heavily sheathed rhizome, 7–9 cm apart, ellipsoid, sometimes almost round, laterally flattened, about 9 × 5 cm. *Leaves* 2, ovate-elliptic, acute, with 5 main nerves, 17–30 × 4–6.5 cm with a 5 cm long, grooved petiole. *Inflorescence* synanthous, peduncle enclosed with bracts and emerging leaves, 5–6 cm long, rachis slender, slightly zig-zag, loosely 4- to 6-flowered, flowers opening simultaneously, floral bracts persistent. *Flowers* about 7.5 cm across, fragrant, whitish- or greenish-yellow, lip white veined with yellow-brown, 4 dark red veins become 5 on side-lobes, similar markings radiate from tuberculae region. *Dorsal sepal* lanceolate, acute, 7-nerved, 2.8 × 0.5cm. *Lateral sepals* oblique, lanceolate, acute, 7-nerved, 2.8 × 0.4 cm. *Petals* linear-lanceolate, broadening from narrow attachment, 5-nerved, 2.7 × 0.4 cm. *Lip* 3-lobed, 2 cm long, 0.9 cm across side-lobes, side-lobes elliptic, front rounded, margin undulate, mid-lobe orbicular, margin erose, undulate, tip bi-lobed (cordate), callus of 3 keels extending from base of lip to base of mid-lobe, median keel slightly shorter, lateral keels elevated, median keel less so, all plate-like, entire, slightly undulate, lateral keels merge with broken tuberculae which cover central region of lip. *Column* straight, 1.9 cm long, widening into winged hood.

DISTRIBUTION. Borneo (Kalimantan, Sarawak, Sabah).

HABITAT. Epiphytic on tree trunks near ground in heath forest and mixed dipterocarp forest; recorded on limestone.

ALTITUDE. Sea level–900 m.

FLOWERING. July–November.

NOTES. The epithet refers to the form of the leaves. Investigations at the National Herbarium of the Netherlands (B. Gravendeel, pers. comm.) indicate that *Coelogyne peltastes* Rchb.f. var. *unguiculata* J.J. Sm. is a small form of *C. pandurata*.

101. COELOGYNE VERRUCOSA

Coelogyne verrucosa S.E.C. Sierra in *Blumea* 45: 309–311 (2000). Type: Borneo, Sabah. *Vermeulen & Lamb* 322 (holo. L; iso. K). **Plate 13.**

DESCRIPTION. Pseudobulbs on a climbing, 0.7–1.2 cm thick, sheathed rhizome, 1.8–5 cm apart, in cross-section very flattened, in outline ovate with margins slightly incurved, pressed against the rhizome and lower part of subsequent pseudobulb, 5–10 × 3.3–5 cm, bracts covering the pseudobulbs. **Leaves** 2, lanceolate, with 7–9 main nerves, 16–38 × 2.5–6 cm with 2–6 cm long petiole. **Inflorescence** proteranthous or synanthous, peduncle partly covered by scales of young shoot, 3.5–14 cm long, rachis 12.5–34 cm long, 6- to 10-flowered, flowers opening simultaneously, floral bracts persistent. **Flowers** with light greenish sepals and petals, lip light greenish with brown patches, at extreme base of lip yellow, keels initially white becoming green and then dark brown; side-lobes with 4–8 brown veins, mid-lobe with light greenish molar-like projections; column light greenish, sometimes with brown lateral lines. **Dorsal sepal** ovate-lanceolate, 9–11-nerved, dorsal prominent, 1.85–4 × 0.6–1.3 cm. **Lateral sepals** falcate, ovate-lanceolate, 7–9-nerved, 2–3.75 × 0.55–0.9 cm. **Petals** lanceolate, 5–7-nerved, rather prominent, 1.75–3.6 × 0.35–0.62 cm. **Lip** 3-lobed, side-lobes in front triangular-ligulate, apex rounded, mid-lobe with rectangular claw 0.25–0.45 cm long, margins straight, mid-lobe becomes irregularly rectangular to quadrangular to ovate, 0.8–1.4 × 0.6–1.2 cm, tip acute, margin broadly and regularly undulating, callus of 3 keels, low at base of lip, median keel extends to base of mid-lobe, lateral keels become more elevated and widening at crest, single or double crested and continuing onto mid-lobe where they become irregular and broken with a few scattered tooth-like projections. **Column** 1.15–1.6 cm long expanding into a distinctly widened hood, tip broadly rounded, emarginate.

DISTRIBUTION. Borneo (Brunei, Sarawak, Sabah).

HABITAT. Epiphytic trunks and large branches of trees, sometimes terrestrial in heath, peat and mixed *Dipterocarp* forest.

ALTITUDE. 10–700 m.

FLOWERING. Once or twice a year in a specific locality but over the whole region effectively all year.

NOTES. The epiphet refers to the large rounded projecting warts on the mid-lobe of the lip.

102. COELOGYNE ZUROWETZII

Coelogyne zurowetzii Carr in *Orchid Rev*. 42: 44 (1934). Type: Borneo, Kalimantan, Sambas, *Zurowetz* s.n. (holo. SING). **Plate 14A.**

DESCRIPTION. Pseudobulbs on a creeping, stout, sheathed rhizome, 3–5 cm apart and making an acute angle with the rhizome, ovate, strongly flattened, somewhat wrinkled, 7.5 × 4 cm, sectioned transversally at 1.5 cm intervals. **Leaves** 2, oblanceolate to narrowly oblong, tapered to acute tip, plicate, with 5 nerves, grooved above, prominent beneath, sulcate, 25 × 6 cm with 3–6 cm long, grooved petiole. **Inflorescence** synanthous, peduncle arched, rachis more or less horizontal, few-flowered, flowers opening simultaneously, floral bracts broadly elliptic, acute, 2.4 × 1.7 cm. **Flowers** with very bright, glistening yellow-green sepals and petals, lip has whitish blade between side-lobes veined with deep chestnut, mid-lobe broader than side-lobes, whitish with suffused pale chestnut at base, mid-lobe warty along middle except at tip with lower warts chestnut, upper warts pure white, keels white tipped with green, column very deep yellow-green. **Dorsal sepal** lanceolate, somewhat acute, 7-nerved, 3.75 × 1.27 cm. **Lateral sepals** curved, lanceolate, acute, 7-nerved, 3.75 × 1.27 cm. **Petals** at base slightly wedge-shaped, then narrowly lanceolate, acute, 3.6 × 0.7 cm. **Lip** 3-lobed, side-lobes curved, triangular, obtuse, mid-lobe clawed, claw 0.28cm long, expanding from sinus, broadly oblong, tip bi-lobed, margins strongly undulate and shortly crenulate, 2 cm long, 1.78 cm across, margins of lip crisped and very finely and evenly fringed, callus of 3 keels simple at base of lip, becoming swollen and extending to sinus of mid-lobe where they become verrucose patch in middle of mid-lobe. **Column** arcuate, 1.65 cm long, expanding to form winged hood, tip bi-lobed, margin entire.

DISTRIBUTION. Borneo (Kalimantan, Sabah).

HABITAT. Epiphytic in heath (Kerangas) forests, freshwater swamp forest, podzolic dipterocarp forest, and on ultramafic substrate.

ALTITUDE. 400–700 m.

FLOWERING. April–November.

NOTES. The species is named after J.E. Zurowetz, the collector of the type specimen.

SECTION *TOMENTOSAE*

Section 12. *Tomentosae* Pfitzer & Kraenzlin in Engler, *Pflanzenr. Orch.-Mon.-Coelog.*: 66 (1907); E.F. de Vogel in *Orchid Monogr.* 6: 1–42 (1992). Type: *Coelogyne tomentosa* Lindl.

Pseudobulbs on a short or long, creeping rhizomes, 2-leaved, leaves herbaceous to (sub) coriaceous. Inflorescence heteranthous or proteranthous to synanthous, peduncle usually short and more or less enclosed by the bracts of the young shoot. Inflorescence few- to many-flowered. In a few species the peduncle longer and with the bracts only at the base. Rachis straight or with a varying degree of zig-zag. Flowers open simultaneously, floral bracts persistent. Mid-lobe of the lip is equal to or less than one-third the length of the lip; if greater than one-third the length of the lip then no papillae on the mid-lobe. A varying number of keels on the lip. Distribution map 13.

Map 13. Distribution of *Coelogyne* Sect. *Tomentosae*.

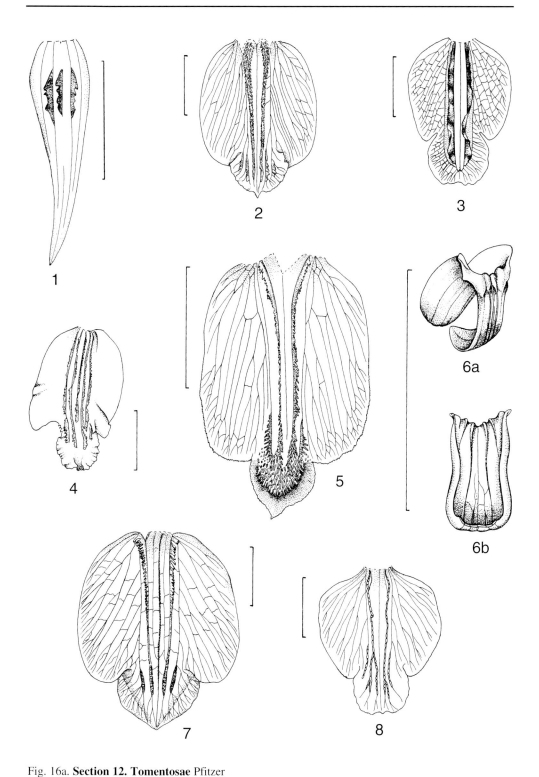

Fig. 16a. **Section 12. Tomentosae** Pfitzer
1. Coelogyne acutilabium de Vogel [after Abdul Satiri]; **2. C. bruneiensis** de Vogel [after C.L. Chan]; **3. C. buennemeyeri** J.J. Sm. [after Abdul Satiri]; **4. C. distans** J.J. Sm. [after C.L. Chan]; **5. C. echinolabium** de Vogel [after C.L. Chan]; **6a–b. C. genuflexa** Ames & C. Schweinf. [after C.L. Chan]; **7. C. hirtella** J.J. Sm. [after C.L. Chan]; **8. C. judithiae** P. Taylor [after Abdul Satiri]. Drawn by Linda Gurr. Scale bar = 1 cm.

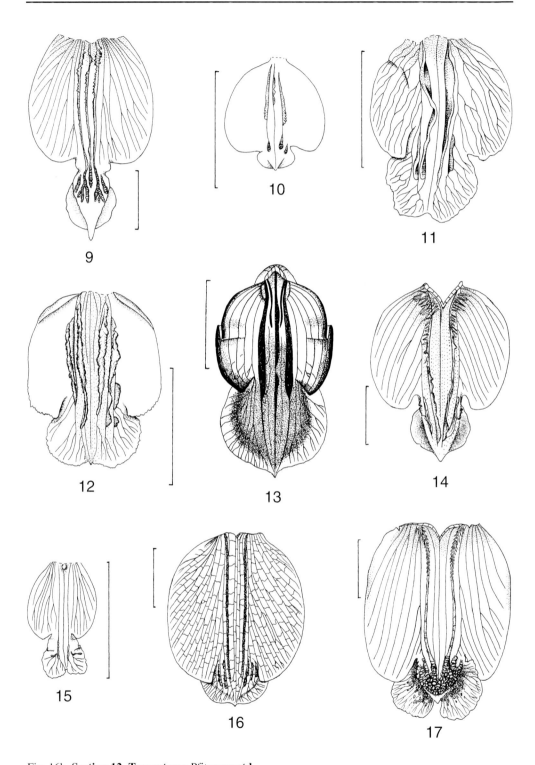

Fig. 16b. **Section 12. Tomentosae** Pfitzer **contd.**
9. C. kaliana P.J. Cribb [after Abdul Satiri]; **10. C. latiloba** de Vogel [after C.L. Chan]; **11. C. longibulbosa** Ames & C. Schweinf. [after C.L. Chan]; **12. C. moultonii** J.J. Sm. [after C.L. Chan]; **13. C. muluensis** J.J. Wood [Kew Coll. *Neilson* 154]; **14. C. odoardi** Schltr. [after C.L. Chan & F.L. Liew]; **15. C. pholidotoides** J.J. Sm. [after C.L. Chan]; **16. C. pulverula** Teijsm & Binn. [after C.L. Chan]; **17. C. radioferens** Ames & C. Schweinf. [after C.L. Chan]. Drawn by Linda Gurr. Scale bar = 1 cm.

SECTION TOMENTOSAE

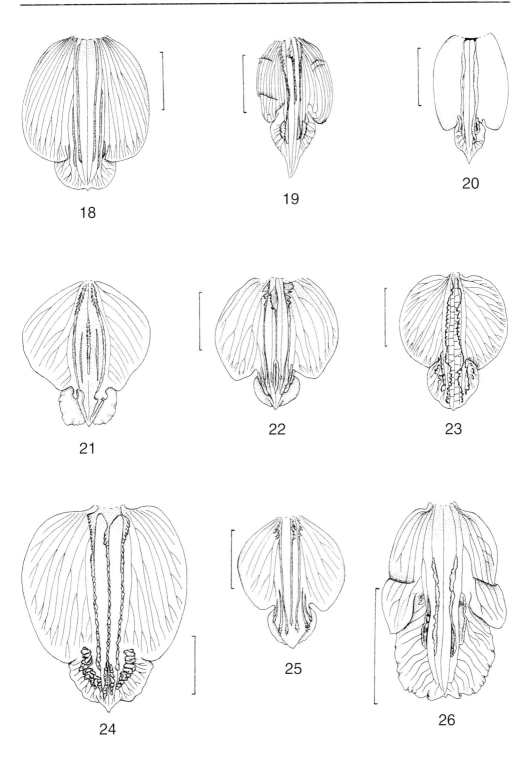

Fig. 16c. **Section 12. Tomentosae** Pfitzer **contd.**
18. C. rhabdobulbon Schltr. [after C.L. Chan]; **19. C. rochussenii** de Vriese [after C.L. Chan]; **20. C. rupicola** Carr [after C.L. Chan]; **21. C. squamulosa** J.J. Sm. [after J.J. Sm.]; **22. C. swaniana** Rolfe [after C.L. Chan]; **23. C. testacea** Lindl. [after Abdul Satiri]; **24. C. tomentosa** Lindl. [after Abdul Satiri]; **25. C. velutina** de Vogel [after Abdul Satiri]; **26. C. venusta** Rolfe [after C.L. Chan]. Drawn by Linda Gurr. Scale bar = 1 cm.

Key to species in section Tomentosae
(after E.F. de Vogel)

1. Flowers rather small, dorsal sepal less than 1.8 cm long 2
   Flowers rather large, dorsal sepal more than 2 cm long 6

2. Pseudobulbs slender-cylindric, more than 11 cm long, if shorter less than 1 cm diam 3
   Pseudobulbs rather shortly fusiform or ovoid, less than 9 cm long 4

3. Inflorescence peduncle more than 12 cm long, floral bracts very large, enclosing the flower (almost) entirely. [Sepals and petals (creamy)-white, lip with a white median band and margins, side-lobes and centre of mid-lobe, including the keels yellowish to brown.] .... *C. moultonii*
   Inflorescence peduncle less than 11cm long, floral bracts not very large, enclosing at most the base of the flowers. [Sepals and petals (creamy)-white, sometimes tinged green, lip (creamy)-white, side-lobes and mid-lobe with a yellow to light brown blotch.] .... *C. longibulbosa*

4. Lip mid-lobe more or less widened from a narrower base, margins undulate, keels extending from the base of the lip to the base of the mid-lobe 5
   Lip mid-lobe ligulate, about equally wide over most of its length, margins not undulate; keels only developed near the base of the mid-lobe. [Flowers cream to greenish-yellow, lip sometimes whitish with a yellowish blotch.] .... *C. genuflexa*

5. Dorsal sepal 0.95–1.15 cm long. Lip mid-lobe broadly spathulate (when flattened), with a broad claw. [Flowers (creamy)-white, side-lobes of the lip with a brown spot, mid-lobe with yellow.] .... *C. pholidotoides*
   Dorsal sepal 1.2–1.5 cm long. Lip mid-lobe obovate to rectangular (when flattened), gradually widening from a broad base. [Sepals, petals and lip (creamy)-white or tinged with pink, side-lobes and mid-lobe of the lip with a yellow blotch or band, keels on the mid-lobe sometimes tipped brown.] .... *C. venusta*

6. Lip clearly 3-lobed, side-lobes pronounced, in front rounded, mid-lobe transversally elliptic to somewhat orbicular or ovate, sometimes with a short and broad claw 7
   Lip not distinctly 3-lobed, sides of the lip low, in front very gradually lowering, 3 keels terminating a third of the way onto mid-lobe. [Colours not known.] .... *C. acutilabium*

7. Pseudobulbs close together, rarely up to 5 cm apart. Floral bracts persistent. Varying number of keels on the lip 8
   Pseudobulbs 5–15 cm apart on the rhizome. Floral bracts deciduous. 5 keels on the lip, sometimes additional lateral keels only partly developed. [Flowers (creamy)-white, inside of side-lobes with brownish stripes, outside with yellow to brown front margin, keels with a brown tip.] .... *C. distans*

8. Lip mid-lobe with keels more or less forming a mass of warts, lobules or papillae 9
   Lip mid-lobe with 2–8 distinctly separate keels 11

9. Hypochile of the lip with 2 keels. On the mid-lobe flattened lobules or rounded papillae 10

Hypochile of the lip with 3 keels. [Lateral keels fleshy, laciniate, extending from base of lip to base of mid-lobe. Median keels continues to tip of mid-lobe, verrucosae, crested plates on outside of keels. Sepals and petals yellowish-brown, side-lobe on outside white with brown/white nerves near base, on inside brown with whitish nerves. Mid-lobe brown with white at base.] .... *C. squamulosa*

10. Pseudobulbs 6–20 cm long. Lip mid-lobe orbicular or broadly ovate with a rounded to acute tip, no radiating rows of papillae or warts from the central mass of slender lobules. [Flowers creamy or dirty white, inside side-lobes of the lip with brownish colouring and paler nerves.] .... *C. echinolabium*
Pseudobulbs 5–10 cm long. Lip mid-lobe butterfly-shaped to transversely elliptic in outline, more or less deeply emarginate or retuse, with radiating rows of papillae and/or elevated warts from the central mass of papillae. [Sepals and petals ochre-yellow, lip (creamy)-white, inside of side-lobes brown with white nerves, outside in front with brown band.] .... *C. radioferens*

11. Keels on the lip hypochile entire or at most with irregularly, more or less deeply incised margins 12
Keels on the lip hypochile with distinct warts, laciniae or teeth 16

12. Inflorescence with a short peduncle, emerging from between the almost entirely developed young leaves. Keels on the hypochile of the lip at most with slightly swollen but not laterally widened margin 13
Inflorescence with a distinctly elongated peduncle, emerging from between the almost entirely developed young leaves. Keels on the hypochile of the lip with a distinctly swollen and laterally widened margin. [Flowers white, sometimes suffused pale salmon, lip with a median bright golden yellow band including the keels and the base of the mid-lobe.] .... *C. rupicola*

13. Inflorescence heteranthous, rachis pendulous, 10- or more-flowered. Lip mid-lobe with plate-like keels 14
Inflorescence heteranthous, rachis stiff, less than 5-flowered. Lip mid-lobe rather broad and swollen, consisting of many elevated cross ridges. [Flowers white, lip median white, inside on the front half orange-brown, side-lobes inside brown with white nerves and margin.] .... *C. latiloba*

14. Lip hypochile with 2 or 3 keels 15
Lip hypochile with 4 keels, rarely a fifth median keel partly developed, lateral keels only develop on the upper half to one quarter length of the hypochile, sometimes part of the keels not developed 16

15. Lip hypochile with 2 keels. [Flowers light pink or (greenish) salmon, lip sometimes lighter.] .... *C. buennemeyeri*
Lip with 3 keels. [Flowers creamy-white with lip reddish-ochre or chestnut except for margins and at the base of the mid-lobe, lateral keels creamy, median keel yellow.] .... *C. muluensis*

16. Top margin of the column incised, distinctly 4–6 lobed. [Flowers white, lip median mustard-coloured to ochre-yellow, keels cream-coloured to the base.] .... *C. judithiae*

Top margin of the column rather shallowly incised, not lobed. [Flowers (creamy)-white sometimes tinged pink or salmon, lip median up to the end of the keels with a yellow to ochrish band of longitudinal lines.] .... ***C. rhabdobulbon***

17. Lip mid-lobe with more or less broad and swollen keels, consisting of molar-like warts or many narrow, elevated cross ridges or with minute tubercles, sometimes partly broken up into low small teeth **18**
    Lip mid-lobe with plate-like keels with irregularly incised or fimbriate to shortly dentate margin **21**

18. Lip hypochile with 3 keels **19**
    Lip hypochile with 2 keels, rarely an additional median keel partly developed **20**

19. Lip mid-lobe with keels truncate, interconnected, molar-like warts, sometime 2 median keels still separate. [Sepals and petals dull light (greenish) yellow to mauve. Lip (creamy)-white, side-lobes inside brown with whitish lines, outside in front with a brown band, median of lip and keels sometimes yellowish, mid-lobe whitish, warts yellowish, between and around brown papillae.] .... ***C. tomentosa***
    Lip mid-lobe with 5–7 keels, all separate, consisting of many elevated cross-ridges. [Flowers white, lip median inside yellow with longitudinal brown lines, side-lobes inside dark brown with white nerves, outside on the junction of the side-lobes and mid-lobe with a small brown spot; mid-lobe base and centre yellow, bordered with brown, margin white.] .... ***C. kaliana***

20. Inflorescence with a stiff, (semi-) erect rachis. Lateral sepals with a pronounced keel. Column broad. [Flowers white, sometimes tinged green, lip outside on the front of the side-lobes with an ochrish to brown band, inside brown with white nerves. Mid-lobe white with a yellow centre which is sometimes bordered with brown, keels in the front dark brown to almost black.] .... ***C. hirtella***
    Inflorescence with a flexible, pendulous rachis. Lateral sepals with a weak keel. Column narrow. [Sepals and petals light ochrish brown to greenish ochre, lip whitish, side-lobes outside in front with a brown band, inside more or less dark brown with whitish nerves, keels white. Mid-lobe white with a broad brown band near the margin and often a narrow brown band across the keels.] .... ***C. pulverula***

21. Keels near the base of the lip not elevated into plate-like projections, at most fimbriate or teeth longer **22**
    Keels near the base of the lip abruptly elevated into high, plate-like, triangular to semi-elliptic, lacerate projections. [Pseudobulbs usually 4 or 5 angled with more or less sharp ridges and hollow sides. Flowers white or rarely tinged pink, side-lobes of the lip outside suffused ochre, brown or with brownish lines, in front with an ochre to brown band, inside brown with white nerves. Mid-lobe with an ochrish to brown blotch over the keels.] .... ***C. swaniana***

22. Lip mid-lobe broadly spathulate, ovate, or almost orbicular to transversely elliptic; tip retuse to broadly acute **23**
    Lip mid-lobe more or less slender ovate, tip rather long acuminate to long acute. [Sepals and petals towards bright (greenish) yellow to greenish, lip whitish, side-lobes outside with brownish stripes, in front with a brown band, inside brown with whitish nerves. Mid-lobe white, base often with a yellow blotch, ends of keels brown; sometimes only the ends of the keels yellow or brown.] .... ***C. rochussenii***

23. Inflorescence with rachis glabrous or sparsely hairy. Lip mid-lobe with 2, 4 or 6 keels **24**
Inflorescence with rachis densely velutinous. Lip mid-lobe with 3, 5 or 7 keels. [Sepals and petals white or pinkish, lip whitish, side-lobes outside in front with a brown band, inside brown with whitish or pinkish nerves, median whitish, to the front yellow with longitudinal brown lines, base of the mid-lobe yellow surrounded by brown.] .... *C. velutina*

24. Lip mid-lobe with keels rather deeply incised into many slender, soft, tooth- or hair-like projections **25**
Lip mid-lobe with keels with irregular to irregularly dentate margin. [Lip white with brown median lines, side-lobes outside in front with a brown band, inside brown with whitish nerves. Mid-lobe yellow to the tip, with brown margin.] .... *C. odoardi*

25. Inflorescence proteranthous to synanthous. Column broad. Lip hypochile with 4, very rarely 2 or 5 keels, the lateral keels rarely absent usually developed only in the front half to one quarter of the hypochile, rarely a median keel developed. [Flowers dull clay- or flesh-coloured or dull (creamy) ochrish. Lip similar coloured or whitish, side-lobes of the lip in- and outside (dark) brown except the margin, without lighter coloured nerves. Mid-lobe with some brown.] .... *C. testacea*
Inflorescence heteranthous. Column narrow. Lip hypochile with 2 keels. [Sepals and petals light yellow, lip white, side-lobes outside in front with a brown band, inside brown with whitish nerves. Mid-lobe white, margins brown.] .... *C. bruneiensis*

103. COELOGYNE ACUTILABIUM

Coelogyne acutilabium de Vogel in *Orchid Monogr.* 6: 11, f.1 (1992). Type: Bogor, cult. Sande*r* s.n. (holo. BO; iso. L).

DESCRIPTION. **Pseudobulbs** no information. **Leaves** no information. ***Inflorescence*** development is not known and peduncle (not defined). Rachis pendulous, slightly zig-zag, 24 cm. long, densely hairy. ***Flowers*** widely open. ***Dorsal sepal*** narrowly elliptic, acute, 9-nerved, 2.8 × 0.53 cm. ***Lateral sepals*** oblique, narrowly ovate, acute, 9-nerved, 2.7 × 0.5 cm. ***Petals*** linear, acute, 3-nerved, dorsal prominent, reflexed, 2.4 × 0.15 cm. ***Lip*** obscurely 3-lobed, shallowly boat shaped, side-lobes not distinct, sides of lip low, front of lip lower, entire, tip reflexed, callus of 3 keels extending from near base of lip to about one third onto mid-lobe, then gradually disappears, deeply incised, broken into triangular and quadrangular elements. ***Column*** curved, 1.5 cm long with a narrow hood, hardly wider at top, tip margin reflexed, rounded, deeply incised to form arms.
  DISTRIBUTION. Not known.
  HABITAT. Not known.
  ALTITUDE. Not known.
  FLOWERING. Not known.
  NOTES. The epithet refers to the shape of lip. Known only from type. The type description was based on a cultivated specimen, now dead, whose origin is unknown. There is every possibility that this species is a perloric form of *C. velutina* de Vogel.

104. COELOGYNE BRUNEIENSIS

Coelogyne bruneiensis de Vogel in *Orchid Monogr.* 6: 11, f.2 (1992). Type: Borneo, Brunei, Belait District, Sungai Ingei, surrounding Batu Melintung, *de Vogel* 27697 (holo. BRUN; iso. K, L). **Plate 14B.**

DESCRIPTION. *Pseudobulbs* on a short rhizome, close together, spindled-shaped, terete to rhomboid, shallowly grooved, 9–14 cm long. *Leaves* 2, rather stiff, narrowly elliptic, narrowly acute, with 3- to 5-main nerves, 26–35 × 2.5–4 cm with 3–6 cm long petiole. *Inflorescence* heteranthous, peduncle enclosed in scales of young shoot, rachis pendulous, somewhat zig-zag, 19 cm long, 12-flowered, flowers opening simultaneously, floral bracts persistent, ovary densely hairy. *Flowers* with light yellow sepals and petals , lip white, side lobes on outside, in front, light brown band, inside brown with white nerves, margin brown, keels in front light brown. *Dorsal sepal* long-elliptic, acute, 7-nerved, 2.5 × 0.9 cm. *Lateral sepals* asymmetric, oblong, acute, 7-nerved, 2.5 × 0.9 cm. *Petals* oblanceolate, acute, 5-nerved, 2.5 × 0.45 cm. *Lip* 3-lobed, side-lobes rounded in front and slightly reflexed, mid-lobe narrowly attached, elliptic, margins reflexed, undulate, tip acute, callus of 2 keels extending from base of lip to near tip of mid-lobe, deeply incised into soft tooth-like projections, a further 4 keels extending from base of mid-lobe to tip of mid-lobe with many more tooth-like projections. *Column* almost straight, narrowly spathulate, about 1.7 cm long, widest in middle, narrowing towards base.

DISTRIBUTION. Borneo (Brunei (Belait District)).
HABITAT. Epiphytic in riverine forest, 2–5 m high on tree trunks, near waterfalls.
ALTITUDE. 100–200 m.
FLOWERING. September.
NOTES. The epithet refers to the country of origin, Brunei, Borneo of the type specimen. Known only from type.

105. COELOGYNE BUENNEMEYERI

Coelogyne buennemeyeri J.J. Sm. in *Bull. Jard. Bot. Buitenzorg* ser. 3, 5: 28 (1922). Type: Sumatra, Sumatera Barat, *Bünnemeijer* 5500 (lecto. BO; iso. lecto. K, L, U).

DESCRIPTION. *Pseudobulbs* on a stout rhizome, 1–2.5 cm apart, ovoid, when dried smooth with sharp longitudinal folds, 3–7.5 cm long. *Leaves* 2, elliptic to ovate, somewhat coriaceous, with 3 to 7 nerves, 13.5–26 × 2–4.5 cm with 2–5 cm long petiole. *Inflorescence* proteranthous to synanthous, peduncle enclosed with bracts, rachis pendulous, straight, 20–40 cm long, 11- to 23-flowered, flowers opening simultaneously, floral bracts persistent, ovary slightly hairy. *Flowers* light pink, salmon-coloured or greenish salmon. *Dorsal sepal* long-elliptic to narrowly ovate, acute, 5-nerved, 2.3–2.7 × 0.7–0.9 cm. *Lateral sepals* narrowly ovate, acute, 5-nerved, 2.3–2.7 × 0.65–0.7 cm. *Petals* linear, acute, 3-nerved, 2.3–2.7 × 0.25–0.28 cm. *Lip* 3-lobed, side-lobes erect, rounded in front, slightly or not recurved, mid-lobe orbicular, tip rounded, margin erose, 2 keels extending from near base of lip to just beyond middle of mid-lobe, with deeply incised margins. *Column* curved, 1.2 cm long, expanding to widest part in middle, then to a large hood, margin reflexed.

DISTRIBUTION. Sumatra (Sumatera Barat).
HABITAT. Epiphytic on trees and shrubs between boulders in upper montane forest or scrub vegetation.

ALTITUDE. 2350–2500 m.
FLOWERING. October–November.
NOTES. The species is dedicated to the collector H.A.B Buennemeijer who was in the employ of the Herbarium at Buitenzorg, Java.

106. COELOGYNE DISTANS

Coelogyne distans J.J. Sm. in *Bull. Dep. Agric. Indes Néerl.* 15: 2 (1908). Type: Borneo, Kalimantan, Pontianak, cult. Bogor s.n. (holo. BO; iso. L).

DESCRIPTION. *Pseudobulbs* on a creeping rhizome 5–15 cm apart, more or less ovoid, slightly grooved, when dried smooth with sharp longitudinal folds, 6.5–12 cm long. *Leaves* 2, obovate to elliptic, with 5–7 main nerves, 25–32 × 2.8–6 cm with 4–7.5 cm long petiole. *Inflorescence* proteranthous, peduncle enclosed with bracts, rachis erect, 1.5–2 cm long, 3- to 4-flowered, flowers opening simultaneously, floral bracts deciduous, ovary slightly hairy. *Flowers* widely open, sepals and petals white or cream-coloured, side-lobes light reddish-brown lines inside, outside yellowish to brown front margin, keels with brown tip. *Dorsal sepal* narrowly elliptic, acute, slightly fleshy, 9–15-nerved, 3–3.5 × 0.5–0.9 cm. *Lateral sepals* somewhat falcate, narrowly elliptic, acute, 9–15-nerved, 2.6–3.2 × 0.7–0.85 cm. *Petals* somewhat falcate, narrowly elliptic to obovate, acute, 7–13-nerved, 2.8–3.2 × 0.45–0.6 cm. *Lip* 3-lobed, side-lobes large, more or less rounded and reflexed, mid-lobe narrowly attached, orbicular, margin slightly undulate, tip acute, callus of 5 keels but sometimes more, extending from base of lip to middle of mid-lobe, median keel terminates at base of mid-lobe, further 2 keels sometimes on outside of existing keels, short, all keels strongly lacerate, often broken up near base, irregular in length. *Column* curved, 2–2.4 cm long, expanding to widest part in middle, then to a large hood, tip margin shallowly notched, irregular dentate, mid part reflexed.
DISTRIBUTION. Borneo (Kalimantan, Sarawak, Sabah).
HABITAT. Epiphytic in riverine forest.
ALTITUDE. Lowlands.
FLOWERING. March–April.
NOTES. The epithet refers to the wide spacing of the pseudobulbs.

107. COELOGYNE ECHINOLABIUM

Coelogyne echinolabium de Vogel in *Orchid Monogr.* 6: 16, f. 6 (1992). Type: Borneo, Sarawak, Gat, Upper Rejang River, *J. & M.S. Clemens* 21639 (holo. L; iso. K, NY). **Plate 14C.**

DESCRIPTION. *Pseudobulbs* on a short rhizome, close together, slender cylindric, when dried with many fine sharp longitudinal folds, 6–20 cm long. *Leaves* 2, thin, narrowly elliptic, acute to acuminate, with 5–7 main nerves, 16–33 × 3–6.5 cm with 3.5–7 cm long petiole. *Inflorescence* heteranthous, peduncle enclosed in bracts, rachis pendulous, about straight to slightly zig-zag, 15–25 cm long, 8- to 18-flowered, flowers opening simultaneously, floral bracts persistent, ovary densely hairy. *Flowers* creamy coloured with brownish colouring on lip, pale nerves. *Dorsal sepal* narrowly elliptic, acute to acuminate, 7–13-nerved, 2.4–2.6 × 0.6–0.85 cm. *Lateral sepals* oblique, narrowly elliptic, acute to acuminate, 7–13-nerved,

2.4–2.6 × 0.6–0.8 cm. **Petals** straight or slightly falcate, narrowly elliptic, acute to acuminate, 5–7-nerved, 2.3–2.6 × 0.4–0.5 cm. **Lip** 3-lobed, side-lobes large, broadly rounded, mid-lobe narrowly attached, orbicular, margins curved upwards, finely erose or dentate, tip rounded to acuminate, callus of 2 strongly dentate keels, extending from base of lip to base of mid-lobe where they become a mass of flattened slender lobules covering the mid-lobe, apart from the margins, a further 2 keels from three quarter point from base and extending to base of mid-lobe. **Column** very slender, straight, 1.5–1.8 cm long, expanding to widest part in middle or near tip, then to a narrow hood, margin of tip dentate.

DISTRIBUTION. Borneo (Sarawak, Brunei).
HABITAT. Epiphytic in lowland dipterocarp forest and riverine forest.
ALTITUDE. 30–50 m.
FLOWERING. January–February, June–July.
NOTES. The epithet refers to the mass of protuberances on the lip of the flower. de Vogel refers to confusion between this species and *C. rhabdobulbon* Schltr. It would seem that the Beccari collection 1868 was a mixture of the two species. A description of the material in Berlin was used by Schlechter to define *C. rhabdobulbon* and Carr's drawing at Kew of the Beccari 1868 type relates to the same species. However, *Beccari* 1868 material in Florence had never been described; de Vogel described the Florence material as *C. echinolabium.*

108. COELOGYNE GENUFLEXA

Coelogyne genuflexa Ames & C. Schweinf., *Orch.* 6: 28 (1920). Type: Borneo, Sabah, Mt. Kinabalu, Marai Parai Spur, *J. Clemens* 251 (holo. AMES). **Plate 14D.**

Coelogyne reflexa J.J. Wood & C.L. Chan in *Lindleyana* 5 (2): 87, f. 5 (1990). Type: Borneo, Sabah, Gunung Alab, *de Vogel* 8666 (holo. K; iso. L).

DESCRIPTION. Pseudobulbs on a thick and woody rhizome, 0.5–2 cm apart, narrowly cylindric, sulcate, dull yellowish-green, 6 × 0.5 cm. **Leaves** 2, oblong-elliptic, rounded at tip, coriaceous, with (3)5–(5)7 nerves, 11–14 × 2.5–3.5 cm with well defined 2–5 cm long petiole. **Inflorescence** synanthous with the halfway to entirely developed leaves, peduncle 10 cm, rachis arcuate, more or less zig-zag, rather densely hairy, about 8 cm long, (14–)20- to 36-flowered, flowers opening progressively, floral bracts persistent, ovary hairy. **Flowers** deep creamy brown, greenish yellow or creamy-green(brown), lip similarly coloured or whitish with a light citron yellow blotch on the hypochile. **Dorsal sepal** lanceolate, acute, 3–5-nerved, 0.9–1.1 x 0.2–0.3 cm. **Lateral sepals** lanceolate, acute, 3–5-nerved, 0.9–1.05 × 0.28–0.35 cm. **Petals** linear, strap shaped, acute, 3-nerved, 0.85–1 × 0.08–0.1 cm. **Lip** 3-lobed, side-lobes in front rounded and reflexed, mid-lobe sharply divided into hypochile and epichile, orbicular, margin curved upwards, tip rounded to acuminate, callus of 4 keels only developed at base of mid-lobe and extending sometimes onto mid-lobe, simple, entire, straight. **Column** arcuate, about 0.8 cm long, broadly winged at tip, tip shallowly tri-lobed with minute lobe on either side of central lobe.

DISTRIBUTION. Borneo (Sarawak, Sabah, Brunei).
HABITAT. Epiphytic on trunks and major branches of trees in low, dense to open mossy upper montane forest, scrub on landslides, sometimes on ultramafic substrate.
ALTITUDE. 1450–2000 m (1200–1900 m).

FLOWERING. February–March, October, December.

NOTES. The epithet refers to the strongly reflexed mid-lobe of the lip of the flower. Examination of the types of this species and that of *C. reflexa* J.J. Wood & C.L. Chan indicate they are conspecific, a judgement that J.J. Wood confirmed in *Sandakania* 12: 22 (1998).

109. COELOGYNE HIRTELLA

Coelogyne hirtella J.J. Sm. in *Bull. Jard. Bot. Buitenzorg* ser. 3, 11: 105 (1931). Type: Borneo, Kalimantan, Gunung Kemoel, *F.H. Endert* 3976 (holo. L, iso. BO). **Plate 14E.**

Coelogyne radioferens sensu J.J. Sm., *Mitt. Inst. Allg. Bot. Hamburg* 7, 32, t. 5, fig. 24 (1927), *non* Ames & C. Schweinf.
C. radiosa J.J. Sm. in *Bull. Jard. Bot. Buitenzorg* ser. 3, 11: 105 (1931). Type: Borneo, *Winkler* 954 (lecto. L).

DESCRIPTION. Pseudobulbs on a stout rhizome, close together, ovoid, with some shallow grooves, when dried finely wrinkled, 2–6 cm long. **Leaves** 2, stiff, lanceolate to linear, with 5–7 main nerves, 14.5–55 × 1.5–7.5 cm with 3.5 cm long petiole. **Inflorescence** proteranthous to synanthous, peduncle enclosed with scales of young shoot, rachis stiff, more or less erect or curved, 7–18 cm long, 5- to 10-flowered, flowers opening in succession, probably 2 at a time, floral bracts persistent, ovary densely hairy. **Flowers** widely opening, sepals and petals white or cream coloured, sometimes tinged pale green, lip white or cream coloured, side-lobes outside in front with an ochrish to brownish blotch, inside brown with white nerves and margin. **Dorsal sepal** narrowly elliptic, acute, 7–11-nerved, 2–3.7 × 0.65–1.2 cm. **Lateral sepals** narrowly elliptic to lanceolate, acute, 7–11-nerved, 2–3.6 × 0.5–1.1 cm. **Petals** narrowly lanceolate, acute, 7–9-nerved, 2–3.6 × 0.5–1 cm. **Lip** 3-lobed, side-lobes rounded and recurved, mid-lobe narrowly attached, convex, margins strongly recurved and then curved up, entire or erose, tip acute or obtuse, callus of 2 keels extending from base of lip to middle of mid-lobe, low and rather broad, initially 0.3 cm high, then diminishing in height, changing from dentate projection into cross-ridged sections, a further 2(4) keels extend from base of mid-lobe to middle of mid-lobe, also cross-ridged sections, all keels become rather swollen papillose. **Column** arcuate, 1.4–1.9 cm long, expanding to widest part in middle, then to a large hood, tip margin dentate, middle part reflexed.

DISTRIBUTION. Borneo (Kalimantan, Sarawak, Sabah, Brunei).

HABITAT. Epiphytic, on lower tree trunks and major branches in dense to open heath forest (Kerangas) and moss forests, sometimes terrestrial or lithophytic on sandstone.

ALTITUDE. 800–2350 m.

FLOWERING. Probably all year around; December–March, July–August.

NOTES. The epithet refers to the hairs on the leaves, petiole and the inflorescence.

110. COELOGYNE JUDITHIAE

Coelogyne judithiae P. Taylor in *Orchid Rev.* 85 (1012): 289, f. 252, 253 (1977). Type: Peninsular Malaysia, Taiping Hills, *Jackman* s.n. (holo. K). **Plate 14F.**

DESCRIPTION. **Pseudobulbs** on rhizome, close together, narrowly ovoid to fusiform, with up to 8 shallow grooves, 10–12 cm long. **Leaves** 2, more or less stiff, narrowly elliptic, acute to acuminate, with 5–7 main nerves, 25–42 × 6–10 cm with 3–7 cm long petiole. **Inflorescence** synanthous, peduncle enclosed in scales of young shoot, rachis pendulous, rather zig-zag, to 15 cm long, 10- to 15-flowered, flowers opening simultaneously, floral bracts persistent, ovary slightly hairy. **Flowers** spreading, white, lip mustard-yellow to ochre-yellow on lower and upper sides, keels cream-coloured. **Dorsal sepal** narrowly elliptic, acute, 7-nerved, 3 × 1 cm. **Lateral sepals** somewhat asymmetric, narrowly elliptic, acute, 11-nerved, 2.7 × 1 cm. **Petals** almost straight, linear, top rounded, irregular, 5–7-nerved, 2.9 × 0.5 cm. **Lip** 3-lobed, side-lobes erect, projecting in front, rounded and hardly reflexed, mid-lobe narrowly attached, convex, semi-orbicular, margins distinctly undulate, tip retuse-obtuse and apiculate, callus of 2 keels sometimes branched, extending from base of lip to middle of mid-lobe, a further 2 keels only develop just before base of mid-lobe and continue to just beyond base of mid-lobe, all keels irregular, slightly undulating and with swollen margin. **Column** arcuate, 2 cm long, expanding to widest part near tip, then to a large hood, tip margin deeply incised, distinctly 6-lobed, middle part not reflexed.

DISTRIBUTION. Peninsular Malaysia (Taiping Hills).
HABITAT. Not known.
ALTITUDE. Not known.
FLOWERING. Not known.
NOTES. The species is dedicated to the collector of the type, Mrs Judith Jackman, a British amateur orchid grower. It is only known from the type.

111. COELOGYNE KALIANA

Coelogyne kaliana P.J. Cribb in *Kew Bull.* 36: 779, f. 1 (1982). Type: Peninsular Malaysia, Selangor/Pahang, Genting Highlands, Ulu Kali, *G. Smith* BE 79 (holo. K). **Plate 15A**.

DESCRIPTION. **Pseudobulbs** on a short rhizome, close together, more or less ovoid, often distinctly 3-ridged, when dried rather finely wrinkled, 1.5–5.5cm long. **Leaves** 2, narrowly elliptic to lanceolate, coriaceous, with 7–9 main nerves, 7–23 × 2–5.5 cm with a 2–8 cm long petiole. **Inflorescence** heteranthous, peduncle enclosed with scales of young shoot, rachis curved, pendulous, more or less zig-zag, 8–30 cm long, 4- to 12-flowered, flowers opening simultaneously, floral bracts persistent, ovary slightly hairy. **Flowers** white, inside of side-lobes of lip brown with white nerves and margin, outside with brown spot at junction of side-lobes and mid-lobe, median part of lip deep yellow with 4 longitudinal brown lines, keels white or yellow. **Dorsal sepal** narrowly elliptic, acute, margin sometime slightly reflexed, 7–11-nerved, 3.2–5 × 0.8–1.7 cm. **Lateral sepals** slightly falcate, narrowly elliptic to lanceolate, acute, 5–9-nerved, 3.2–5 × 0.8–1.7 cm. **Petals** slightly falcate, narrowly elliptic to lanceolate, acute, 5–9 nerved, 3.2–4.9 × 0.9–1.3 cm. **Lip** 3-lobed, side-lobes rounded and reflexed, mid-lobe narrowly attached, orbicular-ovate, margins strongly recurved and curved up again, slightly erose, tip acuminate, callus of 3 keels extending from base of lip to middle of mid-lobe and branching on the mid-lobe, a further 2 keels develop at base of mid-lobe on the outside of existing keels, all keel dentate with median keel a single set of lacerations, other keels a double set of lacerations, on mid-lobe rather broad and swollen with elevated cross-ridges. **Column** arcuate, 4–5.3 cm, expanding to widest part in middle, then to a large hood, tip margin dentate, middle part reflexed, sometimes retuse.

DISTRIBUTION. Peninsular Malaysia (Selangor, Pahang, Perak).

HABITAT. Epiphytic, on shrubs, tree trunks and large branches in montane forest, also terrestrial on moss.

ALTITUDE. 1650–2000 m.

FLOWERING. April–August.

NOTES. The epithet refers to the locality Ulu Kali, Genting Highlands, Peninsular Malaysia where the type specimen was collected.

112. COELOGYNE LATILOBA

Coelogyne latiloba de Vogel in *Orchid Monogr.* 6: 20, f. 10 (1992). Type: Borneo, Sabah, Mt. Kinabalu, Marai Parai spur, *Collenette* A 38 (holo. BM).

DESCRIPTION. *Pseudobulbs* on a short rhizome, close together, ovoid, shallowly grooved, when dried smooth with coarse sharp longitudinal folds, to 4.5 cm long. *Leaves* 2, lanceolate, hard, coriaceous, with 3–5 main nerves, 15.5–32 × 2.3–4.2 cm with 1.5–7.5 cm long petiole. *Inflorescence* heteranthous, peduncle sturdy, short, erect, extending 2.5 cm beyond scales of young shoot, rachis suberect, zig-zag, 2.6–5 cm long, 2- to 3-flowered, flowers opening simultaneously, floral bracts persistent, ovary slightly hairy. *Flowers* pure white, side-lobes inside bright brown with white nerves, lip median white, inside the front half rich bright orange-brown except for a narrow white band in front, keels white. *Dorsal sepal* narrowly elliptic, acute, 5-nerved, dorsal a low rounded keel, 2.5 × 0.7 cm. *Lateral sepals* oblique, narrowly elliptic, acute, 7-nerved, dorsal a low, thick plate-like keel, 2.5 × 0.7 cm. *Petals* almost straight, narrowly elliptic, acute, 5-nerved, dorsal prominent, 2.5 × 0.5 cm. *Lip* 3-lobed, side-lobes high, rounded and reflexed, mid-lobe narrowly attached, convex, elliptic, sometimes a short broad claw, margin undulate, slightly erose, broad triangular tip, callus of 3 keels extends from near base of lip, median keel extends half way to base of the mid-lobe, lateral keels continue to three-quarters way to base of mid-lobe, keels plate-like, irregularly shallow and incised, median keel smaller, on mid-lobe 2 or 3, possibly 4 further keels develop laterally across base of mid-lobe, broad and swollen, comprising cross-ridges. *Column* slightly arcuate, 1.6 cm long, expanding to widest part in middle, then to a large hood, tip margin rounded, slightly irregular, middle part not reflexed.

DISTRIBUTION. Borneo (Sabah (Mount Kinabalu)).

HABITAT. Epiphytic in lower montane forest, sometimes on ultramafic substrate, 1.2 m high on a *Leptospermum* shrub, in light shade.

ALTITUDE. 1700 m.

FLOWERING. September.

NOTES. The epithet refers to the broad side-lobes of the lip to the flowers. Only known from type specimen.

113. COELOGYNE LONGIBULBOSA

Coelogyne longibulbosa Ames & C. Schweinf., *Orch.* 6: 33 (1920). Type: Borneo, Sabah, Mt. Kinabalu, Kiau, *J. Clemens* 79 (holo. BM; iso. AMES, BO, K). **Plate 15B.**

DESCRIPTION. *Pseudobulbs* on a long creeping rhizome, 3–7.5 cm apart, very slender cylindric, when dried smooth with coarse sharp longitudinal fold, 11–25 cm long. *Leaves* 2, narrowly elliptic, acute, with 3–7 main nerves, 15.5–35.5 × 3–7 cm with 1.3–5.7 cm long petiole. *Inflorescence* synanthous, rarely proteranthous, peduncle enclosed in scales of young shoot, rachis limply pendulous, 10–31 cm long, 16- to 25-flowered, flowers opening simultaneously, floral bracts persistent, ovary densely hairy. *Flowers* with white to cream-coloured sepals and petals, sometimes tinged green, lip white or cream-coloured with a yellow, ochre to light brown blotch on side-lobes and mid-lobe. *Dorsal sepal* narrowly ovate, acute to obtuse, 5–7-nerved, dorsal sometimes prominent, 1.25–1.65 × 0.35–0.75 cm. *Lateral sepals* narrowly ovate, acute to obtuse, 5–7-nerved, dorsal prominent, 1.25–1.65 × 0.35–0.75 cm. *Petals* linear-ovate, obtuse to acute, 3-nerved, 1.1–1.65 × 0.15–0.3 cm. *Lip* 3-lobed side-lobes rounded and slightly reflexed, mid-lobe suborbicular, shallowly retuse, and bluntly apiculate, callus of 2 keels extending from base of lip to middle of mid-lobe, plate-like, entire, variable height, partly folded, a further 2 (4) keels develop just before base of mid-lobe and continue to middle of mid-lobe, same form as inner keels. *Column* arcuate, 1 cm long, very broadly winged, sides distinctly projecting forward, broadly rounded, margin irregular or entire, tip acuminate or quadrangular or truncate.

DISTRIBUTION. Borneo (Sarawak, Sabah).

HABITAT. Epiphytic on river-bed boulders and on tree trunks, often near the base of trees in hill and lower montane forest, sometimes on ultramafic substrate.

ALTITUDE. 1000–1900 m.

FLOWERING September–February.

NOTES. The epithet refers to the extremely long, slender pseudobulbs.

114. COELOGYNE MOULTONII

Coelogyne moultonii J.J. Sm. in *Bull. Jard. Bot. Buitenzorg* ser. 2, 3: 54 (1912). Type: Borneo, Sarawak, Bukit Labeng Barisan, Ulu Limbang, *Moulton* 17 (holo. BO). **Plate 15C.**

DESCRIPTION. *Pseudobulbs* on a short rhizome, close together, slender cylindric, when dried with sharp longitudinal folds, 12.5–28 cm long. *Leaves* 2, linear to very narrowly elliptic, with 7–9 main nerves, 34–52 × 5–9.5 cm with 6–11 cm long petiole. *Inflorescence* synanthous, peduncle enclosed within the scales of the young shoot, rachis limply pendulous, straight or very slightly zig-zag, 30–55cm long, 24- to 52-flowered, flowers opening simultaneously, floral bracts persistent, ovate to broadly ovate, ovary rather densely hairy. *Flowers* white to cream-coloured, side-lobes of lip inside and sometimes outside pale yellowish, ochrish to light brown with white margins, between the keels sometimes with similar coloured lines, mid-lobe with a large pale yellowish or light brown spot, also over the front of the keels, and white margins, column whitish, anther with some brown. *Dorsal sepal* narrowly ovate, acute, 7-nerved, dorsal prominent, 1.45–1.8 × 0.35–0.65cm. *Lateral sepals* narrowly ovate, acute, 7-nerved, dorsal prominent, 1.45–1.8 × 0.35–0.65 cm. *Petals* linear, acute, 3-nerved, 1.35–1.7 × 0.15–0.3 cm. *Lip* 3-lobed, side-lobes rounded, mid-lobe shallowly saccate at base, convex, broadly attached, semi-elliptic to broadly obovate, margin irregular, tip broadly acute, callus of 4 keels straight with deeply incised margins, median pair of keels develop near base of lip while lateral keels develop away from base, all extend to just beyond base of mid-lobe, further 2 keels develop only on mid-lobe outside other keels, similar form. *Column* more or less arcuate, 0.9–1.3 cm long, hood widest below the middle, sides distinctly projecting to the front, tip irregular, reflexed.

DISTRIBUTION. Borneo (Sarawak, Sabah).

HABITAT. Epiphytic or lithophytic in rocky areas, landslides and lower montane oak-laurel forest, recorded from ultramafic substrate. On trunks of trees, sometimes lithophytic on mossy rocks.

ALTITUDE. 1100(1400)–2400 m.

FLOWERING. April–June, September–December.

NOTES. The species is dedicated to J.C. Moulton, Curator of the Sarawak Museum (1905–1915), Director of the Raffles Museum and Library (1919–1923), later Chief Secretary to the Government in Sarawak. He made several expeditions in West Borneo.

115. COELOGYNE MULUENSIS

Coelogyne muluensis J.J. Wood in *Kew Bull.* 39(1): 76, f. 2 (1984). Type: Borneo, Sarawak, near Bukit Berar, *Collenette* 2348 (holo. K). **Plate 15D.**

DESCRIPTION. *Pseudobulbs* on a slender-stout, sheathed, climbing rhizome, close together, oblong-cylindric, 2.5–4 × 1.5 cm, fully enclosed at base in large, multiple bracts. *Leaves* 2, narrowly elliptic or oblong-elliptic, acute to acuminate, 23–35 × 4–6 cm with a 3–4 cm long, channelled petiole. *Inflorescence* heteranthous, peduncle completely enveloped in loosely imbricating bracts, about 5 cm long, rachis slender, arcuate, 6 cm long, 3- to 4-flowered, flowers opening in succession from tip of rachis, floral bracts persistent. *Flowers* about 6 cm across, creamy white with lip reddish-ochre or chestnut except on margins and at base of mid-lobe, outer keels creamy and mid keel yellow. *Dorsal sepal* oblong-elliptic, obtuse, hooded, subacute, 3.6 × 1.2 cm. *Lateral sepals* oblong-elliptic, acute, 3-nerved, 3 × 1 cm. *Petals* lanceolate, acute, 3-nerved, 3.5 × 0.5 cm. *Lip* 3-lobed, concave at base side lobes erect, rounded, mid-lobe ovate or oblong-ovate, obtuse to subacute, callus of 3 keels extending from base of lip, keels erect, parallel, median keel prominent, lateral keels reduced and merging into nerves. *Column* arcuate, 1.8 cm long, expanding into winged hood, tip erose.

DISTRIBUTION. Borneo (Sarawak).

HABITAT. Epiphytic in tropical heath forests, alluvial forests and mixed dipterocarp forest with a Kerangas element. Common on slender trees and saplings.

ALTITUDE. Sea level–150 m.

FLOWERING. Not known.

NOTES. The epithet refers to Mount Mulu, Sarawak, Borneo close to where the type was collected. This species was not included by de Vogel (1992) in his revision of the section.

116. COELOGYNE ODOARDI

Coelogyne odoardi Schltr. in *Notizbl. Bot. Gart. Berlin-Dahlem* 8: 14 (1921). Type: Borneo, Sarawak, Gunung Matang, *Beccari* 1678 (holo. FI; iso. K). **Plate 15E.**

DESCRIPTION. *Pseudobulbs* on a short to creeping rhizome, 1.5–3 cm apart, ovoid, shallowly grooved, 1.5–4.5 cm long. *Leaves* 2, elliptic, coriaceous, with 3–5 main nerves, 5–23.5 × 1.8–3.7 cm with 0.8–6.5 cm long petiole. *Inflorescence* synanthous, peduncle enclosed in scales of young shoot or extending 7 cm, rachis curved to pendulous, not to slightly zig-zag, 7–28 cm long, 3- to 13-flowered, flowers opening simultaneously, floral bracts persistent, ovary rather densely hairy. *Flowers* with ochre sepals and petals, lip white, brown band on outside front of side-lobes and brown lines inside, brown lines on mid-lobe. ***Dorsal sepal*** narrowly elliptic, acute, 7–11-nerved, dorsal prominent or a low rounded keel, 2.5–3.6 × 0.8–1.5 cm. ***Lateral sepals*** narrowly ovate, acute to acuminate, 7–9-nerved, dorsal a low rounded keel, 2.5–3.5 × 0.7–1.2 cm. ***Petals*** lingulate, acute, 5–7-nerved, dorsal prominent, 2.2–3.5 × 0.35–0.7 cm. ***Lip*** 3-lobed, side-lobes large, rounded in front and slightly reflexed, mid-lobe convex, ovate-elliptic, broad triangular tip, callus of 2 or 3 keels, median keel absent or only developed half to three quarters the way from base of lip and not extending beyond base of mid-lobe, lateral keels extend from base of lip to three quarters the way onto mid-lobe, near base of lip, keels plate-like with one or two rows of slender dentate projections which gradually lower half way to base of mid-lobe, a further 2–4 keels only on mid-lobe, outside existing keels, extending from base of mid-lobe to middle of mid-lobe, all keels on mid-lobe irregularly dentate or swollen margin. ***Column*** arcuate, 1.8–2.4 cm long, sometimes hairy, expanding to widest part in middle, then to a large hood, tip margin almost entire to irregularly dentate, hardly reflexed.

DISTRIBUTION. Borneo (Sarawak, Sabah, Brunei).
HABITAT. Epiphytic on branches of canopy trees in hill forest and lower montane forest.
ALTITUDE. 200–1900 m.
FLOWERING. May–July.
NOTES. This species is dedicated to Odoardo Beccari (1843–1920), collector of the type specimen. His herbarium collections are in Florence.

117. COELOGYNE PALAWANENSIS

Coelogyne palawanensis Ames, *Orch.* 5: 51 (1915). Type: Philippines, Palawan, Mt. Capoas, *Merrill* 9499 ((holo. AMES).

DESCRIPTION. ***Pseudobulbs*** on a sturdy rhizome, close together, elongate, when dry strongly sulcate, narrowing at tip, 5–7 × 1 cm. ***Leaves*** 2, erect, narrowly elliptic-lanceolate, base and top gradually tapering, to 20 × 3.5 cm with 6 cm long petiole. ***Inflorescence*** synanthous, peduncle enclosed within young scales and developing leaves, 17–20 cm long, 5-flowered, flowers opening probably simultaneously, floral bracts persistent. ***Flowers*** 2.5 cm long, white, side-lobes with yellow veining inside, mid-lobe yellow within. ***Dorsal sepal*** oblong, acute, gradually tapering to the tip, without a keel, 2.7 × 1 cm. ***Lateral sepals*** oblong, acute, gradually tapering to the tip, with a low keel, 2.7 × 1 cm. ***Petals*** narrower, oblong-obtuse, slightly tapering to the tip, 2.9 × 0.6 cm. ***Lip***, 3-lobed, side-lobes one fifth shorter than mid-lobe, broadly semi-cordate, top rounded, mid-lobe 1.2 cm long, no obvious claw, rounded, callus of 2 lightly curved keels, converging at base of lip, at middle of mid-lobe composed of short blunt teeth, ***Column*** curved, 1.5 cm long, at the top broadly winged, minutely denticulate.

DISTRIBUTION. Philippines (Palawan, Mount Capoas).
HABITAT. Not known.
ALTITUDE. 1000 m.
FLOWERING. April.
NOTES. The epithet refers to Palawan Island in the Philippines where the type specimen was collected. de Vogel excluded this species from sect. *Tomentosae* as an 'insufficiently known species'. Ames considered it to be near to *C. swaniana* Rolfe. Based on the descriptions I have seen, I have placed this species in sect. *Tomentosae*.

118. COELOGYNE PHOLIDOTOIDES

Coelogyne pholidotoides J.J. Sm. in *Icon. Bogor.* 2: 24, t. 106b (1903). Types: Borneo, Kalimantan, Damus, *Hallier* s.n.; *Molengraaf* s.n. (syn. L). **Plate 16A.**

DESCRIPTION. *Pseudobulbs* on a short to creeping rhizome, 1.3–3 cm apart, ovoid, shallowly grooved, when dried smooth with coarse sharp longitudinal folds, 4–8.5 cm long. *Leaves* 2, rather stiff, elliptic, acute to rounded, with 5 main nerves, 14.5–22 × 2.5–4.5 cm with 2–5 cm long petiole. *Inflorescence* synanthous, peduncle erect or curved, extending to 29 cm beyond scales of young shoot, rachis pendulous, about straight or sometimes zig-zag, 22–46 cm long, 24- to 62-flowered, flowers open simultaneously, floral bracts persistent, ovary densely hairy. *Flowers* white to cream-coloured, side-lobes of the lip in front with a dark brown spot, mid-lobe with one or two light yellow markings, keels sometimes yellow in the centre with brown edges, column white, stalk in front sometimes yellow. *Dorsal sepal* long elliptic to narrowly ovate, acute, 5–7-nerved, dorsal prominent, 0.95–1.15 × 0.25–0.4 cm. *Lateral sepals* (narrowly) ovate, acute to rounded, 5–7-nerved, dorsal prominent or a low keel, 0.95–1.2 × 0.25–0.4 cm. *Petals* linear, acute to rounded, 1–3-nerved, dorsal more or less prominent, 0.8–1.15 × 0.07–0.13 cm. *Lip* 3-lobed, side-lobes small, in front more or less rounded and recurved, mid-lobe broadly spathulate, distinctly two-lobed, almost truncate to deeply emarginate, tip rounded-acute, lobes rounded with entire undulating margin, callus of 2 entire, low, rounded keels, extending from base of lip to just beyond middle of mid-lobe, sometimes a further 2 short keels outside existing keels on mid-lobe only, all keels on mid-lobe, low, swollen, either undulate or broken into warty projections. *Column* slightly arcuate, 7.5–9 cm long, expanding slightly into narrow hood, tip margin notched, irregularly dentate.

DISTRIBUTION. Borneo (Kalimantan, Sarawak, Sabah).

HABITAT. Epiphytic, sometimes lithophytic in hill forest, on trunks and branches of trees or exposed to the sun on sandstone or on rock outcrops.

ALTITUDE. 800–1500 m.

FLOWERING. September–October, December.

NOTES. The epithet refers to the resemblance of the flowering plant to *Pholidota* species.

119. COELOGYNE PULVERULA

Coelogyne pulverula Teijsm. & Binn. in *Natuurk Tijdschr. Ned.-Indie* 24: 306 (1862). Type: Sumatra, *Teijsmann & Binnendijk* s.n. (holo. L). **Plate 16B.**

Coelogyne dayana Rchb.f. in *Gard. Chron.* ser. 2, 21: 826 (1884). Type: Borneo, Veitch, *Curtis* s.n. (holo. W; iso. K).

DESCRIPTION. *Pseudobulbs* close together on a short rhizome, spindled-shaped to long-ovoid, rather shallowly grooved, 7–25 cm long. *Leaves* 2, rather stiff, oblanceolate to long elliptic, with 7–9 nerves, 25–65 × 4.5–11 cm with 6–15 cm long petiole. *Inflorescence* proteranthous to synanthous with less than halfway developed leaves, peduncle enclosed in bracts, rachis pendulous straight to zig-zag 25–110 cm long, 10- to 55-flowered, flowers opening simultaneously, floral bracts persistent, ovary densely hairy. *Flowers* with pale to light yellowish brown to greenish ochre sepals and petals, lip outside white, front of side-lobes with brown band, side-lobes inside brown to dark brown with white nerves and white margin, mid-lobe white including the keels; a broad band near the margin and often a narrow brown cross band over the keels. *Dorsal sepal* narrowly elliptic, acute, margin more or less reflexed, 5–13-nerved, 2.3–3.3 × 0.5–0.95 cm. *Lateral sepals* narrowly elliptic, acute, 5–9-nerved, 2.3–3.3 × 0.55–0.9 cm. *Petals* narrowly elliptic, rounded to acute, 5–7-nerved, 2.3–3.3 × 0.22–0.7 cm. *Lip* 3-lobed, 2–3 cm long, side-lobes large, broadly rounded in front, reflexed, mid-lobe narrowly attached, orbicular, margins strongly reflexed, tip acute, callus of 2 low keels extending from base of lip to middle of mid-lobe, wart-like or tooth-like projections, further 4–6 keels on outside, on mid-lobe extending from base of mid-lobe to varying distances along mid-lobe but nominally centred at the middle of the mid-lobe. *Column* slightly arcuate, 1.5–2 cm long, expands to narrow hood, widest just below tip, tip margin dentate, sometimes deeply notched.

DISTRIBUTION. Borneo (Sarawak, Sabah, Brunei), Java, Peninsular Malaysia, Thailand, Sumatra (West coast).

HABITAT. Epiphytic on trunks and major branches of trees in lowland rain forest and lower montane forest, sometimes lithophytic on rocks, often near or over streams and rivers.

ALTITUDE. 275–1900 m.

FLOWERING. Nearly all year.

NOTES. The epithet refers to the powdery appearance of the flower. *C. pulverula* Teijsm & Binn. and *C. dayana* Rchb.f. have been judged to be conspecific by earlier authors although the type specimens were judged to have a size difference. In recent history, *C. dayana* was given precedence by de Vogel whilst G. Seidenfaden gave *C. pulverula* precedence. There is no doubt that the name *C. dayana* has been favoured by the orchid-growing community but priority dictates that *C. pulverula* Teijsm & Binn. is the correct name and *C. dayana* Rchb.f. is a later synonym. *Coelogyne dayana was* dedicated to the 19th Century orchid grower, John Day of Tottenham, London, whose well illustrated notebooks recording his active interest in orchids are in the possession of Kew. Curtis collected the type specimen of *C. dayana* for Veitch & Sons and it flowered for the first time in cultivation in 1884. Examples of this species, as *C. dayana*, are to be found in many collections.

120. COELOGYNE RADIOFERENS

Coelogyne radioferens Ames & C. Schweinf., *Orch.* 6: 38, pl. 81 (1920). Type: Borneo, Sabah, Mt. Kinabalu, Paka-Paka Cave, *J. Clemens* 200 (holo. AMES). **Plate 16C**.

DESCRIPTION. *Pseudobulbs* on a long creeping rhizome, 2–7 cm apart, more or less ovoid, shallowly grooved, 5–10 cm long. *Leaves* 2, more or less stiff, long elliptic, acute, with 5–9 main nerves, 26–47 × 2.5–7.5 cm with 6.5–19 cm long petiole. *Inflorescence* proteranthous, peduncle enclosed in scales of young shoot, rachis curved at the base, top part pendulous, about straight to more or less zig-zag, 9.5–25 cm, 6- to 15-flowered, flowers opening simultaneously, floral bracts persistent, ovary densely hairy. *Flowers* with light to deep ochre yellow sepals and petals, lip white or cream-coloured, side-lobes outside in front with a brown band with often whitish lines, inside light to deep brown with white lines and white margins, mid-lobe whitish with often brown lines between the keels. *Dorsal sepal* narrowly elliptic, acute, reflexed, 7–11-nerved, dorsal low rounded keel, 3.2–4 × 0.8–1.2 cm. *Lateral sepals* oblique, narrowly elliptic to lanceolate, acute, reflexed, 7–11-nerved, dorsal low rounded keel, 3.2–4 × 0.75–1.2 cm. *Petals* almost straight, narrowly elliptic to linear, acute to rounded, 5–7-nerved, dorsal prominent or low rounded keel, 3.2–4 × 0.35–0.65 cm. *Lip* 3-

lobed, side-lobes rounded and reflexed, mid-lobe narrowly attached, elliptic, margin erose, reflexed, bi-lobed, tip retuse, broadly acute, callus of 2 keels extending from base of lip, with hair-like projections, and terminating at base of mid-lobe and become an expanded mass of low, rounded, separate papillae which extend to the tip of the mid-lobe, from central mass, papillae or elongated warts radiate over the mid-lobe. *Column* arcuate, 1.6–1.8 cm long, expanding to widest part in middle, then into large hood, tip margin truncate, notched, irregularly dentate.

DISTRIBUTION. Borneo (Kalimantan, Sarawak, Sabah, Brunei).

HABITAT. Epiphytic on mossy trunks and big branches of trees in lower and upper montane forest, sometimes lithophytic on mossy rocks.

ALTITUDE. 1300–3000 m.

FLOWERING. All year.

NOTES. The epithet refers to the rows of papillose and warts radiating from the central mass of papillose on the mid-lobe.

121. COELOGYNE RHABDOBULBON

Coelogyne rhabdobulbon Schltr. in *Notizbl. Bot. Gart. Berlin-Dahlem* 8: 15 (1921). Type: Borneo, Sarawak, *Beccari* 1868 (holo. B†; lecto. drawing by Carr (K); not *Beccari* 1868 in FI, see *C. echinolabium*). **Plate 16D.**

Coelogyne pulverula sensu Lamb & C.L Chan in Luping, Chin & Dingley (eds.), *Kinabalu, Summit of Borneo* 238, pl. 28 (1978), *non* Teijsm. & Binn.

DESCRIPTION. *Pseudobulbs* on a short rhizome close together, slender, spindle-shaped to long-cylindric, with few shallow grooves, 7–28 cm long. *Leaves* 2, long-obovate to long-elliptic, acute, tip often slightly acuminate, with 5–9 main nerves, 13–49 × 4–14 cm with 2.5–17 cm long petiole. *Inflorescence* synanthous, peduncle enclosed in scales of young shoot, rachis limply pendulous, about straight to slightly zig-zag, 16–63 cm long, 10- to 15-flowered, flowers opening simultaneously, floral bracts persistent, ovary sparsely hairy. *Flowers* white or cream-coloured, sometimes tinged pale violet or pink, a broad yellow to ochrish band or longitudinal lines on mid-lobe, sometimes the side-lobes with similar lines. *Dorsal sepal* narrowly ovate to oblong or elliptic, acute, 7–9-nerved, dorsal prominent or low rounded keel, 2.2–3 × 0.6–1.1 cm. *Lateral sepals* somewhat oblique, narrowly ovate to oblong or elliptic, acute, 7–9-nerved, dorsal low rounded keel, 2.2–2.8 × 0.7–1.15 cm. *Petals* almost straight to almost falcate, narrowly lanceolate to linear, acute, 5–7-nerved, 2.2–2.8 × 0.2–0.5 cm. *Lip* 3-lobed, side-lobes large, in front rounded, mid-lobe narrowly attached, concave, orbicular, margin curved upwards, undulate, erose, top retuse, acute, callus of 2 keels extend from near base of lip to just before middle of the mid-lobe, a further 2 outer keels develop half to three quarters the way from base of lip to base of mid-lobe, all keels plate-like, entire or erose, further 2–4 keels develop only on the outer side of existing keels on mid-lobe only, on mid-lobe keels undulate and with irregularly incised margin. *Column* arcuate, 1.8–2.3 cm long, expanding to large hood, tip margin notched, dentate.

DISTRIBUTION. Borneo (Sarawak, Sabah).

HABITAT. Epiphytic on base of tree trunks in lower montane oak-laurel forest, sometimes lithophytic on rocks.

ALTITUDE. 900–2000 m (800–1500 m).

FLOWERING. February, May–June, September–October.

NOTES. The epithet refers to the slender, cylindrical, stick-like pseudobulbs. See also under *C. echinolabium* de Vogel.

122. COELOGYNE ROCHUSSENII

Coelogyne rochussenii de Vriese in *Ill. Orch.* t. 2; t. 11, f. 6 (1854). Type: Java, Gunung Salak, cult. Bogor (lecto. de Vriese illustration above). **Plate 17A.**

Coelogyne plantaginea Lindl. in *Gard. Chron.* 1855: 20 (1855). Type: cult. Bishop of
   Winchester s.n. (holo. K).
C. macrobulbon Hook.f., *Fl. Brit. India* 5: 830 (1890). Types: Peninsular Malaysia, Penang,
   Wallich s.n.; Perak, King' collector s.n., *Scortechini* s.n. (syn. K).
C. stellaris Rchb.f. in *Gard. Chron.* ser. 2, 26: 8 (1886). Type: Borneo, *Lobb* s.n. (holo. W).
C. steffensii Schltr. in *Feddes Repert. Sp. Nov. Regni Veg. Beih.* 21: 130 (1925). Type: Consul
   Steffen s.n. (holo. B†).
Pleione rochussenii (De Vries) Kuntze, *Rev. Gen. Pl.* 2: 680 (1891).
Pleione plantaginea (Lindl.) Kuntze, *Rev. Gen. Pl.* 2: 680 (1891).
Pleione macrobulbon (Hook.f.) Kuntze, *Rev. Gen. Pl.* 2: 680 (1891).

DESCRIPTION. **Pseudobulbs** on a creeping rhizome, 2–4 cm apart, narrowly conical, ribbed except when very young, to 20 cm long. **Leaves** 2, ovate, apex rounded but slightly acuminate, with 5–7 main nerves, about 25 × 11 cm, widest in the apical half and gradually narrowed at base to a 5 cm long petiole. **Inflorescence** heteranthous, peduncle enclosed in scales of young shoot, rachis curved initially, then pendulous, to 70 cm long and bearing 20- to 35-flowerd, flowers opening simultaneously, floral bracts persistent, ovary slightly hairy to glabrous. **Flowers** about 5 cm across, sepals and petals pale green, lip side-lobes dark brown with white veins on the inside, white outside, mid-lobe tip pure white. **Dorsal sepal** narrowly elliptic, acute, 5–9-nerved, dorsal prominent or low rounded keel, 2–3.2 × 0.4–0.8 cm. **Lateral sepals** oblique, narrowly ovate, acute, 5–9-nerved, dorsal low rounded keel, 1.8–3.1 × 0.35–0.7 cm. **Petals** narrowly elliptic, acute, 3–7-nerved, 1.8–3.1 × 0.18–0.5 cm. **Lip** 3-lobed, side-lobes relatively low, front broadly rounded, mid-lobe narrowly attached, convex, slender ovate, sessile or with broad claw, margin erose, reflexed and curved upwards, top acuminate-acute, callus of 3 fimbriate keels extend from the base of lip to base of mid-lobe, lateral keels continue to middle of mid-lobe, further 2 outer keels extend from base of mid-lobe to middle of mid-lobe, all with fimbriate margins. **Column** arcuate, 1.2–1.8 cm long, expanding to widest part in middle, then into a rather pronounced hood, top margin truncate-retuse, with small teeth, sometimes notched, middle part slightly reflexed.

DISTRIBUTION. Thailand, Peninsular Malaysia, Sumatra (East & West coasts), Java (West), Borneo (Kalimantan, Sarawak, Sabah, Brunei), Philippines, Sulawesi (Minahassa), Maluku.

HABITAT. Epiphytic, sometimes lithophytic in riverine and hill forest, freshwater and peat swamp forest, podzolic dipterocarp forest on very wet sandy soil; on isolated trees in wet regions.

ALTITUDE. Sea level–1500 m.

FLOWERING. Nearly all year.

NOTES. This species was dedicated to J.J. Rochussen who was governor-general of the Dutch East Indies in the mid nineteenth Century. He encouraged the scientific investigations of Teijsmann and Binnendijk whose names occur fairly regularly associated with the descriptions of *Coelogyne* species. de Vogel (1992) did not locate a type specimen and judged that it was probably not collected. If that is so, then the de Vriese illustration is the holotype.

123. COELOGYNE RUPICOLA

Coelogyne rupicola Carr in *Gard. Bull. Straits Settlem.* 8: 210 (1935). Type: Borneo, Sabah, Mt. Kinabalu, main spur above Kamborangah, *Carr* 3552 in SFN 27793 (holo. SING; iso. K). **Plate 17B.**

DESCRIPTION. *Pseudobulbs* on a short to creeping rhizome, close together to 3 cm apart, ovoid to slender fusiform, grooved, 4.5–16.5 cm long. *Leaves* 2, narrowly obovate to slender elliptic, top acute, herbaceous to hard coriaceous, with 3–9 main nerves, 12–29 × 1.5–4.5 cm with 4–14.5 cm long petiole. *Inflorescence* synanthous, peduncle erect or curved, extending to 28.5 cm beyond scales of young shoot, rachis suberect to curved with top pendulous, zig-zag, 6.5–22.5 cm long, 5- to 9-flowered, flowers opening simultaneously, floral bracts persistent, ovary densely hairy. *Flowers* white, sometimes tinged with green; apart from side-lobes and margin of mid-lobe, lip rather deep yolk-yellow to ochrish, keels to the base sometimes white, column white with sometimes additional colours under the hood ranging from pink to orange-red, brown. *Dorsal sepal* narrowly elliptic, sometimes reflexed, acute, 5–11-nerved, dorsal prominent or a low rounded keel, 2.2–2.4 × 0.65–1 cm. *Lateral sepals* slightly oblique, narrowly elliptic to long ovate, acute, 7–11-nerved, dorsal low rounded keel, 2.2–3.4 × 0.7–1 cm. *Petals* almost straight, narrowly elliptic to obovate, acute, 5–9-nerved, dorsal prominent, lateral nerves branched, 2.2–3.4 × 0.45–0.7 cm. *Lip* narrowly attached, elliptic, 3-lobed, side-lobes rounded and sometimes slightly reflexed, mid-lobe convex, almost quadrangular, margin undulate, broad triangular tip, callus of 2 distinctly swollen, laterally widened and smooth to warty margined, plate-like keels, extending from base of lip to just beyond middle of mid-lobe, a further 2 or 3 similar keels on mid-lobe, a very short median keel and 2 outer keels which merge with the existing keels. *Column* arcuate, 1.7–2 cm long, expanding to widest part in middle, then into large hood, tip margin rounded, shallowly dentate, middle slightly reflexed.

DISTRIBUTION. Borneo (Sabah (Mount Kinabalu)).

HABITAT. Epiphytic, on the base of mossy shrubs, and terrestrial or lithophytic on rocks, sometimes in very damp places, in shade to rather exposed places.

ALTITUDE. 1400–3500 m.

FLOWERING. Nearly all year.

NOTES. The epithet refers to the usual lithophytic nature of the type.

124. COELOGYNE SQUAMULOSA

Coelogyne squamulosa J.J. Sm. in *Bull. Dep. Agric. Indes Néerl.* 15: 3 (1908). Type: Borneo, Kalimantan, Semedum, *Hallier* s.n. (holo. BO). **Plate 17C.**

DESCRIPTION. *Pseudobulbs* on a thick, creeping rhizome, forming an acute angle with the rhizome, 1.5 cm apart, ovoid, transversely elliptic in cross-section, 2.75 × 2 cm, initially with dark, coriaceous and minutely black hair covered bracts enclosing base. *Leaves* 2, spreading, lanceolate, obtuse, mucronate, slightly undulate, with 7 main nerves, darker colour, scarcely green beneath, 18 × 3.6 cm with 4.5 cm long, grooved petiole which has a tinge of brown. *Inflorescence* synanthous or hysteranthous, peduncle erect, slender, 2.75–6.75 cm long, 2 enveloping bracts at base, rachis dark and minutely scaly, zig-zag, 1.85–2.8 cm long, 1- to 2-flowered, flowers opening simultaneously, floral bracts persistent. *Flowers* widely opening, 5 cm across, sepals and petals shining yellowish-brown, side-lobe on outside white with brown/white nerves near base, on inside brown with white nerves, mid-lobe brown with white at base, column yellow-brown and green base, keels brown. *Dorsal sepal* oblong, obtuse, apiculate, strongly grooved and concave in middle, on both sides convex, 9-nerved, 2.9 × 1.1 cm. *Lateral sepals* oblique, oblong, acute, margin rolled back at edge, mid-nerve grooved and with dorsal keel, 2.8 × 0.85 cm. *Petals* linear, slightly spathulate, acute, convex, mid nerve sulcate, dorsal slightly thickened, 2.8 × 0.35 cm. *Lip* concave, 2.55 × 2.45 cm, 3-lobed, side-lobes erect, above column and curved inwards, concave, tip free, rounded, mid-lobe deflexed, transversally quadrangular-ovate, emarginate, apiculate, crenulate, convex, triangular section in middle thickened, 0.75 × 1.1 cm, callus of 3 verrucose keels, the median keel extending from base of lip and vanishing towards the tip of the mid-lobe, the lateral keels slightly fleshy and slashed with many incisions, extending from base of lip and abruptly terminating at base of mid-lobe, in addition, on both sides, a crested plate. *Column* slender, arcuate, 2 cm long, at tip winged, wings apiculate, cut deeply and sharply acute-angled away from the wings, tip finely serrate.

DISTRIBUTION. Borneo (Kalimantan, Sarawak).
HABITAT. Epiphytic in hill dipterocarp forest.
ALTITUDE. 800 m.
FLOWERING. September.
NOTES. The epithet refers to the covering of minute scales on the inflorescence. This species was not included by de Vogel (1992) in his treatment of the section.

125. COELOGYNE SWANIANA

Coelogyne swaniana Rolfe in *Bull. Misc. Inf., Kew* 183 (1894). Type: ?Philippines, cult. Sander s.n. (holo. K). **Plate 17D.**

Coelogyne quadrangularis Ridl. in *J. Linn. Soc. Bot.* 32: 323 (1896). Type: Peninsular Malaysia, Gunung Hijan, *Ridley* s.n. (holo. SING).

DESCRIPTION. *Pseudobulbs* on a long creeping rhizome, 2–5 cm apart, rather fat to slender ovoid, distinctly angular, usually with sharp ridges and hollow sides, 3.5–14.5 cm long. *Leaves* 2, broadly elliptic to obovate, acute-shortly acuminate, subcoriaceous, with 3–5(–7) main nerves, 7.5–40 × 1.8–10 cm with 3.5–14.5 cm long petiole. *Inflorescence* heteranthous, peduncle emerging from within scales covering base of pseudobulb, rachis curved at the base, sometimes zig-zag, 5–30 cm long, 5- to 17-flowered, flowers opening simultaneously, floral bracts persistent, ovary slightly to densely hairy. *Flowers* with white to cream-coloured sepals and petals, lip whitish, side-lobes ochre or brown and an ochre to brown band in front, inside ochre to dark brown with white veins and median white with 2 brownish lines between the keels, keels brown or white with brown apex. *Dorsal sepal* narrowly elliptic, acute, margin reflexed, 7–9-nerved, dorsal prominent, 2.5–3.3 × 0.55–1.05 cm. *Lateral sepals* oblique, narrowly ovate, acute, 5–7-nerved, dorsal low and rounded, 2.4–3.3 cm × 0.8–0.95

cm. ***Petals*** narrowly elliptic, acute, 5-nerved, 2.4–3.3 × 0.35–0.75 cm. ***Lip*** 3-lobed, side-lobes large, front broadly rounded and sometimes reflexed, mid-lobe narrowly attached, convex, saccate at base, ovate, margin entire, reflexed or curved upwards, top acute, callus of 3 keels extending from base of lip to just beyond middle of mid-lobe, initially elevated, plate-like and with lacerated projections then diminish in size before thickening and becoming larger with dentate margins, median keel slightly shorter, similar, further 2–4 keels develop across base of mid-lobe and merge with the existing keels. ***Column*** arcuate, 1.6–2.1 cm long, expanding to widest part in middle, then into large hood, tip margin slightly dentate, mid part lightly reflexed.

DISTRIBUTION. Peninsular Malaysia, Sumatra, Borneo (Kalimantan, Sarawak, Sabah, Brunei).

HABITAT. Epiphytic in crowns and on major branches of trees in dense to open forest, also often terrestrial in humus, moss or lithophytic on limestone and sandstone.

ALTITUDE. Sea level–2000 m.

FLOWERING. Almost all year.

NOTES. The species is dedicated to Mr. J.M. Swan, a British artist. The type specimen was supposed to come from the Philippines but there is some doubt about this as it is felt the wrong provenance was given to Rolfe when he made his description.

126. COELOGYNE TESTACEA

Coelogyne testacea Lindl. in *Bot. Reg.* 28: Misc. 38 (1842). Type: Singapore, cult. Loddiges s.n. (holo. K). **Plate 17E.**

Coelogyne sumatrana J.J. Sm. in *Bull. Dep. Agric. Indes Néerl.* 25, 3 (1908). Type: Bogor, cult. Gravesande s.n. (holo. BO; iso. AMES, L, SING).

DESCRIPTION. ***Pseudobulbs*** on a short rhizome, close together to 2.5 cm apart, compressed, narrowly ovoid, shallowly grooved, 4–11 cm long. ***Leaves*** 2, oblanceolate to narrowly elliptic, with 5–7 main nerves, 10.5–45 × 2.5–6 cm with 2–16.5 cm long petiole. ***Inflorescence*** proteranthous or synanthous, peduncle enclosed in scales of young shoot, rachis pendulous, almost straight to more or less zig-zag, 8–26 cm long, 6- to 18-flowered, flowers opening simultaneously, floral bracts persistent, ovary slightly hairy to glabrous. ***Flowers*** dull clay- or flesh-coloured or dull creamy-ochre, lip similarly coloured or whitish, side-lobes of the lip both in and outside, dark brown except the margin, mid-lobe with some brown. ***Dorsal sepal*** long ovate to lanceolate, acute, 5–9-nerved, 2.1–2.7 × 0.65–0.9 cm. ***Lateral sepals*** oblique, lanceolate, acute, 5–9-nerved, dorsal a low, rounded or plate-like keel, 2.1–2.7 × 0.65–0.9 cm. ***Petals*** oblanceolate to narrowly elliptic, rounded to acute, 3–5-nerved, 2.1–2.6 × 0.25–0.55 cm. ***Lip*** 3-lobed, side-lobes large, front rounded and reflexed, mid-lobe broadly attached, slightly saccate at base, convex, broadly spathulate-elliptic, margin undulate and erose to dentate, top retuse, callus of 2 slender, slightly incised keels, extending from base of lip to near tip of mid-lobe, becoming irregular, soft, with dentate or hair-like projections, similar further 2–4 keels develop on mid-lobe, extending from about base of mid-lobe on the outside of existing keels. ***Column*** arcuate, 1.5–1.9 cm long, expanding into hood, top margin irregularly dentate, truncate, sometimes retuse or deeply notched, mid part reflexed.

DISTRIBUTION. Peninsular Malaysia (Pahang), Singapore, Sumatra, Borneo (Kalimantan, Sabah).

HABITAT. Epiphytic on trunks and branches of trees, often near the ground, in lowland rain forest and riverine forest, also terrestrial, often growing in large clumps.

ALTITUDE. Sea level–1400 m.

FLOWERING. June–July, October–March.

NOTES. The epithet refers to the dull ochre colour of the sepals, which is reminiscent of some types of baked pottery and tiles. de Vogel (1992) indicated that he was unsure whether this species should be in sect. *Tomentosae* because it has some similarity to *C. trinervis*. I feel he made the right decision to place it in sect. *Tomentosae* as my examination of a wide range of sect. *Flaccidae* material, including a number of *C. trinervis* specimens, suggests to me that they show characteristics common to the species in that section which *C. testacea* does not have.

127. COELOGYNE TOMENTOSA

Coelogyne tomentosa Lindl., *Fol. Orch.-Coelog.* 3 (1854). Type: Borneo, locality unknown, *Lobb* 187 (holo. K; iso. W). **Plate 18A.**

Coelogyne massangeana Rchb.f. in *Gard. Chron.* ser. 2, 10: 684 (1878). Type: cult. Masssange s.n. (holo. W).
Pleione tomentosa (Lindl.) Kuntze, *Rev. Gen. Pl.* 2: 680 (1891).
Coelogyne densiflora Ridl. in *J. Roy. Asiat. Soc. Straits Branch* 39: 81 (1903). Type: Peninsular Malaysia, *Kelsall* s.n. (holo. K).
C. dayana Rchb.f. var. *massangeana* Ridl., *Mat. Fl. Mal. Pen.* 1: 127 (1907).
C. tomentosa Lindl. var. *massangeana* Ridl., *Mat. Fl. Mal. Pen.* 4: 130 (1907).
C. cymbidioides sensu Ridl. in *J. Linn. Soc. Bot.* 38: 329 (1908), *non* Rchb.f.

DESCRIPTION. *Pseudobulbs* on a short rhizome, close together, slightly flattened, more or less ovoid, shallowly grooved, 4–15 cm long. *Leaves* 2, ovate to longly elliptic, abruptly acuminate, sometimes acute, with 5–7 main nerves, 13.5–70 × 2.5–10 cm with 6–17 cm long petiole. *Inflorescence* heteranthous, peduncle enclosed in scales of young shoot, rachis pendulous, about straight to more or less zig-zag, 28–55 cm long, 10- to 30-flowered, flowers opening simultaneously, floral bracts persistent, ovary slightly to densely hairy. *Flowers* with dull light greenish yellow to mauve-yellow or whitish sepals and petals, lip whitish sepals and petals, keels similar coloured or yellowish, often separated by brown lines, side-lobes outside brown, inside brown with whitish veins, warts yellow. *Dorsal sepal* narrowly elliptic, margins sometimes reflexed, acute, 9–11-nerved, 2.3–4.2 × 0.75–1.1 cm. *Lateral sepals* oblique, narrowly elliptic, acute, 7–9-nerved, dorsal low rounded to plate-like keel, 2.3–4 × 0.7–1.1 cm. *Petals* almost straight, narrowly elliptic to oblanceolate, acuminate, 5–7-nerved, 2.4–3.9 × 0.4–0.8 cm. *Lip* 3-lobed, side-lobes large, front broadly rounded and more or less reflexed, mid-lobe narrowly attached, convex, elliptical-orbicular, margin irregular, reflexed, top margin notched, irregularly dentate, mid part hardly recurved, callus of 3 keels extending from base of lip, swollen with dentate projections, becoming less elevated and with blunt projections, on mid-lobe keels expand into a mass of truncate, molar-like projections covering the mid-lobe with the exception of the margin region. *Column* arcuate, 1.5–2 cm long, expanding to widest part in middle, then into large hood, tip margin notched, irregularly dentate, tip margin notched, mid part slightly reflexed.

DISTRIBUTION. Peninsular Malaysia (Kedah, Pahang, Perak, Penang), Sumatra, Java.

HABITAT. Epiphytic on tree trunks and major branches in lower and upper montane forest, lithophytic on steep limestone cliffs.

ALTITUDE. 1500–2100 m.

FLOWERING. All year but mostly March–October.

NOTES. The epithet refers to the tomentose ovary. de Vogel (1992) explained why *C. massangeana* Rchb.f. is deemed to be conspecific with *C. tomentosa*. He also indicated why it was necessary to define a new type *C. velutina*. This species is still usually cultivated under the name *C. massangeana*.

128. COELOGYNE VELUTINA

Coelogyne velutina de Vogel in *Orchid Monogr.* 6: 38, f. 23, pl. 4d (1992). Type: Origin unknown, *van Beusekom & Phengklai* 957 (holo. L; iso. BKF, C, E, K, P). **Plate 18B.**

?*Coelogyne tomentosa* Lindl. var. *penangensis* Hook.f., *Fl. Brit. India* 5. 830 (1885). Type: *Maingay* 2238 (1633) (holo. K).

DESCRIPTION. *Pseudobulbs* on a short rhizome, close together, more or less ovoid, 4.5–10 cm long. *Leaves* 2, sometimes quite stiff, long elliptic to obovate, acuminate, with 5–7 main nerves, 16–38 × 3.5–10.5 cm with 3–13 cm long petiole. *Inflorescence* heteranthous, peduncle enclosed in scales of young shoot, rachis pendulous, about straight, very densely hairy with (red)brown velutinous hairs, 23–59 cm long, 12- to 26-flowered, flowers opening simultaneously, floral bracts persistent, ovary very densely hairy. *Flowers* with white or pinkish sepals and petals, lip whitish, side-lobes outside with a brown band, inside dark brown with whitish or pinkish nerves and a narrow white margin, keels bright golden yellow. *Dorsal sepal* narrowly elliptic, acute, 7–11-nerved, dorsal a low rounded keel, 2.5–3.3 × 0.6–1.05 cm. *Lateral sepals* oblique, narrowly elliptic, acute, 8–9-nerved, dorsal low, rounded to plate-like keel, 2.5–3.3 × 0.65–0.95 cm. *Petals* almost straight, narrowly elliptic, acute, 3–5-nerved, keel prominent, 2.5–3.2 × 0.35–0.75 cm. *Lip* narrowly attached, strongly reflexed, strongly convex, 3-lobed, side-lobes large, in front broadly rounded and only at the junction with the mid-lobe reflexed, mid-lobe margin entire to irregular, curved upwards, top broadly acute with tip projecting, callus of 3 low, plate-like keels extending from near base of lip where they are more elevated with dentate margin, then smaller in size with no dentate margin, continuing to beyond middle of mid-lobe with median keel much shorter and terminating near base of mid-lobe, a further 4 short keels across the base of mid-lobe only, all keels on mid-lobe more elevated with dentate margins. *Column* arcuate, 1.6–2.1 cm long, expanding into large hood, tip margin rounded, irregularly dentate, mid part hardly reflexed.

DISTRIBUTION. Thailand, Peninsular Malaysia.

HABITAT. Epiphytic on trees and shrubs in lower montane forest and open forest, lithophytic on rocks.

ALTITUDE. 800–1950 m.

FLOWERING. February–May.

NOTES. The epithet refers to the velutinous (velvety) covering of the rachis and ovary.

129. COELOGYNE VENUSTA

Coelogyne venusta Rolfe in *Gard. Chron.* ser. 3, 35: 259 (1904). Type: Provenance unknown, cult. Glasnevin B.G., ex Sander (holo. K, iso. BM). **Plate 18C.**

DESCRIPTION. *Pseudobulbs* on a long creeping rhizome, 2.5–5.5 cm apart, straight or curved, slender ovoid, shallowly grooved, 3.5–7.5 cm long. *Leaves* 2, sometimes rather stiff, elliptic or ovate to linear, 10.5–25 × 1.5–6 cm with 1.2–5.5 cm long petiole. *Inflorescence* synanthous, peduncle enclosed in scales of young shoot or extended to 7 cm, rachis pendulous, slightly zig-zag, 14.5–42 cm long, 16- to 35-flowered, flowers opening simultaneously, floral bracts persistent, ovary very densely hairy. *Flowers* white or cream-coloured, sometimes tinged pink, lip white with a yellow to ochre spot on the side-lobes and a similar spot or band at the base of the lip across the keels, keels sometimes tipped with brown. *Dorsal sepal* narrowly ovate, acute, 5–7-nerved, dorsal prominent or low rounded keel, 1.35–1.7 × 0.35–0.58 cm. *Lateral sepals* narrowly ovate, acute, 5–7-nerved, low rounded keel, 1.35–1.7 × 0.35–0.55 cm. *Petals* linear, acute, 1–3-nerved, dorsal prominent, 1.32 × 0.12–0.3 cm. *Lip* 3-lobed, side-lobes triangular, in front projecting and distinctly reflexed, rounded to acute, mid-lobe convex, obovate-rectangular, tapering to the base, top rounded to emarginate, erose, with broad triangular tip, callus of 2 keels extending from half way along hypochile to middle of mid-lobe, slender, low, irregular dentate margin, similar, further 2–4 keels on outside of existing keels, extending from base of mid-lobe and slightly shorter than main keels. *Column* arcuate, 0.9–1.3 cm long, expanding into large hood, top margin truncate to retuse, almost entire to irregularly dentate, slightly reflexed, often notched.

DISTRIBUTION. Borneo (Sarawak, Sabah, Brunei).

HABITAT. Epiphytic on trunks and branches of trees in lower montane ridge forest, sometimes terrestrial on rocks; sometimes very exposed to the sun.

ALTITUDE. 800–2350 m.

FLOWERING. October–June, but no record in April.

NOTES. The epithet refers to the graceful nature of the plant. The type description was based on material provided by Sander & Sons said to be from Yunnan in China. As stated by de Vogel (1992), such a provenance is unlikely and it probably reflects the custom amongst traders to give misleading information. The specimen at BM does not mention Sander & Sons; it is recorded as an isotype.

SECTION *HOLOGYNE*

Section 13. *Hologyne* (Pfitzer & Kraenzlin) D.A. Clayton **stat. nov.** Type: *Coelogyne miniata* (Blume) Lindl.

Coelogyne subgenus *Hologyne* Pfitzer & Kraenzlin in Engler, *Pflanzenr. Orch.-Mon.-Coelog.*: 20 (1907).

Inflorescence proteranthous or synanthous, peduncle bare. Flowers fleshly, opening simultaneously, lip more or less entire, generally without distinct side-lobes, knee-shaped or S-bend half way along the lip, basal part of the lip embraces the column, keels extending from base of lip to base of mid-lobe. Distribution map 14.

Key to species in Section Hologyne

Map 14. Distribution of *Coelogyne* Sect. *Hologyne*; (1) *C. miniata*, (2) *C. obtusifolia*, (3) *C. malipoensis*.

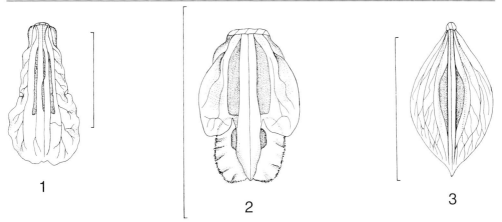

Fig. 17. **Section 13. Hologyne** (Pfitzer) D.A. Clayton
1. Coelogyne malipoensis Z.H. Tsi [after L. Averyanov]; **2. C. obtusifolia** Carr [*Carr* 3149]; **3. C. miniata** (Blume) Lindl. [Cult Kew A 1923]. Drawn by Linda Gurr. Scale bar = 1 cm.

Key to species in Section Hologyne

1. Pseudobulbs with 2 leaves 2
   Pseudobulb with 1 leaf. [Lip spathulate with no evident side-lobes, 3 smooth keels.] .... ***C. malipoensis***

2. Pseudobulbs cylindric, lip indistinctly 3-lobed. [Few-flowered, flowers pale orange-red, keels deep red, 3 keels, median keel less distinct than lateral keels.] .... ***C. miniata***
   Pseudobulbs narrowly ovoid, lip 3-lobed. [Lip side-lobes rounded, mid-lobe subquadrate, papillae on mid-lobe, margin undulate, sepals and petals translucent pale yellow, lip pale salmon, white on mid-lobe, 2 fleshy keels.] .... ***C. obtusifolia***

130. COELOGYNE MALIPOENSIS

Coelogyne malipoensis Z.H. Tsi in *Inst. Bot., Acad. Sin., Bull. Bot. Res.* 15: 14–15 (1995). Type: China, Yunnan, Malipo, *Z.H. Tsi 92-129* (holo. PE). **Plate 18D & E.**

DESCRIPTION. ***Pseudobulbs*** on a thick rhizome, about 2 cm apart, elongate-ovate, 2.5–4.5 cm long, about 1 cm dia. at tip. ***Leaf*** 1, elliptic to obovate-elliptic, obtuse to acuminate, coriaceous, 13–15 × 5–5.5 cm with slender, 5–6.5 cm long, abruptly contracted petiole. ***Inflorescence*** synanthous or hysteranthous, peduncle sheathed at base, with rachis 20 cm long, lax, 1.5 cm long, few-flowered, flowers opening simultaneously, at least two at a time, bracts persistent, narrowly oblong. ***Flowers*** with white sepals and petals, lip white with yellow on the mid-lobe. ***Dorsal sepal*** and ***lateral sepals*** somewhat similar, ovate-lanceolate, acuminate, 1-nerved, 2 × 0.7 cm. ***Petals*** narrowly ovate, acute, tip somewhat spathulate, 1.5 × 0.6 cm. ***Lip*** somewhat spathulate, undivided, 1.8 cm long, near tip 0.9 cm across, tip obtuse, callus of 3 smooth keels extending from near base of lip to base of mid-lobe. ***Column*** broad and flat, tip bi-lobed, winged, tip somewhat truncate, about 0.5 cm across.

DISTRIBUTION. China (Yunnan (Malipo)), North Vietnam.
HABITAT. Epiphytic on trees over limestone.
ALTITUDE. 1400 m.
FLOWERING. October–December.
NOTES. The epithet refers to the locality, Malipo, Yunnan, China where the type specimen was collected.

131. COELOGYNE MINIATA

Coelogyne miniata (Blume) Lindl., *Gen. Sp. Orch. Pl.* 42 (1895). Type: Java, Gunung Salak, *Blume* s.n. (holo. BO; iso. K, L). **Plate 19A.**

Chelonanthera miniata Blume, *Bijdr.* 385 (1825).
Coelogyne simplex Lindl., *Fol. Orch. Coelog.* 13 (1854). Type: Java, *Lobb* s.n. (holo. K-LINDL).
Pleione miniata (Blume) Kuntze, *Rev. Gen. Pl.* 2: 680 (1891).
Coelogyne lauterbachiana Kraenzl. in *Notizbl. Bot. Gart. Berlin*, 1: 113 (1896). Type: New Guinea, northeast, *Lauterbach* s.n. (holo. B†).
Hologyne miniata (Blume) Pfitzer in Engler, *Pflanzenr. Orch.-Mon.-Coelog.* 132 (1907).
Hologyne lauterbachiana (Kraenzl.) Pfitzer in Engler, *Pflanzenr. Orch.-Mon.-Coelog.* 133 (1907).

DESCRIPTION. ***Pseudobulbs*** on a slender, creeping, branching, lightly sheathed, nodular rhizome, 2.5 cm apart, cylindric, 4 × 0.2–0.4 cm (when dried), enclosed in bracts at base when young. ***Leaves*** 2, elliptic, acute, rather thin texture, with 7 nerves, 7–13 × 1.8–3 cm with 1.7 cm long, grooved petiole. ***Inflorescence*** synanthous or proteranthous, peduncle terete, stiff, 5.5 cm long, rachis becoming more slender, zig-zag, 2–3 cm long, 4-flowered, flowers opening simultaneously, floral bracts deciduous. ***Flowers*** about 2.5 cm across, pale to deep orange-red, keels deep red. ***Sepals*** broadly ovate, acute, 1.5 cm long. ***Petals*** narrowly triangular-linear, 1.5 cm long. ***Lip*** indistinctly 3-lobed, side-lobes not very prominent, narrowly semi-elliptic, mid-lobe ovate, acuminate, callus of 3 keels, lateral keels extending from base of lip to base of mid-lobe, median keel less distinct, shorter and only on mid-lobe. ***Column*** graceful.

DISTRIBUTION. Sumatra (West coast), Bali, Java, Lesser Sunda Islands.
HABITAT. Epiphytic on branches of trees in lower and upper montane forest.
ALTITUDE. 1000–2400 m.
FLOWERING. June–?September.
NOTES. The epithet refers to the reddish colour of the flower.

132. COELOGYNE OBTUSIFOLIA

Coelogyne obtusifolia Carr in *Gard. Bull. Straits Settlem.* 8: 205 (1935). Type: Borneo, Sabah, Mt. Kinabalu, below Bundu Tuhan, *Carr* 3149 in SFN 27897 (holo. SING; iso. AMES, K).

DESCRIPTION. *Pseudobulbs* on creeping, branched, stout rhizome covered with fleshy tubular bracts, 3.7 cm apart, narrowly ovoid, to 6 cm long. *Leaves* 2, narrowly oblong, broadly obtuse, rather thin in texture, keel prominent beneath, 9.5–17.5 × 1–1.8 cm with 1.5–2.5 cm long, grooved petiole. *Inflorescence* synanthous, peduncle bare, hardly exists, but elongating to 9.5 cm in fruit, rachis somewhat zig-zag to 4.5 cm long, 10- to 15-flowered, flowers opening rapidly in sequence, nearly simultaneously with 5 to 6 open together, floral bracts deciduous. *Flowers* with transparently pale yellow sepals and petals with a pale salmon tint, lip pale salmon with white on mid-lobe, converging keels pale salmon, column brown. *Dorsal sepals* triangular-ovate, acuminate, acute, 5 nerves, 0.9 × 0.45 cm. *Lateral sepals* ovate-oblong, acute, somewhat S-shaped, 5-nerved, keeled on back, 0.9 × 0.4 cm. *Petals* linear, subacute, slightly narrowed in middle, margins irregular towards tip, 1-nerved, 0.8 × 0.07 cm. *Lip* 3-lobed, 0.8 × 0.6 cm, side-lobes rounded, mid-lobe 0.8 × 0.8 cm, right-angular recurved, subquadrate, very shortly bi-lobed, with minute tooth in sinus, papillose, margins undulate and minutely erose, inner nerves inconspicuously elevated, callus of 2 fleshy keels from base of lip to base of mid-lobe, converging at tip. *Column* 0.6 cm long, not winged, hood transversally oblong, upper margins with 4 teeth.
DISTRIBUTION. Borneo (Sabah).
HABITAT. Epiphytic or terrestrial in hill and lower montane forest.
ALTITUDE. 600–1500 m.
FLOWERING. July–August.
NOTES. The epithet refers to the obtuse form of the leaves.

SECTION *RIGIDIFORMES*

Section 14. *Rigidiformes* Carr in *Gard. Bull. Straits Settlem.* 8: 213 (1935). Type: *Coelogyne rigidiformis* Ames & C. Schweinf.

Pseudobulbs close together or a short distance apart on a stout, sheathed rhizome, either ovoid, pyriform, fusiform to cylindric with 1-leaf at the apex. The inflorescence is heteranthous, with a peduncle, usually bare, short and erect; the rachis is arcuate with some closely imbricating bract-like sheaths. Large persistent bracts involute over the base of the flower. Flowers are fleshy and open simultaneously. The lip is more or less entire without distinct side-lobes, the basal part of the lip embraces the column. Distribution map 15.

The Genus Coelogyne

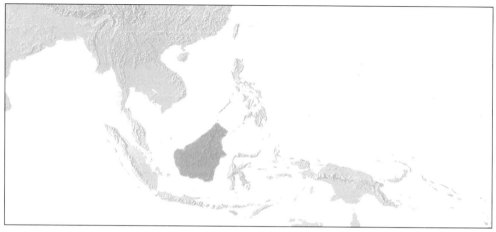

Map 15. Distribution of *Coelogyne* Sect. *Rigidiformes*.

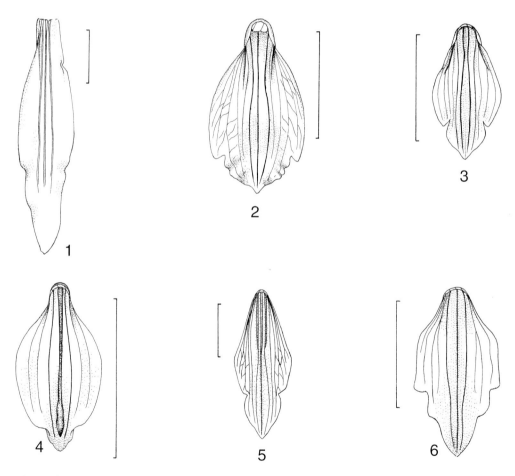

Fig. 18. **Section 14. Rigidiformes** Carr
1. Coelogyne albobrunnea J.J. Sm. [after C.L. Chan]; **2. C. clemensii** Ames & C. Schweinf. var. **clemensii** [*J & M.S. Clemens* 50.408]; **3. C. clemensii** Ames & C. Schweinf. var. **angustifolia** Carr [*Carr* 3091]; **4. C. craticulaelabris** Carr [*W.L. Chew & E.J.H. Corner* 4453, after Eleanor Catherine]; **5. C. exalata** Ridl. [Type (K)]; **6. C. plicatissima** Ames & C. Schweinf. [*Clemens* 204]. Drawn by Linda Gurr. Scale bar = 1 cm.

KEY TO SPECIES IN SECTION RIGIDIFORMES

1. Pseudobulbs ovoid **2**
    Pseudobulbs fusiform to cylindric or pyriform **4**

2. Pseudobulbs close together **3**
    Pseudobulbs more or less 3 cm apart. [Inflorescence synanthous. Lip entire, without side-lobes, a fold about three quarters the way along the lip; 3 simple keels joined in the middle by a transverse keel.] .... *C. craticulaelabris*

3. Inflorescence proteranthous. Lip entire, without side-lobes, at the base convolute; 3 distinct nerves, not keels .... *C. exalata*
    Inflorescence probably synanthous. Lip with tooth-like side-lobes, mid-lobe triangular, 3 keels terminating at the tip of the mid-lobe .... *C. albobrunnea*

4. Pseudobulbs fusiform to cylindric **5**
    Pseudobulbs pyriform. [Inflorescence heteranthous. Lip indistinctly 3-lobed, tip truncated, a fold in the middle; 3 low, simple keels terminating near the tip of the mid-lobe]:
    Leaves linear-oblong to oblong, 35–53 cm × 4–5 cm. Floral bracts persistent. .... *C. clemensii* var. *clemensii*
    Leaves linear-lanceolate, 15–33 cm × 0.5–2 cm. Floral bracts persistent. .... *C. clemensii* var. *angustifolia*
    Leaves linear-oblong to oblong, 35–53 cm × 4–5 cm. Floral bracts convolute. .... *C. clemensii* var. *longiscapa*

5. Dorsal sepal 1.6 cm long. Lip convolute for half its length, then lanceolate, a third of the distance from the tip emarginate; 3 low, simple keels terminating at the tip of the mid-lobe .... *C. plicatissima*
    Dorsal sepal 2.2 cm long. Lip tubular-involute for about two thirds its length, surrounding the column, a third of the distance from the tip slightly emarginate, slightly thickened tip; 3 low, simple keels terminating at the middle of the mid-lobe .... *C. rigidiformis*

133. COELOGYNE ALBOBRUNNEA

Coelogyne albobrunnea J.J. Sm. in *Bull. Jard. Bot. Buitenzorg* ser. 3, 11: 103 (1931). Type: Borneo, Kalimantan, Gunung Kemoel, *Endert* 3868 (holo. BO; iso. L).

DESCRIPTION. *Pseudobulbs* on a short, stout rhizome, close together, ovoid, 6 cm long, when dry, strongly wrinkled longitudinally. *Leaf* 1, erect, lanceolate, longitudinally grooved, acuminate-acute, plicate, coriaceous, with 7–9 nerves prominent beneath, also minute dots along the underside, 50 × 13 cm, base narrows into long petiole. *Inflorescence* lax, drooping, rachis 24 cm long, to 13-flowered, flowers opening simultaneously. *Flowers* large, fleshy, brown with a white lip and column. *Dorsal sepal*, narrowly lanceolate, shortly obtuse, 5-nerved, 4.25 × 0.9 cm. *Lateral sepals* obtuse angled-concave, distinct groove and laterally convex, oblique, narrowly lanceolate, obtuse, 5-nerved, 4.2 × 0.9 cm. *Petals* oblique, lanceolate, narrowly obtuse, 3-nerved, 4 × 0.65 cm. *Lip* grooved and surrounds the column, scarcely 3-lobed, elliptic-lanceolate, base truncated, many nerved, side-lobes obtuse angled, tooth-like, mid-lobe triangular, obtuse and thickened, conical shaped tip to mid-lobe, callus of 3 keels virtually undivided extending from base of lip to tip of mid-lobe. *Column* slender, slightly arcuate, 2.3 cm. long, winged, apex gently broadens.

DISTRIBUTION. Borneo (Kalimantan, Sarawak).
HABITAT. Epiphytic in lower montane forest on steep slope of ravine.
ALTITUDE. 700–1200 m.
FLOWERING. October.
NOTES. The epithet refers to the brown sepals and petals combined with a white lip and column. Some authors refer it to sect. *Hologyne* Pfitzer but as it is a near ally to C. *rigidiformes* Ames & C. Schweinf., I place it in sect. *Rigidiformes* Carr.

134. COELOGYNE CLEMENSII

Coelogyne clemensii Ames & C. Schweinf., *Orch.* 6: 23 (1920). Type: Borneo, Sabah, Mt. Kinabalu, Marai Parai Spur, *J. Clemens* 227 (holo. AMES; iso. BM, K). **Plate 19B.**

var. **clemensii**

DESCRIPTION. *Pseudobulbs* on a woody rhizome, slender, pyriform to cylindric, 4–5 cm long. *Leaf* 1, linear-oblong to oblong, abruptly acuminate, erect, rigid, strongly plicate, coriaceous, many nerved with 7 nerves prominent beneath, 35–53 × 3.8–5.3 cm, gradually narrowing into (8)12–25 cm long, deeply grooved petiole. *Inflorescence* heteranthous, peduncle short, including rachis up to 17.5 cm long, rachis suberect, zig-zag, loosely 5- to 7-flowered, flowers open simultaneously, floral bracts persistent. *Flowers* white and fragrant. *Dorsal sepal* elliptic-oblong, strongly cymbiform, rounded at tip, strongly concave, 1.6 × 0.8 cm. *Lateral sepals* elliptic-oblong, strongly cymbiform, complicate-mucronate at tip, hooded, 5-nerved, 1.6–1.8 × 0.8–0.9 cm. *Petals* narrowly elliptic, acute, 3-nerved, 1.5 × 0.6 cm. *Lip* tubular-convolute about the column for about half its length, then somewhat reflexed with the tip gradually spreading and becoming undulate, lip oval, about 1.75 × 1.1 cm, saccate at the base, indistinctly 3-lobed, side-lobes longitudinally semi-orbicular, apex porrect, rounded, overlapping with large fold in the middle of each margin, tip broadly truncate and more or less retuse-apiculate, callus of 3 low, simple, fleshy keels extending from base of lip to near tip of mid-lobe. *Column* slender, straight, about 1.3 cm long, winged.

DISTRIBUTION. Borneo (Sabah (Mount Kinabalu)).
HABITAT. Epiphytic in hill forests, lower and upper montane forest, sometimes on ultramafic substrate.
ALTITUDE. 900–2200 m.
FLOWERING. June.
NOTES. This species is dedicated to J. Clemens the collector of the type specimen.

var. **angustifolia** Carr in *Gard. Bull. Straits Settlem.* 8: 212 (1935). Type: Borneo, Sabah, Penibukan ridge, *Carr* 3091 in SFN 26453 (holo. SING; iso. K).

DESCRIPTION. *Pseudobulbs* on a woody rhizome, 1–1.5 cm apart, slender, pyriform to cylindric, 4–5 × 1.5 cm. *Leaf* 1, linear-lanceolate, acuminate, conduplicate acute, erect, rigid, strongly plicate, coriaceous, many nerved with 7 nerves prominent beneath, 15–33.5 × 0.5–2 cm with 8 cm long, grooved petiole. *Inflorescence* heteranthous, peduncle erect, slender, to 8 cm long, rachis zig-zag, to 4 cm long, loosely 6-flowered, flowers open simultaneously, floral bracts persistent. *Flowers* with bright yellow-green or olive-green sepals and petals, the sepals often and the petals sometimes suffused dull red towards tip or all dull red, lip similar, column pale green or pale reddish, hood white. *Dorsal sepal* elliptic-oblong, strongly cymbiform, rounded at tip, strongly concave, 1.25 × 0.52 cm. *Lateral sepals* elliptic-oblong, strongly cymbiform, hooded, complicate-mucronate at tip, 5-nerved, 1.37 × 0.52 cm. *Petals* narrowly elliptic, acute, 3-nerved 1.17 × 0.37 cm. *Lip* tubular-convolute about the column for about half its length, then somewhat reflexed with the tip gradually spreading and becoming undulate, lip oval, about 1.3 × 0.75 cm, saccate at the base, indistinctly 3-lobed, side-lobes longitudinally semi-orbicular, apex porrect, rounded, overlapping with large fold in the middle of each margin, apex minutely apiculate, callus of 3 low simple fleshy keels run from base to near tip of mid-lobe. *Column* slender and straight, 0.8 cm long, winged, hood extending from middle, rounded, very shortly emarginate with an apical tooth.

DISTRIBUTION. Borneo (Sabah, Sarawak).

HABITAT. Epiphytic in lower and upper montane forest; mossy forest; on ultramafic substrate.

ALTITUDE. 800–2400 m.

FLOWERING. July–August.

NOTES. The epithet refers to the very narrow leaves.

var. **longiscapa** Ames & C. Schweinf., *Orch.* 6: 25 (1920). Type: Borneo, Sabah, Mt. Kinabalu, *Haslam* s.n. (holo. AMES).

DESCRIPTION. *Pseudobulbs* on woody rhizome, slender, pyriform to cylindric, up to 3 cm long. *Leaf* 1, linear-oblong to oblong, abruptly acuminate, erect, rigid, strongly plicate, coriaceous, many-nerved, none prominent, 35–53 × 3.8–5.3 cm, tapering into about 8 cm long, deeply grooved petiole. *Inflorescence* heteranthous, peduncle short, rachis, rigidly arcuate, zig-zag, overall about 16 cm long, 5- to 7-flowered, flowers open simultaneously, floral bracts tightly convolute. *Flowers* white and fragrant. *Dorsal sepal* elliptic-oblong, strongly cymbiform, rounded at tip, strongly concave, about 5-nerved, 1.6 × 0.8 cm. *Lateral sepals* elliptic-oblong, strongly cymbiform, hooded, complicate-mucronate at tip, about 5-nerved, 1.6–1.8 × 0.8–0.9 cm. *Petals* lanceolate, acute, about 3-nerved, 1.6 × 0.45 cm. *Lip* tubular-convolute about the column for about half its length, then slightly reflexed with the tip gradually spreading and becoming undulate, lip oval, about 1.75 × 1.1 cm, saccate at the base, indistinctly 3-lobed, side-lobes longitudinally semi-orbicular, front porrect, rounded, overlapping with large fold in the middle of each margin, apex broadly truncate and more or less retuse-apiculate, mid-lobe obtusely narrowed, callus of 3 low simple fleshy keels extending from base of lip to near tip of mid-lobe. *Column* slender and straight, about 1.3 cm long, winged.

DISTRIBUTION. Borneo (Sabah (Mount Kinabalu)).

HABITAT. Epiphytic in hill and lower montane forest.

ALTITUDE. ?900–1600 m.

FLOWERING. July–August.

NOTES. The epithet refers to the differences in this variation when compared with the type specimen. Known only from the type.

135. COELOGYNE CRATICULAELABRIS

Coelogyne craticulaelabris Carr in *Gard. Bull. Straits Settlem.* 8: 214 (1935). Type: Borneo, Sabah, Mt. Kinabalu, Lumu-Lumu, *Carr* 2665 in SFN 27965 (holo. SING). **Plate 19C**.

DESCRIPTION. *Pseudobulbs* on a stout, pendulous, much branched rhizome covered initially in imbricating sheaths, 21 cm apart on primary rhizome, forming acute angle with rhizome, 2.75 cm apart on the branching rhizome, ovoid, more or less 4-sided, 1.3–3 cm long. *Leaf* 1, lanceolate, tip conduplicate, obtuse, rigid, plicate, with 5 nerves, 7.5–17.5 × 1.1–3.1 cm with 0.7–1.7 cm long, grooved petiole. *Inflorescence* proteranthous, peduncle short, nodding, bare, 1 cm long, rachis zig-zag, to 6 cm long, lax, few-flowered, flowers open ?simultaneously, floral bracts persistent, large, involute and embracing lower portion of flowers. *Flowers* rather fleshy, hardly open, sepals and petals pale yellowish suffused salmon towards fleshy tip. *Dorsal sepal* oblong-lanceolate, subacute, 5-nerved, 1.05 × 0.4 cm. *Lateral sepals* subfalcate, 5-nerved, 1.15 × 0.38 cm. *Petals* subfalcate, obovate, subacute, 3-nerved, outer nerves branched from base, 0.95 × 0.45 cm. *Lip* entire, swollen, sides erect and closely embracing column, mid-lobe oblong-oblanceolate, tip obtuse, very fleshy, sides incurved in a short sharp fold at about a quarter the way from tip, callus of 3 simple keels joined at about middle by transverse keel and in upper half with 2 simple keels not reaching tip of mid-lobe and joined together by transverse keel which creates an elevated median keel. *Column* short, 0.57 cm long, hood obovate, tip minutely erose.

DISTRIBUTION. Borneo (Kalimantan, Sarawak, Sabah, Brunei).

HABITAT. Epiphytic, sometimes lithophytic in lower and upper montane forest, montane ericaceous scrub, ridge forest; recorded on sandstone.

ALTITUDE. 900–2400 m.

FLOWERING. January, August.

NOTES. The epithet seems to refer to the lip being in the form of a crater or an excavated or hollow portion.

136. COELOGYNE EXALATA

Coelogyne exalata Ridl. in *J. Roy. Asiat. Soc. Straits Branch* 49: 29 (1908). Type: Borneo, Sarawak, Serapi, Matang, *Ridley* 124170 (holo. SING; iso. K). **Plate 19D**.

Coelogyne subintegra J.J. Sm. in *Bull. Dep. Agric. Indes Néerl.* 22: 12 (1909). Type: Borneo, Kalimantan, Gunung Kenepai, *Hallier* s.n. (holo. BO).

DESCRIPTION. *Pseudobulbs* on a stout, sheathed rhizome, close together, subglabrous, rounded, 2.5–3.2 × 1 cm (when dried). *Leaf* 1, lanceolate, acuminate-mucronate, coriaceous, with 7 nerves, 15–41 × 4.4–7.6 cm with 5.5–10 cm long, smooth and rounded petiole. *Inflorescence* heteranthous, peduncle erect, stiff, 18–20 cm long, rachis erect, thick, nodes pronounced, 20 cm long, 6- to 8-flowered, flowers opening in ?succession, floral bracts persistent. *Flowers* with green sepals tinted brown or light brown, petals green, lip green, column white. *Dorsal sepal* lanceolate, spathulate, acute, 2.5 × 0.6 cm. *Lateral sepals* oblong, acute, 2.5 × 0.6 cm. *Petals* lanceolate, spathulate, acute, shorter than sepals. *Lip* spathulate, entire with side-lobes, at base convolute, tip acute, with 3 distinct nerves but no obvious keels. *Column* straight, 1.35 cm long, hood flattened and broadly winged, wings reflexed.

DISTRIBUTION. Borneo (Sarawak, Sabah, Brunei).
HABITAT. Epiphytic, in mixed hill-dipterocarp forest, lower and upper montane forest, oak-laurel forest, terrestrial in open scrub under *Leptospermum*.
ALTITUDE. 600–2700 m.
FLOWERING. August.
NOTES. The epithet presumably refers to the small stature of the pseudobulbs.

137. COELOGYNE PLICATISSIMA

Coelogyne plicatissima Ames & C. Schweinf. *Orch.* 6: 35 (1920). Type: Borneo, Sabah, Mt. Kinabalu, Paka-Paka Cave, *J. Clemens* 204 (holo. AMES; iso. BM, K). **Plate 19E.**

DESCRIPTION. *Pseudobulbs* fusiform-cylindric to broadly pyriform on a stout rhizome, 2.5 × 1.5 cm (when dried), enclosed with bracts at base. *Leaf* 1, linear, oblong-linear or oblong-elliptic, acute, narrows to base, very strongly plicate, rigid, erect, coriaceous, 21.4 × 2.14 cm with about 6.2 cm long, cylindric, stout, sulcate petiole. *Inflorescence* heteranthous, peduncle suberect to arcuate, 24 cm, rachis arcuate or nodding, more or less zig-zag, about 8 cm long, loosely few-flowered. *Flowers* with deep brownish-salmon sepals, petals olive green suffused pale brownish-salmon towards tip, lip similar, column pale yellowish, hood and wings white. *Dorsal sepal* lanceolate-oblong, acute, carinate, shallowly cymbiform, 1.6 × 0.6 cm. *Lateral sepals* lanceolate-oblong, acute, dorsally carinate, shallowly cymbiform, 1.8 × 0.6 cm. *Petals* narrowly elliptic, subacute, obtuse, 3-nerved, 1.4 × 0.5 cm. *Lip* 1.7 × 0.9 cm, in natural position convolute for about half its length, when extended broadly lanceolate, anterior portion fleshy-thickened, obtuse, slightly reflexed near the base, at about one third distance from apex slightly emarginate on each side, callus of 3 low, simple, fleshy keels extending from base of lip to the tip of the mid-lobe. *Column* short surrounded by tube of the lip, about 0.85 cm long, including wing, very broad at the truncate, entire tip.
DISTRIBUTION. Borneo (Sabah, Sarawak).
HABITAT. Epiphytic in lower and upper montane forest, oak-laurel forest, montane ericaceous scrub, frequently on ultramafic substrate.
ALTITUDE. 1500–3000 m.
FLOWERING. June.
NOTES. The epithet refers to the strongly plicate form of the leaves.

138. COELOGYNE RIGIDIFORMIS

Coelogyne rigidiformis Ames & C. Schweinf., *Orch.* 6: 40 (1920). Type: Borneo, Sabah, Mt. Kinabalu, Kiau, *J. Clemens* 71 (holo. AMES).

DESCRIPTION. *Pseudobulbs* clothed with fibrous remnants of sheaths, cylindric to narrowly pyriform, 3–4 cm long. *Leaf* 1, oblong-linear, gradually narrowed at complicate base, strongly plicate, rigid, erect, coriaceous, 20–28.5 × 1.8–2.3 cm with 5.5–7.5 cm long, cylindric petiole. *Inflorescence* heteranthous, peduncle erect, 9.4–21 cm long, rachis arcuate or zig-zag, to 12 cm long, loosely few-flowered. *Flowers* greenish with a purple end to column. *Dorsal sepal* linear-oblanceolate, acute, obtuse, indistinct 5-nerved, 2.1 × 0.5 cm. *Lateral sepals* linear-oblanceolate, acute, more or less obtusely cymbiform, indistinctly 5-

nerved, 2.2 × 0.5 cm. ***Petals*** linear, obtuse or unequally subtruncate, 3-nerved at base, 2 × 0.35 cm. ***Lip*** 2.2 × 0.75 cm, in natural position, tubular-involute for two thirds of the length surrounding the column, when extended, narrowly elliptic, at about one third distance from tip slightly emarginate on each side, thereafter narrowed to an obtuse, slightly thickened tip, callus of 3 low, simple keels extending from base of lip to centre of mid-lobe with the median keel extending almost to the tip of the mid-lobe. ***Column*** slender, about 1.4 cm long, broadly winged at apex, wing crenate-undulate.

DISTRIBUTION. Borneo (Sabah (Mount Kinabalu)).
HABITAT. Epiphytic in hill and lower montane forest.
ALTITUDE. 900–1600m.
FLOWERING. Not known.
NOTES. The epithet refers to the rigid form of the species. C.E. Carr defined sect. *Rigidiformes* to characterise similar species.

SECTION *VEITCHIAE*

Section 15. ***Veitchiae*** D.A. Clayton **sect. nov.**

Pseudobulbi grandes, oblongato-fusiformes vel pyriformes, apice bifoliati, arcte siti. Inflorescentia heterantha, pedunculo nudo gracili pendulo, rachidi pendenti multifloro, floribus eodem tempore aperientibus, bracteis floralibus persistentibus. Flores parvi in niveis dispositi. Typus: *Coelogyne veitchii* Rolfe.

Pseudobulbs large, close together, fusiform-oblong to pyriform, 2 leaves at the apex. Inflorescence heteranthous, peduncle bare, slender, pendulous, rachis pendulous, many-flowered, floral bracts persistent. Flowers opening simultaneously, small and in snow-white clusters. Distribution map 16.

A monotypic section.

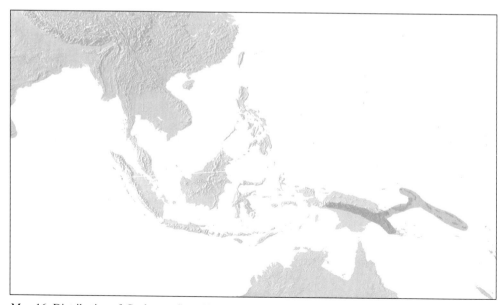

Map 16. Distribution of *Coelogyne* Sect. *Veitchiae*.

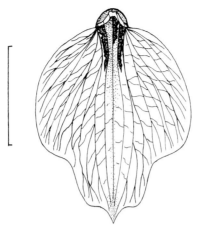

Fig. 19. **Section 15. Veitchiae** D.A. Clayton
1. Coelogyne veitchii Rolfe [Kew Coll.]. Drawn by Linda Gurr. Scale bar = 1 cm.

139. COELOGYNE VEITCHII

Coelogyne veitchii Rolfe in *Bull. Misc. Inf., Kew* 1895: 282 (1895). Types: New Guinea, cult. *Veitch* s.n. (syn. K); *Burke* s.n. (syn. K). **Plate 19F**.

DESCRIPTION. ***Pseudobulbs*** on a stout, sheathed rhizome, fusiform-oblong to pyriform, longitudinally wrinkled, to 9 × 1.7 cm when dry, base enclosed with long bracts, which become deciduous. ***Leaves*** 2, elliptic-lanceolate, acuminate, with 5 main nerves, many other nerves, 16–40 × 3.5–6 cm with 4 cm long, grooved petiole. ***Inflorescence*** heteranthous, peduncle slender, pendulous, 6.3 cm long, rachis pendulous, not zig-zag, 32.5–68 cm long, 34-flowered, flowers opening simultaneously, floral bracts persistent. ***Flowers*** small, snow-white clusters. ***Dorsal sepal*** ovate, acuminate, 5-nerved, 1.8 × 0.7 cm. ***Lateral sepals*** ovate, acute, 3-nerved, 1.8 × 0.7 cm. ***Petals*** linear, acute, 1.8 × 0.2 cm. ***Lip*** obscurely 3-lobed, 0.8 × 0.65 cm, side-lobes erect, oblong, front rounded, margin erose, mid-lobe cordate, tip reflexed, margins undulate, erose, tip with mucronate. ***Column*** slender, straight, 1 cm long, expanding into hood with acute tipped wings, margin erose.

DISTRIBUTION. Irian Jaya, Papua New Guinea (Torricelli, Bismarck, Finisterre Ranges), Solomon Islands (Guadacanal).

HABITAT. Epiphytic on old moss covered trees in lower montane forests.

ALTITUDE. 800–1100 m.

FLOWERING. September–October.

NOTES. The species is dedicated to Peter Veitch, the collector of the type specimen. It is said to resemble the genus *Pholidota* Lindl. Many authors place this species in sect. *Tomentosae* but de Vogel (1992) excluded it on the grounds that the lip has keels which are only developed at the base of the lip and the structure of the lip is quite different from sect. *Tomentosae* species. It has more in common with sect. *Rigidiformes* species in the structure of the lip, particularly the small side-lobes. It requires a section of its own which I have chosen to call *Veitchiae*.

SECTION *PTYCHOGYNE*

Section 16. ***Ptychogyne*** (Pfitzer & Kraenzlin) D.A. Clayton **stat. nov.** Type: *Coelogyne flexuosa* Rolfe.

Coelogyne subgenus *Ptychogyne* Pfitzer & Kraenzlin in Engler, *Pflanzenr. Orch.-Mon.-Coelog.*: 20 (1907).

Pseudobulbs close together. Inflorescence hysteranthous, peduncle bare, many-flowered, floral bracts deciduous, arranged in 2-rows on a distinctly zig-zag rachis. Lip distinctly 3-lobed, transverse fold at the base of the lip. Flowers opening simultaneously, almost entirely white, slight amount of yellow on lip. Distribution map 17.

A monotypic section.

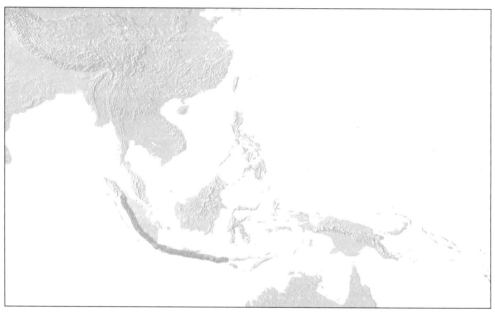

Map 17. Distribution of *Coelogyne* Sect. *Ptychogyne*.

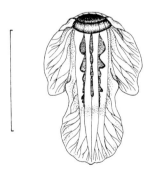

Fig. 20. **Section 16. Ptychogyne** (Pfitzer) D.A. Clayton
1. Coelogyne flexuosa Rolfe [Cult. D.A. Clayton DAC 32]. Drawn by Linda Gurr. Scale bar = 1 cm.

140. COELOGYNE FLEXUOSA

Coelogyne flexuosa Rolfe in *Bull. Misc. Inf., Kew* 1892: 209 (1892). Type: Java, cult. Glasnevin B.G. (holo. K). **Plate 20A.**

Coelogyne bimaculata Ridl. in *J. Linn. Soc.* 32: 327 (1896). Type: Peninsular Malaysia, Perak, Larut Hills, *Ridley* s.n. (holo. SING).
Ptychogyne flexuosa (Rolfe) Pfitzer in Engler, *Pflanzenr. Orch.-Mon.-Coelog.* 18 (1907).
Ptychogyne bimaculata (Ridl.) Pfitzer in Engler, *Pflanzenr. Orch.-Mon.-Coelog.* 20 (1907).

DESCRIPTION. **Pseudobulbs** large, close together, ovoid-conical with some rounded angles but otherwise smooth, to 15 × 5 cm. **Leaves** 2, lanceolate, acute, margins slightly undulate, 9-nerved, 30 × 5 cm with a 2 cm long petiole. **Inflorescence** hysteranthous, peduncle slender, erect, becomes arched, about 20 cm long, rachis zig-zag, 15 cm long, about 18-flowered, flowers open simultaneously, floral bracts deciduous. **Flowers** almost entirely white with only a little yellow on lip, sweet scented and placed in two rows. **Sepals** 1.7 × 0.7 cm. **Petals** 1.4 × 0.4 cm. **Lip** 3-lobed, transverse fold at base of lip, side-lobes, erect and rounded, mid-lobe triangular above a narrowed base, margins undulate, tip more or less acute, callus of 3 parallel keels from the saccate base but only the median keel continues as far as the tip. **Column** slender, 0.7 cm long, narrow wings at tip.
DISTRIBUTION. Peninsular Malaysia (Larut Hills), Sumatra (West coast), Java, Bali.
HABITAT. Epiphytic and high on isolated tree trunks and branches where light levels are high in lower montane forest.
ALTITUDE. 900–1520 m.
FLOWERING. May.
NOTES. The epithet refers to the distinctly zig-zag nature of the inflorescence.

SECTION *LAWRENCEANAE*

Section 17. *Lawrenceanae* D.A. Clayton **sect. nov.**

Pseudobulbi bifoliati in rhizomate crasso arcte siti. Inflorescentia hysterantha, pauciflora, bracteis paucis grandibus imbricatis inter pedunculum atque florem primum rhachidi sitis, floribus deinceps aperientibus sed longaevis. Flores pergrandes, sepalis 7 cm longis vel longioribus late patentibus, labello multicarinato, carinis tribus prominentibus distincte laciniatis. Typus: *Coelogyne lawrenceana* Rolfe.

Pseudobulbs close together on stout rhizome, 2-leaves. Inflorescence hysteranthous, a few large imbricate bracts at the interface between the peduncle and the first flower on the rachis, few-flowered. Flowers opening in succession but long-lasting, very large with sepals 7 cm or longer, wide spreading, lip with many keels, 3 prominent, distinctly laciniate keels. Distribution map 18.

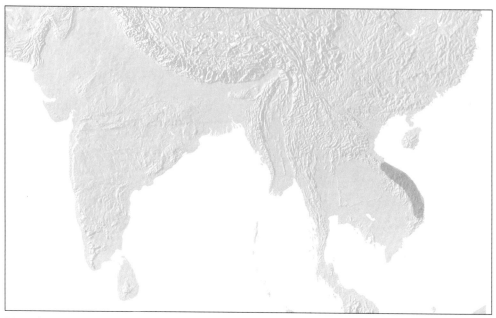

Map 18. Distribution of *Coelogyne* Sect. *Lawrenceanae*.

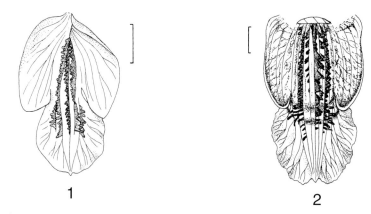

Fig. 21. **Section 17. Lawrenceanae** D.A. Clayton
1. Coelogyne eberhardtii Gagnep. [after Seidenfaden (1992)]; **2.** **C. lawrenceana** Rolfe [Cult. D.A. Clayton DAC 48]. Drawn by Linda Gurr. Scale bar = 1 cm.

KEY TO SPECIES IN SECTION LAWRENCEANAE

Pseudobulbs ovoid to pyriform. Inflorescence hysteranthous, peduncle bare, erect, rachis with persistent bracts at the base, 1- to 3-flowered. Lip large, mid-lobe orbicular, 3 prominent lacinate keels, 4 additional keel-like filaments .... ***C. lawrenceana***
Pseudobulbs cylindrical. Inflorescence hysteranthous, rachis with no persistent bracts at the base, 1-flower. Lip ovate-oblong, mid-lobe ovate, obtuse, shortly clawed, 3 prominent lacinate keels .... ***C. eberhardtii***

141. COELOGYNE EBERHARDTII

Coelogyne eberhardtii Gagnep. in *Bull. Mus. Nat. Hist. Paris* sér. 2, 2: 423 (1930). Type: Vietnam, Langbian, *Eberhardt* 1887 (holo. P; iso. C). **Plate 20B.**

DESCRIPTION. *Pseudobulbs* cylindrical. *Leaves* 2, oblong-lanceolate, tip attenuated-obtuse, with 5 nerves, 12–16 × 2.5 cm with short petiole. *Inflorescence* hysteranthous, peduncle 2–3 cm long, rachis 2.5–3 cm long, single flower. *Flowers* white with brown striations. *Dorsal sepal* oblong, acuminate-acute, with 7 branched nerves, 4.6 × 1.3 cm. *Lateral sepals* similar, somewhat falcate, with 7 branched nerves, 4.8 × 1.3 cm. *Petals* linear, acuminate, with 5 branched nerves, 4.7 × 0.5 cm. *Lip* ovate-oblong, 4 cm long, widening from base to 2.5 cm at two thirds along lip, side-lobes not stretched, rounded, 0.8 × 0.7 cm, mid-lobe ovate, obtuse, shortly clawed, callus of 3 prominent keels from base of lip to middle of mid-lobe, obtuse lacerations. *Column* 2 cm long, broadly obtuse tip.

DISTRIBUTION. Vietnam (Da Lat).
HABITAT. Not known.
ALTITUDE. Not known.
FLOWERING. Not known.
NOTES. The species named after P. Eberhardt, the collector of the type specimen.

142. COELOGYNE LAWRENCEANA

Coelogyne lawrenceana Rolfe in *Gard. Chron.* ser. 3, 37: 227 (1905). Type: Vietnam, cult. Sander, *Micholitz* s.n. (lecto. icon. Bot. Mag. 133: t. 8164). **Plate 20C.**

Coelogyne fleuryi Gagnep. in *Bull. Mus. Hist. Nat. Paris* sér. 2, 2: 424 (1930). Types: Vietnam, Langbian, *Chevalier* 30900 (syn. P), *Poilane* 5975 (syn. NY, P).

DESCRIPTION. *Pseudobulbs* on a stout, sheathed rhizome, 2.5–4 cm apart, ovoid to pyriform, grooved longitudinally, 9– 11.5 × 2.5–4 cm, enclosed with large, 13 cm long bracts at base which become brown and papery. *Leaves* 2, lanceolate, acuminate, with 8 main nerves, coriaceous, glossy, to 25–40 × 3.5–5.2 cm with 2.2–4 cm long, open channelled petiole. *Inflorescence* hysteranthous, a long time in developing, peduncle bare, erect, 13–20 cm long, rachis initially erect with large convolute bracts which become brown, to 8 cm long, then arching, stiff, to 10 cm long, 1- to 3-flowered, flowers opening in succession, floral bracts large, convolute, deciduous. *Flowers* about 10–13 cm across, widely spreading, waxy, fragrant, sepals and petals greenish-yellowish-brownish, lip with brown-veined side-lobes, a large brown patch and golden-yellow crescent patch near base of lip, further crescent shaped yellow patch, surrounded with brown margin centred on the claw of the lip, keels whitish with brown tips, column pale yellow. *Dorsal sepal* ovate-oblong, acute, 9-nerved, 7.1 × 2.5 cm. *Lateral sepals* ovate-lanceolate, acute, 7-nerved, 7 × 2.1 cm. *Petals* linear-oblong, obtuse, 3-nerved, 7 × 0.6 cm. *Lip* 7.8 cm long, 3-lobed, side-lobes erect, level with base of column, rectangular, rounded in front, 3.8 cm long, 1 cm deep, margins entire, mid-lobe large, 3.6 × 3.6 cm, orbicular with up-turned side margins, tip acute, porrect, margins deeply undulate, minutely erose, callus of 7 keels, the 3 main keels extending from base of the lip to one third way onto mid-lobe, straight, deeply laciniate to crested, lateral keels more prominent than median keel, further 2 keel-like filaments between median and lateral keels, additional 2 keel-like filaments outside lateral keels at the interface of the claw with the mid-lobe. *Column* stout, arcuate, 3.7 cm long, expanding from 0.6 cm across to 1.4 cm at winged hood, tip distinctly incurved, front slightly rounded, dentate at each end, slightly bi-lobed.

DISTRIBUTION. Vietnam.
HABITAT. Not known.
ALTITUDE. Not known.
FLOWERING. February–March.

NOTES. The species is dedicated to Sir James John Trevor Lawrence (1831–1913), born in London and who had a private collection of orchids at Burford, near Dorking, Surrey, England. He grew and exhibited the species at the Royal Horticultural Society in 1905 and was awarded an AM. He had obtained the plant from Messrs. Sander & Sons, for whom Micholitz had collected the original specimen. Lawrence was President of the Royal Horticultural Society, 1885–1913. Messrs. Sander & Sons presented a plant to Kew and this was illustrated by J.N. Fitch is in *Botanical Magazine* 133: t. 8164 (1907). The illustration must be judged to be the holotype. A painting of this species by Dr. L. Simond in Paris (no. 187 P) is misidentified as *Coelogyne speciosa* Lindl. I agree with Seidenfaden, that *C. fleuryi* Gagnep. is a synonym of this species.

SECTION *COELOGYNE*

Section 18. **Coelogyne**. Type: *Coelogyne cristata* Lindl.

Section *Cristatae* Pfitzer & Kraenzlin in Engler, *Pflanzenr. Orch.-Mon.-Coelog.*: 60 (1907).

Pseudobulbs broadly ovoid or fusiform to cylindric, close together or some distance apart on the rhizome. Inflorescence either heteranthous, synanthous or hysteranthous. Flowers generally large and mainly white with yellow markings on the lip, some with brownish marking on the lip. Lip with several parallel keels which may be crisped, crenulate, fimbriate, lacerate or laciniate. Distribution map 19.

Map 19. Distribution of *Coelogyne* Sect. *Coelogyne*.

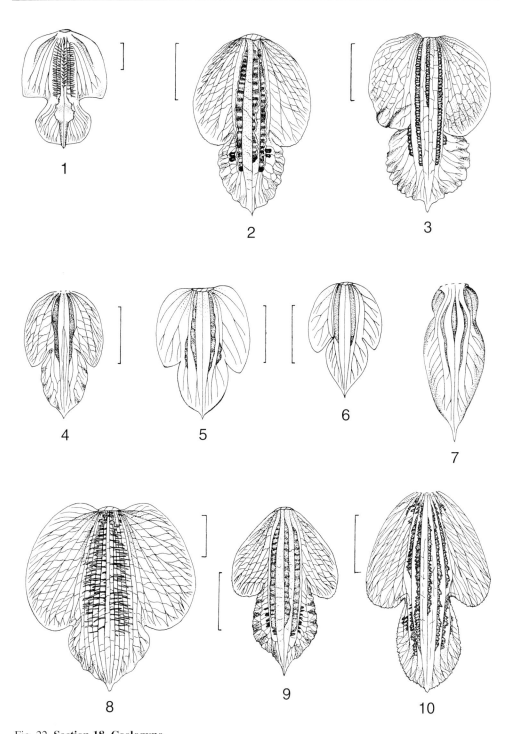

Fig. 22. **Section 18. Coelogyne**
1. Coelogyne cristata Lindl. [Cult. D.A. Clayton DAC 65]; **2. C. cumingii** Lindl. [Kew Spirit Coll. 18902]; **3. C. foerstermannii** Rchb.f. [after C.L. Chan]; **4. C. glandulosa** Lindl. var. **glandulosa** [*Bourne* 1898]; **5. C. glandulosa** Lindl. var. **bournei** S. Das & S.K. Jain [after Das & Jain (1976)]; **6. C. glandulosa** Lindl. var. **sathyanarayanae** S. Das & S.K. Jain [after Das & Jain (1976)]; **7. C. malintangensis** J.J. Sm. [after J.J. Smith]; **8. C. mooreana** Sand. ex Rolfe [Cult. D.A. Clayton, DAC 9]; **9. C. nervosa** A. Rich [*Wight* 2963]; **10. C. sanderiana** Rchb.f. [*Wood* 958]. Drawn by Linda Gurr. Scale bar = 1 cm.

Key to species in section Coelogyne

1. Inflorescence heteranthous **2**
   Inflorescence synanthous or hysteranthous **3**

2. Pseudobulbs distant on the rhizome, narrowly cylindrical. Lip mid-lobe obovate, with 3 crenulate keels. [Flowers white with yellow marking on the lip.] .... *C. foerstermannii*
   Pseudobulbs close together or distant on the rhizome, oblong-ovoid. Lip mid-lobe suborbicular, with 4 or 5 fimbriate keels. [Flowers pure white, keels yellow between side-lobes and 2 yellow crenulate plates on the mid-lobe.] .... *C. cristata*

3. Inflorescence synanthous **4**
   Inflorescence hysteranthous **8**

4. Floral bracts persistent **5**
   Floral bracts deciduous **6**

5. Pseudobulbs rugose, obovoid, young leaves about fully formed. Inflorescence with bracts on the peduncle. [Lip with 3 crisped keels.] .... *C. nervosa*
   Pseudobulbs smooth, ovoid, young leaves only partly developed. Inflorescence with peduncle enclosed within bracts. [Flowers white, lip tinged with yellow]:
   Lip with 3 keels .... *C. glandulosa* var. *glandulosa*
   Lip with 2 keels .... *C. glandulosa* var. *sathyanarayanae*
   Lip with 4 keels .... *C. glandulosa* var. *bournei*

6. Pseudobulbs ovoid **7**
   Pseudobulbs cylindric **9**

7. Lip mid-lobe elliptic, with 5 crenulate keels. [Flowers white, lip with yellow or orange veins, margin crisped.] .... *C. cumingii*
   Lip mid-lobe ovate, with 3 fimbriate keels, thickened at the base of the lip. [Flowers white, lip with yellow markings.] .... *C. mooreana*

8. Lip mid-lobe ovate, with 3 low, slender, crenulate, undulate keels. [Flowers white, lip with brown markings.] .... *C. kemiriensis*
   Lip mid-lobe ovate, with 3 simple keels. [Flowers yellowish.] .... *C. malintangensis*

9. Pseudobulbs fusiform, lip mid-lobe oblong, rounded, acute, reflexed, with 5 or more keels, margin undulate. [Flowers large, white, lip with brown veins.] .... *C. sanderiana*
   Pseudobulbs ovoid, lip mid-lobe oblong, with 3 keels, 2 keels papillose. [Flowers greenish-white to white, median keel orange.] .... *C. concinna*

143. COELOGYNE CONCINNA

Coelogyne concinna Ridl. in *J. Roy. Asiat. Soc. Straits Branch* 61: 40 (1912). Type: Sumatra, Dolok Baros, Belawan, *Moisseniac* s.n. (holo. not located).

DESCRIPTION. *Pseudobulbs* close together, ovoid, 4-angled, apex blunt, deep green, appressed, 1.9 × 1.4 cm. *Leaf* 1, elliptic-lanceolate, acuminate, spathulate, thinly coriaceous, with 5 nerves, 15 × 5 cm with 1.4 cm long petiole. *Inflorescence* hysteranthous, peduncle enclosed in bracts, 2.5 cm, rachis not prominent, 1- to 2-flowered. *Flowers* with greenish-white sepals, petals white, lip white with orange markings on median keel, column white. *Sepals* lanceolate-oblong, subacute, 3.8 × 0.65 cm. *Petals* narrowly linear, 3.8 cm long. *Lip* 2.5 cm long, 3-lobed, side-lobes rounded, mid-lobe oblong, rounded, callus of 3 keels, lateral papillose keels from base of lip to tip of mid-lobe, with minute papillae near base, median keel not papillose, terminating in middle of mid-lobe. *Column* 1.25 cm long, hood rounded, slightly notched.

DISTRIBUTION. Sumatra (Dolok Boros).

HABITAT. Epiphytic on trees in hill forests.

ALTITUDE. Not known.

FLOWERING. Not known.

NOTES. The epithet refers to the small, neat nature of the species.

144. COELOGYNE CRISTATA

Coelogyne cristata Lindl., *Coll. Bot.* sub. t. 33 (1821). Type: Nepal, Toka, *Wallich* s.n. (Wall. Cat. no. 1958) (holo. K; iso. CAL). **Plate 20D.**

Cymbidium speciosissimum D. Don in *Prodr. Fl. Nepal.* 35 (1825). Type: Nepal, *Wallich* (holo. BM).

DESCRIPTION. *Pseudobulbs* on a stout, sheathed rhizome, closely spaced to 5–6 cm apart, oblong-ovoid, smooth, 5–7.5 × 4 cm. *Leaves* 2, linear-lanceolate, acute, slightly undulate, 15–30 × 2–2.7 cm with no distinct petiole. *Inflorescence* heteranthous, peduncle sheathed, rachis pendulous, overall 15–30 cm long, 3- to 10- flowered, flowers open simultaneously, floral bracts up to 5 cm long, persistent. *Flowers* very large, about 8 cm across; sepals and petals white, lip white with yellow keels between side-lobes and 2 crenulate yellow plates on mid-lobe, column white. *Sepals* narrowly elliptic-oblong, subacute, undulate, dorsal nerve prominent, many other nerves indistinct, 3.7–5 × 1.7 cm. *Petals* narrowly elliptic-oblong, acute, undulate, dorsal nerve prominent, many other nerves indistinct, 4.5 × 1.7 cm. *Lip* 3-lobed, up to 4 × 3.5 cm, side-lobes large, rounded, front angled, reflexed, entire, mid-lobe suborbicular, tip tri-lobed, margin reflexed, entire, callus of 4 or 5 fimbriate keels from base of lip to base of mid-lobe, 3 keels continue to tip of mid-lobe, lateral keels elevated and plate-like, median keel a prominent nerve line. *Column* arcuate, 2.6–3.2 cm long, broadly winged.

DISTRIBUTION. Tibet, Nepal, Bhutan, Northeast India (Himalayas (Kumaon to Sikkim, Khasia Hills)), China.

HABITAT. Epiphytic on trees, sometimes lithophytic on rocks in shaded, lower and upper montane forest.

ALTITUDE. 1000–2600 m.

FLOWERING. February–April depending on altitude.

NOTES. The epiphet refers to the crested form of the keels on the lip of the flowers. There have been numerous variations cited in the literature but these are, in the main, just colour variations. The main horticulatural varieties are:

cv. Arrigadh, *Gard. Chron.* ser. 3, 3: 462 (1888). Sepals and petals plain, not crisped, keels of lip orange-yellow.

cv. Chatsworth brought by Gibson in 1837; larger than normal form, regular in form, about 9 cm across; sepals and petals white; lip white with distinctly deep yellow keels between side lobes and 2 crenulate deep yellow plates on mid-lobe. Sepals narrowly elliptic-oblong, subacute, undulate, 3.7–5 × 1.7 cm. Petals similar to sepals, acute, 4.5 × 1.7 cm. Lip 3-lobed, up to 4 × 3.5 cm, side-lobes large, rounded, mid-lobe suborbicular with 4 fimbriate keels from base of lip to tip of mid-lobe. It is said to flower in the autumn.

var. *citrina* Williams in Godfroy's *Orchidophile* : 212 (1881). The epiphet for this variation refers to the paler form of the yellow/orange colour on the lip of the flowers. There is a tendency for orchid growers to call this variety *C. cristata* var. *lemoniana* because this name was used by Sir Charles Lemon of Carclew, Cornwall, England, probably in the late 1870s.

var. *hololeuca* Rchb.f. in *Gard. Chron.* ser. 2, 16: 563 (1881). There is *C. hololeuca* Hort. and this name was judged to be a synonym of *C. cristata* var. *alba*. The true epiphet should be *C. cristata* var. *hololeuca* Rchb.f. and the characteristics of this variation are the all-white form of the flower.

var. *maxima* Sander in *Reichenbachia* 1, t. 6 in *The Garden* 31, t. 585 (1887). Flowers larger in all their parts.

cv. Trentham. Flowers produced 6–8 weeks later than the other forms.

var. *duthiei* Rchb.f. in Engler, *Pflanzenr. Orch.-Mon.-Coelog.* (1907). Type: India, Tehni Garwal, *Duthie* 23941 (holo. K). Very large, about 8 cm across; virtually the same as *C. cristata* Lindl.

var. *albina;* var. Woodland's and var. *intermedia* are further varieties.

145. COELOGYNE CUMINGII

Coelogyne cumingii Lindl. in *Bot. Reg.* 26: misc. 187: 76 (1840). Type: Singapore, cult. Loddiges s.n. (holo. K). **Plate 20E.**

Coelogyne longebracteata Hook.f., *Fl. Brit. India* 6: 194 (1890). Type: Peninsular Malaysia, Perak, *Kunstler* s.n. (holo. CAL).
C. casta Ridl. in *J. Linn. Soc., Bot.* 32: 322 (1896). Type: Peninsular Malaysia, Selangor, Hitai, *Kelshall* s.n. (holo. SING).
Pleione cumingii (Lindl.) Kuntze, *Rev. Gen. Pl.* 2: 680 (1891).

DESCRIPTION. **Pseudobulbs** on a stout, creeping, sheathed rhizome, 2.5–3 cm apart, ovoid, shiny, becoming wrinkled with age, often yellowish, 5–7.5 × 1.5–4 cm, enclosed with persistent bracts at base. **Leaves** 2, linear-lanceolate, acute-acuminate, glossy, with 5 nerves, 15–19 × 2–4 cm with 2.5–5 cm long, grooved petiole. **Inflorescence** synanthous, peduncle stiff, erect, enclosed with loosely imbricating bracts and new developing leaves at base, 10 cm long, rachis erect or arching, to 15 cm long, loosely 3- to 5-flowered, flowers opening simultaneously, floral bracts persistent, yellowish, shiny. **Flowers** fragrant, about 6–7 cm across, lasting well. Sepals and petals white, lip veined with yellow or orange, with 5 yellow or orange keels. **Dorsal sepal** elliptic-lanceolate, acute-acuminate, 5-nerved, 3.5 × 0.6 cm. **Lateral sepals** elliptic, acute-acuminate, 5-nerved, 3 × 0.5 cm. **Petals** linear, acute, 3 × 0.3 cm. **Lip** 3-lobed, side-lobes erect, elliptic, front obtuse, 1.5 cm long, mid-lobe broadly elliptic, margins crisped, callus of 5 keels, 3 extending from base of lip to middle of mid-lobe, 2 further keels on outside of lateral keels only present on mid-lobe, keels crenulate. **Column** slender, arcuate, 1.9–2.8 cm long, expanding into winged hood, tip tri-lobed.

DISTRIBUTION. Laos, Riau Archipelego, Thailand (Khao Kheo, Songkla), Peninsular Malaysia, including Singapore, Borneo (Kalimantan, Sabah, Sarawak), Sumatra (Palembang).

HABITAT. Epiphytic on trees, lithophytic on rocks, sometimes in large masses in hill forest on ultramafic substrate.

ALTITUDE. 900–2130 m.

FLOWERING. April, May, July–August, later at higher altitude.

NOTES. Dedicated to Hugh Cuming (1791–1865), who was born in West Alvington, Devonshire, England. He collected orchids for Messrs Loddiges of Hackney and is best known for his expedition to the Philippines from 1836 to 1840 and his early work on the shipping of live orchids from Manila.

146. COELOGYNE FOERSTERMANNII

Coelogyne foerstermannii Rchb.f. in *Gard. Chron.* ser. 2, 26: 262 (1886). Type: Sumatra, *Förstermann s.n.* (holo. W). **Plate 20F.**

Coelogyne maingayi Hook.f., *Fl. Brit. India* 5: 831 (1890). Type: Peninsular Malaysia, Melaka, *Maingay 1636* (holo. K; iso. CAL).
Pleione foerstermannii (Rchb.f.) Kuntze, *Rev. Gen. Pl.* 2: 680 (1891).
Pleione maingayi (Hook.f.) Kuntze, *Rev. Gen. Pl.* 2: 680 (1891).
Coelogyne kingii Hook.f. in *Ann. Bot. Gard. Calcutta* 6: 25, t. 38 (1895). Type: Peninsular Malaysia, Perak, *Kunstler s.n.* (holo. K; iso. CAL).

DESCRIPTION. **Pseudobulbs** on a creeping, darkly sheathed rhizome, about 11 cm apart, narrowly cylindrical, grooved with age, to 11 × 1.5 cm, base enclosed in bracts. **Leaves** 2, lanceolate, acute, stiff, rather heavy-textured, closely folded, to 26 × 4 cm with 5 cm long, grooved petiole. **Inflorescence** heteranthous, peduncle erect or arching, sheathed along its length, 18–24 cm long, rachis zig-zag, to 15-flowered, flowers opening simultaneously, floral bracts persistent. **Flowers** fragrant, almost 8 cm across, white, lip with central yellow area near apex. **Dorsal sepal** lanceolate, acute, 7-nerved, 3.2 × 0.5 cm. **Lateral sepals** lanceolate, acuminate, 5-nerved, 3.5 × 0.6 cm. **Petals** oblong, obtuse, 2.5 × 0.4 cm. **Lip** 3-lobed, side-lobes large, semi-orbicular, margins entire but erose in front, mid-lobe obovate, crenate, undulate, callus of 3 crenulate keels, prominent and extending from base of lip to tip of mid-lobe. **Column** slender, slightly arcuate, 1.5 cm long, expanding from base to form hood, tip crenulate.

DISTRIBUTION. Borneo (Kalimantan, Sarawak, Sabah), Maluku, Peninsular Malaysia, including Singapore, Sumatra (East & West coasts).

HABITAT. Epiphytic in lowland rain forest on tall trees and in hill forest.

ALTITUDE. Sea level–500 m.

FLOWERING. After 3 months dry weather, several times a year.

NOTES. The species dedicated to the collector Ignatz F. Förstermann (1854–1895) who was born at Koblenz, Germany. He was a collector for Messrs. Sander and Sons of St Albans from 1880–86 and was later manager of their branch in Summit, New Jersey, USA.

147. COELOGYNE GLANDULOSA

Coelogyne glandulosa Lindl., *Fol. Orch.-Coelog.* 6 (1854). Type: South India, Nilgiri Hills, Pykara, *R. Wight* s.n. (holo. K).

var. **glandulosa**

DESCRIPTION. *Pseudobulbs* on a stout rhizome, 0.5–1.2 cm apart, ovoid, 3–8 cm × 1.2 cm, slightly curved when old, ridged and furrowed, enclosed with coriaceous bracts at base. *Leaves* 2, elliptic-oblong to oblong-lanceolate, acute, tapering towards base, entire, slightly undulate, coriaceous, green, with 7–11 nerves, 9–20 × 2.3–4.6 cm with a 0.5–1.5 cm petiole. *Inflorescence* synanthous, peduncle with 2 sheathing bracts at base, erect 13 cm long, rachis erect, not zig-zag, 6 cm long, 2- to 8-flowered, flowers opening simultaneously, floral bracts persistent. *Flowers* white with a white lip tinged with yellow or orange in the middle. *Sepals* oblong-lanceolate, acute, entire, glabrous, subcarnose, 7-nerved, dorsal keeled, 2.3–3 × 0.9–1.3 cm. *Petals* elliptic, acute, entire, glabrous, subcarnose, 7-nerved, not keeled, 2.2–2.8 × 1–1.4 cm. *Lip* 1.8–2.2 × 1–1.3 cm, 3-lobed, side-lobes oblong, obtuse, entire, clasping the column, mid-lobe elliptic, acute, entire, callus of 3 keels extending from base of lip to base of mid-lobe, almost entire. *Column* slightly arcuate 1.4–1.6 cm long, broadly winged, slightly bi-lobed at apex.

DISTRIBUTION. South India (Nilgiri and Palni (Pulney) Hills).
HABITAT. Epiphytic on trees and lithophytic on rocks in lower montane forest.
ALTITUDE. 1000–2000 m.
FLOWERING. June.
NOTES. The epithet refers to the gland-like keels on the lip to the flower.

var. **bournei** S. Das & S.K. Jain in *Bull. Bot. Surv. India* 18 (1–4): 244 (1976, publ. 1979). Type: South India, Palni (Pulney) Hills, Poombarai, *Bourne* 2941 (holo. CAL).

This variety differs from the type only in the flowers having a lip mid-lobe with 4 keels, almost entire lateral keels very short.

DISTRIBUTION. South India (Palni (Pulney) Hills).
HABITAT. Epiphytic in lower montane forest.
ALTITUDE. 1000 m.
FLOWERING. May.
NOTES. The variation is named after its collector Bourne. See notes under sect. *Lentiginosae, C. mossiae* Rolfe.

var. **sathyanarayanae** S. Das & S.K. Jain in *Bull. Bot. Surv. India* 18 (1–4): 242 (1976, publ.1979). Type: South India, Kodaikanal, *Saldanha* 5211 (holo. CAL).

?*Coelogyne mossiae* Rolfe in *Kew Bull.* 1894: 156 (1894), *pro parte*. Type: South India, Nilgiri Hills, *S.N. Moss* s.n. (holo. K).

This variety differs from the type only in the flowers having a lip mid-lobe with 2 keels with entire glands of equal length.

DISTRIBUTION. South India (Kodaikanal, Palni (Pulney) Hills).
HABITAT. Epiphytic lower montane forest.

ALTITUDE. 1000 m.
FLOWERING. March.
NOTES. The variation is dedicated to Sathranaray. See Notes under sect. *Lentiginosae, C. mossiae* Rolfe. I have seen a photograph of the sheet determined by Das & Jain as the holotype of *C. glandulosa* var. *sathyanarayanae*, which was originally labelled *C. mossiae* Lindl. It comes from the Blatter Herbarium, St. Xavier's College, Bombay (No. CS 5211), from Kodaikanal, Palni Hills and was collected on 10th March 1960. The specimen does seem to be a variation of *C. glandulosa* and not of *C. mossiae*.

148. COELOGYNE KEMIRIENSIS

Coelogyne kemiriensis J.J. Sm. in *Blumea* 5: 298 (1943). Type: Sumatra, Aceh, Gunung Kemiri (eastern slope), *van Steenis* 9636 (holo. L; iso. BO). **Plate 21A**.

DESCRIPTION. ***Pseudobulbs*** tubular, tip acuminate, 6.75 cm long, when new enclosed with bracts. ***Leaves*** 2, lanceolate, acute-acuminate, with 5 nerves prominent beneath, 17.5 × 2.2 cm, base gently narrows into 4 cm long, grooved petiole. ***Inflorescence*** synanthous, erect, peduncle 9.5 cm long, rachis zig-zag, 4.5–5 cm long, 5- to 6-flowered, short lived, flowers opening in succession. ***Flowers*** white with brown markings on keels and in throat of lip. ***Dorsal sepal*** narrowly oblong, narrowly obtuse, apiculate, 5-nerved, dorsal prominent, 3.8 × 1.1 cm. ***Lateral sepals*** narrowly oblique, oblong, somewhat falcate, acute-apiculate, 5-nerved, 3.7 × 0.9 cm. ***Petals*** oblique, lanceolate, acute, 3.6 × 0.64 cm. ***Lip*** 3-lobed, side-lobes free, stretch outwards and forwards, oblique, ovate, obtuse, mid-lobe longer than side-lobes, ovate, rounded-obtuse, apiculate, margin erose, undulate, callus of 3 slender, low, crenulate-undulate keels, from base of lip with median keel extending to near tip of mid-lobe, lateral keels elevated but vanishing in middle of mid-lobe. ***Column*** slender, lightly arcuate, 2.9 cm long, slightly winged, wings irregularly angled.
DISTRIBUTION. Sumatra (Aceh (Gunung Kemiri)).
HABITAT. Terrestrial in mossy forest on mountain ridge and on plateau.
ALTITUDE. 2900–3300 m.
FLOWERING. Not known.
NOTES. The epiphet refers to the origin of the type specimen and probably the only locality where the species is found, Gunung Kemiri, Aceh, Sumatra.

149. COELOGYNE MALINTANGENSIS

Coelogyne malintangensis J.J. Sm. in *Bull. Jard. Bot. Buitenzorg* ser. 3, 5: 26 (1922). Type: Sumatra, Gunung Malintang, *Büennemeyer* 4050, 4216, 4220 (syn. BO, isosyn. L).

DESCRIPTION. ***Pseudobulbs*** on a strong, elongated, branched, initially sheathed rhizome, 6–7 cm apart, somewhat cylindric, 3.4–5.7 cm long, initially sheathed with imbricating bracts. ***Leaves*** 2, erect, elliptic to lanceolate-elliptic, slightly acuminate to acute, undulate, with 5–7 major nerves, 5.75–11 × 2.1–3.6 cm with 0.6–1.3 cm long, grooved petiole. ***Inflorescence*** synanthous, erect, peduncle bare, thickening towards tip, 4.7–10 cm long, rachis zig-zag, 2–5.5 cm long, lax 4- to 6-flowered, flowers opening simultaneously. ***Flowers*** yellowish-green or flesh-coloured. ***Dorsal sepal*** incurved, tip recurved, ovate, long triangle-acuminate or acute with mucronate, grooved, 5-nerved, 2.6 × 1 cm. *Lateral sepals* obtuse-angled, tip recurved, oblique, ovate-lanceolate, acute, grooved, fleshy, 5-nerved, 2.5 × 0.7 cm. *Petals* oblique, somewhat linear, gently narrowing towards tip, acute, grooved, 3-nerved, 2.7 × 0.26 cm. *Lip* nearly parallel with column then recurved, 2.1 cm long when flattened, contracting gently from base with a nominal mid-lobe being somewhat ovate, acute with mucronate tip, undulate, tip recurved, at base of lip laterally erect, rounded on the sides, on inside hollowed, Y-shaped nerves, callus of 3 simple keels in a smooth curve outwards, thickening from base of lip and extending to tip of mid-lobe. *Column* slender, arcuate, 0.9 cm long, narrowing into wings, 3-cornered, tip truncate, somewhat 3-lobed.

DISTRIBUTION. Sumatra (West coast (Gunung Malintang, Lake Toba).

HABITAT. Epiphytic on trees and brushwood in upper montane forest.

ALTITUDE. 1500–2200 m.

FLOWERING. July–August.

NOTES. The epiphet refers to the locality Gunung Malintang, Sumatra, where the type specimen was collected.

150. COELOGYNE MOOREANA

Coelogyne mooreana Sand. ex Rolfe in *Bull. Misc. Inf., Kew* 1907: 129 (1907). Type: Vietnam, Lanbian, *Micholitz* s.n. (holo. K; iso. BM, C, P). **Plate 21B.**

C. psectrantha Gagnep. in *Bull. Mus. Hist. Nat. Paris* sér. 2, 2: 425 (1930). Types: Vietnam, Da Lat, *Chevalier* 38631, 38843, 38645; Nhatang, *Poilane* 11222, 4406, 5088; Nhatang, *Evrard* 1032, 2531 (all syn. P).

DESCRIPTION. *Pseudobulbs* on a sheathed rhizome, 1 cm apart or clustered, ovoid, bluntly angulate, slightly furrowed, 6.5 × 2 cm. *Leaves* 2, narrowly linear-oblanceolate, acute, with 7 nerves, 25–40 × 2.5–3.5 cm with 7.5 cm long, grooved petiole. *Inflorescence* synanthous or proteranthous, peduncle erect or arching, to 9 cm long, rachis slender, straight, 11 cm long, (can be up to 50 cm long overall), 4- to 8-flowered, flowers opening simultaneously, floral bracts deciduous. *Flowers* about 10 cm across, sepals and petals pure white, lip white with a dark orange or ochreous at base of lip, column pale yellow. *Dorsal sepal* elliptic-lanceolate, acuminate, 7–9-nerved, 4.3–5 × 1.7 cm. *Lateral sepals* broadly lanceolate, acute-acuminate, 7-nerved, slightly keeled, 3.5–5 × 1.3–1.5 cm. *Petals* elliptic-lanceolate, acute-acuminate, 7–9-nerved, 4 × 1.6 cm. *Lip* 3-lobed, 4 × 3 cm, side-lobes erect, embrace column, broadly rounded, mid-lobe ovate, acute, callus of 3 fimbriate keels extending from base of lip to base of mid-lobe. *Column* 2.5 cm long, expanding into hood.

DISTRIBUTION. Vietnam (Da Lat, Dunkia-Mi-Ba, Nhalrang, Mount Dango-pot).

HABITAT. Not known.

ALTITUDE. 1300 m.

FLOWERING. Spring to early summer.

NOTES. Dedicated to F.W. Moore, Director of the Glasnevin Botanical Gardens, Dublin.

151. COELOGYNE NERVOSA

Coelogyne nervosa A. Rich in *Ann. Sci. Nat. Paris* sér. 2, 15: 16 (1841). Type: South India, Nilgiri Hills, Neddoubetta, *Perrottet* 1840 (holo. P). **Plate 21C.**

C. corrugata Wight in *Bot. Reg.* 5(1): 5, t. 1639 (1851). Type: South India, *Wight* s.n. (holo. K).

DESCRIPTION. ***Pseudobulbs*** on rhizome serially arranged, close together, angled, broadly oblong-ovoid, 3–6 × 2–3 cm, old pseudobulbs covered in dry scales and fibres. ***Leaves*** 2, elliptic, acute, slightly undulate, with 9 nerves, 5–27 × 1.6–4.5 cm, with short, indistinct petiole. ***Inflorescence*** synanthous, peduncle erect, bare, extending to 15 cm long, rachis short, zig-zag, loosely 7-to 12-flowered, flowers opening simultaneously, floral bracts persistent. ***Flowers*** about 5 cm across, fragrant, white, lip veined with yellow, side-lobes veined inside with orange-yellow, slightly wavy brownish-orange keels. ***Dorsal sepal*** elliptic-oblong, acute, with many nerves, median nerve prominent, 2.7 × 0.85 cm. ***Lateral sepals***, elliptic, oblique at base, acute, with many nerves, median nerve prominent, 2.7 × 0.8 cm. ***Petals*** elliptic, acute, 3-nerved, median nerve prominent, 2.6 × 0.95 cm. ***Lip*** 2.3 × 1.5 cm, 3-lobed, side-lobes erect, oblong, obtuse, embracing the column, mid-lobe lanceolate, margin obscurely crenulate, callus of 3 erect, crisped keels. ***Column*** about 1.7 cm long, erect, winged, hooded.

DISTRIBUTION. South India (Karnataka, Kerala, Tamil Nadu).
HABITAT. Epiphytic in lower and upper montane forest.
ALTITUDE. 1000–2300 m.
FLOWERING. August–September.
NOTES. The epithet refers to the many veins on the floral segments and the veining on the side-lobes of the lip.

152. COELOGYNE SANDERIANA

Coelogyne sanderiana Rchb.f. in *Gard. Chron.* ser. 3, 1: 764 (1887). Type: Provenance unknown, *Förstermann* s.n. (holo. W). **Plate 21D**.

DESCRIPTION. ***Pseudobulbs*** on a sheathed rhizome, close together, spindle-shaped 5–13 × 1.5–3 cm, enclosed with bracts at base. ***Leaves*** 2, oblong-lanceolate, acute, with 5 prominent nerves beneath, 38–54 × 5–6.2 cm with 15–20 cm long, grooved petiole. ***Inflorescence*** hysteranthous, peduncle erect, 20 cm long, rachis stiff, erect to pendulous, 30 cm long, 5- to 9-flowered, flowers opening simultaneously, floral bracts deciduous. ***Flowers*** about 9 cm across, fragrant, very showy, white, lip with brown-veined side-lobes, yellow hypochile, yellow keels. ***Sepals*** lanceolate, acuminate, keeled beneath. ***Petals*** narrower. ***Lip*** 3-lobed, side-lobes oblong, erect, crisped at front margin, mid-lobe oblong, acute, reflexed and undulate, callus of 5 strong keels, extending from base of lip to base of mid-lobe, median keel slightly shorter, 4 further keels on outside of lateral keels near base of lip only. ***Column*** slender, broadly winged hood, wing tips membraneous, tri-lobed, mid-lobe semi ovate, side-lobes angled.

DISTRIBUTION. Sumatra, Sunda Islands, Borneo (Kalimantan, Sarawak, Sabah, Brunei).
HABITAT. Epiphytic in secondary hill forest, moss forest on sandstone ridges or terrestrial on limestone rocks in shade.
ALTITUDE. 800–1300 m.
FLOWERING. April–May.
NOTES. Another species dedicated to Frederick Sander of F. Sander & Sons of St Albans.

SECTION *FUSCESCENTES*

Section 19. ***Fuscescentes*** Pfitzer & Kraenzlin in Engler, *Pflanzenr. Orch.-Mon.-Coelog.*: 42 (1907). Type: *Coelogyne fuscescens* Lindl.

Pseudobulbs close together or a small distance apart on the rhizome, ovoid, or fusiform to cylindric. Inflorescence proteranthous to synanthous. Peduncle enclosed in emerging young leaves and a few bracts, rachis generally few flowered, flowers open simultaneously and floral bracts deciduous or persistent. Flowers membraneous, dorsal sepal larger than lateral sepals and petals, forming a deep, enclosing hood over the column. Keels on the lip generally entire or erose. Distribution map 20.

Map 20. Distribution of *Coelogyne* Sect. *Fuscescentes*.

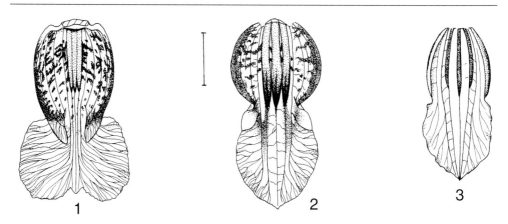

Fig. 23. **Section 19. Fuscescentes** Pfitzer
1. Coelogyne assamica Linden & Rchb.f. [Cult. D.A. Clayton DAC 5]; **2. C. fuscescens** Lindl. var. **fuscescens** [Cult. C. Howe DAC 6]; **3. C. fuscescens** Lindl. var. **integrilabia** (Pfitzer) Schltr. [after Seidenfaden (1975)]. Drawn by Linda Gurr. Scale bar = 1 cm.

KEY TO SPECIES IN SECTION FUSCESCENTES

1. Pseudobulbs fusiform to cylindric **2**
   Pseudobulbs ovoid. [Flowers yellow, lip sepia brown and ochre. Lip mid-lobe transverse ovate, 3 keels, clearly thickening, floral bracts persistent.] .... *C. picta*

2. Lip with side-lobes triangular, acute in front, mid-lobe orbicular or quadrangular **3**
   Lip with side-lobes rounded in front, mid-lobe broadly ovate to cordate:
       Flowers generally yellow or pale yellowish green, 3 fleshy, undulate keels .... *C. fuscescens* var. *fuscescens*
       Flowers pale yellow, pale yellow-green or pale yellow-brown, 3 fleshy, simple, straight keels .... *C. fuscescens* var. *brunnea*
       Flowers pale brown, lip partly sepia, side-lobes of lip rudimentary, 3 fleshy keels .... *C. fuscescens* var. *integrilabia*
       Flowers pale apple green, no trace of brown, 3 fleshy keels .... *C. fuscescens* var. *viridiflora*

3. Lip with mid-lobe orbicular, 3 keels, median keel lacking shape. [Flowers pale yellow, lip darker yellow with dark veins on the side-lobes, veins on the mid-lobe orange-brown.] .... *C. assamica*
   Lip with mid-lobe quadrangular, 3 keels, plate-like with an interrupted margin. [Sepals and petals pale green, lip light green outside, white inside, with yellowish-purple margin and yellowish keels.] .... *C. dichroantha*

153. COELOGYNE ASSAMICA

Coelogyne assamica Linden & Rchb.f. in *Allg. Gartenzeitung* 25: 403 (1857). Type: India, Assam, cult. Linden s.n. (holo. W; iso. C). **Plate 21E.**

C. fuscescens Lindl. var. *assamica* (Linden & Rchb.f.) Pfitzer & Kraenzl. in Engler, *Pflanzenr. Orch.-Mon.-Coelog.* 43 (1907).
C. annamensis Rolfe in *Bull. Misc. Inf., Kew* 1914: 214 (1914). Type: Vietnam, Da Lat, cult. Glasnevin B.G. (holo. K).
C. siamensis Rolfe in *Bull. Misc. Inf., Kew* 1914: 373 (1914). Type: Thailand, Bangkok, *Roebelin* 292-12 (holo. K).
C. dalatensis Gagnep. in *Bull. Mus. Hist. Nat. Paris* sér. 2, 2: 423 (1930). Type: Vietnam, Da Lat, *Evrard* 1262 (holo. P).
Cymbidium evrardii Guill. in *Bull. Soc. Bot. Fr.* 77: 339 (1930). Type: Vietnam, Da Lat, *Evrard* 1283 (holo. P).
Coelogyne saigonensis Gagnep. in *Bull. Mus. Hist. Nat. Paris* sér. 2, 10: 435 (1938). Type: Vietnam, cult. Saigon B.G. (holo. P).

DESCRIPTION. *Pseudobulbs* on a sturdy rhizome, close together, spindled-shaped, 6–7 × 1.5–2 cm in middle. *Leaves* 2, elliptic-lanceolate, tip barely attenuated, acute, with 3–5 nerves, prominent below, 16–25 × 3.5–6 cm with 3 cm long petiole. *Inflorescence* proteranthous, peduncle enclosed with imbricating bracts at base, 10–12 cm long, rachis slightly zig-zag, thickened becoming slender, arching, 19–25 cm long, usually 8-flowered, flowers opening simultaneously, floral bracts deciduous. *Flowers* more than 7 cm across, pale yellow, lip darker yellow with intricate brown nerves on side-lobes, veins on mid-lob, orange-brown. *Dorsal sepal* oblong, acuminate to slightly acute, with 7–9 branched nerves, 3.2 × 0.9 cm. *Lateral sepals* linear, acute, falcate, with branched nerves, 2.8 × 0.6 cm. *Petals* narrowly linear, with 3 branched nerves, 3 × 0.25 cm. *Lip* obovate with deep recess on each side, 2.6 × 1.2 cm, 3-lobed, conspicuously veined, side-lobes not spread out, triangular, mid-lobe somewhat orbicular, shortly clawed, acute or subacute, 1.2 × 1.3 cm, callus of 3 keels but median keel lacking in shape. *Column* clavate, arcuate, 2 cm long, tip 0.6 cm across, acuminate.

DISTRIBUTION. Bhutan, Northeast India (Assam), Burma, China (Yunnan), Laos, Vietnam (Da Lat), Thailand.

HABITAT. Epiphytic in riverine forest.

ALTITUDE. 700 m.

FLOWERING. January.

NOTES. The epithet refers to Assam, India where the type specimen was collected.

154. COELOGYNE DICHROANTHA

Coelogyne dichroantha Gagnep. in *Bull. Mus. Hist. Nat. Paris* sér. 2, 22: 506 (1950). Type: Indo-china, locality unknown, icon. *Eberhardt* (holo. P).

DESCRIPTION. *Pseudobulbs* on a slender, sheathed rhizome, to 1.2 cm apart, cylindric, 4-angled, 5 cm long. *Leaves* 2, lanceolate, acute, with 7 nerves, 15–18 × c. 4 cm with c. 1.5 cm long, grooved petiole. *Inflorescence* proteranthous or synanthous with partially to entirely developed leaves, peduncle during flowering enclosed at base by young shoot, peduncle c. 3.8 cm long, rachis curved, zig-zag, c. 1.8 cm long, 3-flowered, flowers opening in succession, floral bracts ovate-lanceolate, acute, persistent. *Flowers* with pale green sepals and petals, lip light green outside, white inside, with yellowish-purple margin and yellowish keels. *Dorsal sepal* oblong, acute, 2.5 × 1 cm. *Lateral sepals* falcate, oblong, acute, 2.2 × 0.7 cm. *Petals* recurved, linear, acute, 2.2–2.4 × 0.1 cm. *Lip* 1.8 × 1.3 cm, 3-lobed, side-lobes in front acute, diverging, front margin entire, with acute sinus, mid-lobe quadrangular, with broad, short claw, tip acute, margin entire, recurved, sides of mid-lobe not pronounced as side-lobes, callus of 3 keels extending from base of lip to 0.8–1.1 cm from tip of mid-lobe, median keel shorter, keels plate-like with an interrupted margin. *Column* hood with acute apical margin.

DISTRIBUTION. Vietnam.

HABITAT. Not known.

ALTITUDE. Not known.

FLOWERING. Not known.

NOTES. The epithet refers to the two colours (yellowish-green and brown) of the flowers. Only the type description and an assessment of the watercolour by Eberhardt are available. There are no known specimens available for examination.

155. COELOGYNE FUSCESCENS

Coelogyne fuscescens Lindl., *Gen. Sp. Orch. Pl.* 41 (1830). Type: Nepal, Toka, *Wallich* s.n. (Wall. Cat. no. 1962) (holo. K-LINDL; iso. CAL).

var. **fuscescens. Plate 21F.**

DESCRIPTION. *Pseudobulbs* on a stout, creeping rhizome, 1–2.5 cm apart, erect, cylindric-fusiform, curved and deeply grooved with age, 8–14.5 × 1–3.2 cm. *Leaves* 2, oblanceolate to oblong-elliptic, acute, with 6 prominent nerves beneath, 15–25 × 5–10 cm with short petiole. *Inflorescence* proteranthous, rarely synanthous, peduncle erect, sheathed with 4–5 short, imbricating bracts, to 9 cm long, rachis erect or suberect, slightly zig-zag, to 8 cm long, 2- to 10-flowered, flowers opening simultaneously, floral bract deciduous. *Flowers* 6 cm across, sepals and petals pale yellow or pale yellow-green, flushed brown towards apex; lip whitish with a pale yellow-green central stripe and marked with brown, keels vermilion red in form of a flash, also narrow band of same colour at base of lip, side-lobes whitish with brown band just inside margin. *Dorsal sepal* oblong-lanceolate, acute, 5–7-nerved, 3.5–4 × 1.25–1.5cm. *Lateral sepals* oblong-lanceolate, acute, 7-nerved, 3–4 × 0.9 cm. *Petals* linear, acute, 1-nerved, 2.7–3.5 × 0.3 cm. *Lip* narrowly elliptic-oblong, acute, 4 × 1.5 cm, 3-lobed, side-lobes short, narrowly oblong, rounded, reflexed, mid-lobe broadly ovate to cordate, elliptic, membraneous, margins entire, undulate, callus of 3 fleshy, undulate keels extending from base of lip to base of mid-lobe then continuing as branched nerves which converge at tip of mid-lobe. *Column* arcuate, 2 cm long, winged, apex mucronate.

DISTRIBUTION. Nepal, Northeast India (Sikkim, Khasia Hills, Lushai Hills), Bhutan, Lower Burma (Tenasserim), China, Northeast Thailand.

HABITAT. Epiphyte on mossy branches in damp, shady hill forest and lower montane forest.

ALTITUDE. 600–1800 m.

FLOWERING. October–December.

NOTES. The epithet refers to the brownish colouring of the flowers. A very variable species both in shape and size of the lip and the colour of the flowers.

var. **brunnea** (Lindl.) Lindl., *Fol. Orch.-Coelog.* 11 (1854). Type: 'East Indies', cult. Syon Park, 1844 (holo. K-LINDL). **Plate 22A.**

Coelogyne brunnea Lindl. in *Gard. Chron.* 1848: 71 (1848).
C. cynoches C.S.P. Parish & Rchb.f. in *Trans. Linn. Soc. London* 30: 147 (1874). Type: Burma, *Parish* 195 (holo. W; iso. K).

DESCRIPTION. *Pseudobulbs* on a creeping, sheathed rhizome, 1.5–2.5 cm apart, erect, cylindric-fusiform, curved and wrinkled with age, 6–11 × 1–2 cm. *Leaves* 2, oblanceolate to oblong-elliptic, acute, with 5 prominent nerves, 22–28 × 4–6 cm with 3–4 cm long, grooved petiole. *Inflorescence* synanthous, peduncle erect, sheathed with loosely imbricating bracts at base, 3.5–6 cm long, rachis erect, becoming slender, slightly zig-zag, 13.5 cm long, 2- to 7-flowered, flowers opening in succession, maybe a 1 to 3 at a time, floral bracts deciduous. *Flowers* 3.5–5 cm across, sepals and petals pale yellow or pale yellow-green or pale yellow-brown, flushed brown towards apex; lip whitish, marked with brown on mid-lobe, side-lobes paler, keels vermilion red in form of a flash, also narrow band of same colour at base of lip. *Dorsal sepal* oblanceolate, acute, 5–7-nerved, 4 × 0.9 cm. *Lateral sepals* oblong-lanceolate, acute, 7-nerved, 3–4 × 0.6–0.9 cm. *Petals* linear-lanceolate, acute, 1-nerved, 2.7–3.8 × 0.2–0.3 cm. *Lip* narrowly elliptic-oblong, 3-lobed, 4 × 0.5 cm, side-lobes erect, oblong, front rounded but not distinct, mid-lobe ovate, tip acute, margin entire, crenulate, callus of 3 fleshy, simple, straight keels extending from base of lip to base of mid-lobe, continuing as branching nerves over mid-lobe, 3 keels continue as nerve lines and converge at tip of mid-lobe. **Column** slender, 2.2 cm long, expands to form hood.

DISTRIBUTION. Northeast India (Himalayas), Lower Burma (Tenasserim, Moulmein), China, Thailand (Doi Angka, Mae Kaw), Laos, Vietnam.

HABITAT. Epiphytic on trees in evergreen hill forest and shady valleys.

ALTITUDE. 910–1200 m.

FLOWERING. October–December.

NOTES. The distinction in flower colour between the typical variety and this one is not clear-cut.

var. **integrilabia** (Schltr.) Pfitzer in Engler, *Pflanzenr. Orch.-Mon.-Coelog.* 43 (1907). Type: Peninsular Malaysia, Penang, coll. not cited (holo. K).

C. integrilabia Schltr. in *Orchis* 9: 13 (1915).

DESCRIPTION. No information on **Pseudobulbs** or **Leaves**. **Inflorescence** synanthous, few-flowered, (no further information). **Flowers** colour could be pink. **Sepals** and **Petals** pale brown, lip sepia coloured along margins in part, keels reddish-brown. **Lip** 3 lobed, side-lobes rudimentary (no further information).

DISTRIBUTION. Nepal.

HABITAT. Not known.

ALTITUDE. Not known.

FLOWERING. Not known.

NOTES. The epithet for this variation reflects the supposedly entire form of the lip to the flower where there are virtually no side-lobes. There is very little information on this variation and it could be a pink form.

var. **viridiflora** U.C. Pradhan, *Indian Orchids: Guide Indentif. & Cult.* 2: 268 (1979). Type: India, Kalimpong, collected for *U.C. Pradhan* 56 (holo. CAL).

DESCRIPTION. **Pseudobulbs** on a stout, creeping rhizome, about 2.5 cm apart, erect, cylindric-fusiform, 8–10 × 1–2 cm. **Leaves** 2, oblanceolate to oblong-elliptic, acute, 15–28 × 5–10 cm with petiole. **Inflorescence** synanthous, peduncle and rachis erect or suberect, enclosed at base with 4–5 short bracts, 16 cm long, 2- to 10-flowered. **Flowers** 3.5–5 cm across, sepals and petals pale apple-green with no trace of brown, mid-lobe pale green with brown markings, side-lobes white with a few brown veins, keels vermilion red (brick-red) forming a flash, also narrow band of same colour at base of lip. **Sepals** oblong-lanceolate, 3–4 × 0.9 cm. **Petals** linear, acute, 2.7–3.5 × 0.3 cm. **Lip** narrowly elliptic-oblong, 4 × 1.5 cm, 3-lobed, side-lobes short, rounded, reflexed, mid-lobe broadly ovate, elliptic, margins undulate, 3 fleshy keels from base of lip.

DISTRIBUTION. Northeast India (Himalayas (Kalimpong)).

HABITAT. Epiphyte on mossy branches in hill forest and lower montane forest.

ALTITUDE. 1300 m.

FLOWERING. October.

NOTES. The epithet refers to its green flowers. Although this variety was first described in 1979, in the Lindley Library of the Royal Horticultural Society, London, there is a painting of it made my Capt. G. Bailey of the 104th Bombay Rifles which was published in his *Orchids Collected in the Lushia Hills* (17/32).

156. COELOGYNE PICTA

Coelogyne picta Schltr. in *Notizbl. Bot. Gart. Berlin* 8: 118 (1922). Type: Upper Burma, cult. Dahlem-Berlin B.G., *Hennis* s.n. (holo. B†).

DESCRIPTION. **Pseudobulbs** on a short, strong moderately thick rhizome, close together, erect, oval to ovoid, not angular, 4–4.5 × 1.8–2.2 cm. **Leaves** 2, erect-widely spreading or somewhat erect, lanceolate-linear, acute, 10–14 × 2–2.5 cm, base gradually narrowing into petiole. **Inflorescence** synanthous, peduncle bare, arcuate, 5 cm long, rachis lax few-flowered, floral bracts persistent, ovary glabrous. **Flowers** overall yellow with a sepia-brown and ochre lip. **Dorsal sepal** oblong, somewhat acute, 3 cm long. **Lateral sepals** oblique, oblong-lanceolate, acute, dorsal keel, 3 cm long. **Petals** oblique, narrowly linear, acute, reflexed, 3 cm long. **Lip** broadly oblong, about 3 cm long, concave, 3-lobed, side-lobes erect, shortly triangular, obtuse, mid-lobe and isthmus short, transversely ovate, tip slightly rounded with shallow notched ends, 1.2 cm long, in middle 1.4 cm across, callus of 3 keels clearly thickening as they extend from base of lip to middle of mid-lobe. **Column** lightly arcuate, 2.2 cm long, apex somewhat bi-lobed.

DISTRIBUTION. Upper Burma.
HABITAT. Epiphytic.
ALTITUDE. Not known.
FLOWERING. September.
NOTES. The epithet refers to the coloured form of the lip of the flower.

SECTION *OCELLATAE*

Section 20. ***Ocellatae*** Pfitzer & Kraenzlin in Engler, *Pflanzenr. Orch.-Mon.-Coelog.*: 56 (1907). Type: Type: *Coelogyne ocellata* Lindl. (= *Coelogyne punctulata* Lindl.).

Pseudobulbs close together or a very short distance apart on the rhizome, ovoid, oblong-ovoid or pyriform. Inflorescence hysteranthous, proteranthous, synanthous or heteranthous. Scape may be bare or have a few bracts on the peduncle, 1- to many-flowered, floral bracts deciduous. Lip with eye-shaped spots on the side-lobes and/or mid-lobe, dorsal sepal, lateral sepals and petals almost of equal length. Keels crenate, crenulate or with papillae, rarely undulate, conspicuous or entire. Distribution map 21.

KEY TO SPECIES IN SECTION OCELLATAE

1. Inflorescence hysteranthous 2
   Inflorescence proteranthous to synanthous 3

2. Pseudobulbs obpyriform, inflorescence peduncle enclosed with bracts to rachis, lower flowers concealed by bracts. [Flowers white, streaked dull red, 3 keels, median keel short, lateral keels terminating on the mid-lobe, crenate, thickened.] .... ***C. occulata***
   Pseudobulbs pyriform or ovoid, inflorescence bare. [Flowers with bright orange-yellow spots on side-lobes, two similar spots on mid-lobe, near base, brown veining, 3 dentate keels converge towards the base of mid-lobe then diverge and disappear on mid-lobe, further dentate, curled keels appear on outside of existing keels.] .... ***C. punctulata***

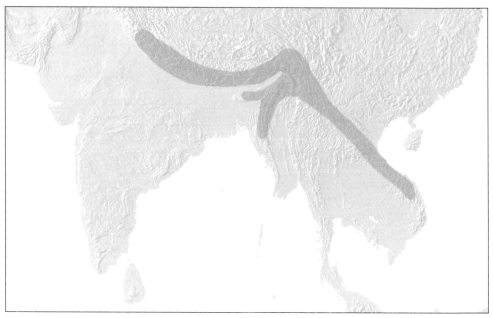

Map 21. Distribution of *Coelogyne* Sect. *Ocellatae*.

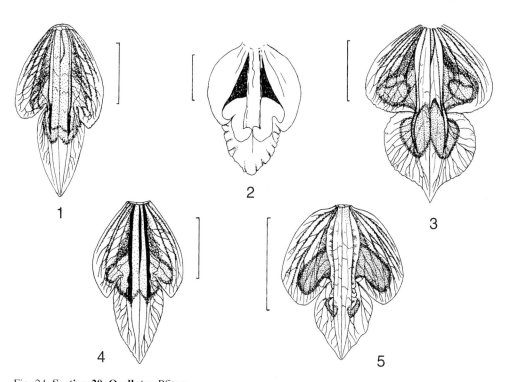

Fig. 24. **Section 20. Ocellatae** Pfitzer
1. Coelogyne corymbosa Lindl. [Type]; **2. C. hitendrae** S. Das & S.K. Jain [after Orchid Review (1978)]; **3. C. nitida** (Wall. ex D. Don) Lindl. [Type]; **4. C. occulata** Hook.f. var. **occulata** [Type]; **5. C. punctulata** Lindl. [Type]. Drawn by Linda Gurr. Scale bar = 1 cm.

3. Pseudobulbs 2.5–4 cm long **4**
   Pseudobulbs 7.5–10 cm long. [Inflorescence bare, 3- to 8-flowered. Flowers white, 4 yellow eyes on the side-lobes, 2 yellow patches bordered with red on mid-lobe, 3 keels terminating at the base of mid-lobe.] .... ***C. nitida***

4. Inflorescence synanthous with bracts enclosing the base of the peduncle, 2- to 4-flowered **5**
   Inflorescence proteranthous, 1- to 2-flowered. [Flowers white, lip with continuous golden yellow band bordered with reddish-brown on side-lobes and mid-lobe, also lip streaked reddish-brown, 3 keels, lateral keels terminate at base of mid-lobe, median keel at the middle of mid-lobe, all keels crenulate, slightly undulate.] .... ***C. hitendrae***

5. Sepals lanceolate, 2.5–3.8 × 0.9 cm, petals narrowly lanceolate, 3.3 × 0.8 cm. [Flowers white, 4 large yellow eyes bordered with orange-red on lip, 3 low keels terminating at the base of mid-lobe, crenulate.] .... ***C. corymbosa***
   Sepals subelliptic, 3.2–3.5 × 1.5 cm, petals subrhombic-elliptic, 3.2–3.5 × 1.7 cm. [Flowers cream, with brown veins on the side-lobes and deep orange keels.] .... ***C. gongshanensis***

157. COELOGYNE CORYMBOSA

Coelogyne corymbosa Lindl., *Fol. Orch.-Coelog.* 7 (1854). Type: Northeast India, Sikkim, *J.D. Hooker* 136 (holo. K; iso. CAL). **Plate 22B.**

DESCRIPTION. ***Pseudobulbs*** on a stout rhizome, clustered, developing serially, ovoid to subrhomboidal, 2.5–4 × 1.8 cm with dark brown sheathing bracts at base. ***Leaves*** 2, erect, elliptic-lanceolate, acute, towards coriaceous, with 5 nerves, 10–20 × 2–3.5 cm with 3 cm long, grooved petiole. ***Inflorescence*** synanthous or proteranthous, peduncle erect or drooping, covered in sheaths to rachis, 8–15 cm long, rachis slender, 2–3 cm long, 2- to 4-flowered, flowers opening simultaneously, floral bracts deciduous. ***Flowers*** fragrant, heavy in texture, up to 7 cm across, white with 4 large yellow eyes bordered with orange-red on lip. ***Sepals*** lanceolate, acute, with 1 prominent nerve, 2.5–3.8 × 0.9 cm. ***Petals*** narrowly lanceolate, acute, with 1 prominent nerve, 3.3 × 0.8 cm. ***Lip*** 3-lobed, 3 × 1.7 cm, side-lobes rounded, mid-lobe ovate to ovate-lanceolate, acute to acuminate, callus of 3 low keels, crenulate, extending from base of lip to base of mid-lobe. ***Column*** arcuate, 1.8 cm long, broadly winged, tip 3-lobed.

DISTRIBUTION. Nepal, Bhutan, Northeast India (Himalayas from Sikkim eastwards, Khasia Hills), China.

HABITAT. Epiphytic on moss-covered branches and tree trunks in damp, shady upper montane forest.

ALTITUDE. 1400–3000 m.

FLOWERING. April–July depending on altitude.

NOTES. The epithet refers to the clustered inflorescence. *C. corymbosa* var. *heteroglossa* Rchb.f. in *Gard. Chron.* 8 (1878) was mentioned in a letter to Gustav Mann (in pencil) dated 20th April 1901 where he discussed differences between *Bot. Mag.* t. 6955 and Pantling's no. 185 in the following terms: 'Flowers larger than *C. corymbosa* but of the same colour. Side-lobes overlap base of very broad mid-lobe, 3 membraneous, denticulate keels extend from base of mid-lobe, 2 broader brown areas occur on each side of the epichile between the side-lobes which have brown veins; each brown area ends in a yellow one and there is a 4-lobed angulate, deep yellow area on the base of the broad mid-lobe, with a dark narrow border and with short petioles'. Reichenbach asked the question—Can it be a hybrid between *C. corymbosa* and *C. brevifolia* or *C. ocellata*? Firstly, *C. brevifolia* is not a recognised name but it has been used to define a form of *C. punctulata*. Secondly, *C. ocellata* Lindl. is a synonym of *C. punctulata* Lindl. Therefore, there is the possibility of natural hybridisation since *C. corymbosa* and *C. punctulata* are sect. Ocellatae Pfitzer species and they are to be found in similar habitats in Nepal, Northeast India and Burma and their flowering times slightly overlap.

158. COELOGYNE GONGSHANENSIS

Coelogyne gongshanensis H. Li ex S.C. Chen in *Fl. Rep. Pop. Sin.* 18:345, 412 (2) (1999). Type: China, Yunnan, Gong Shan County, Dulong River, *Nan Shui Bei Diao Exp. Team* 8516 (holo. KUN).

DESCRIPTION. *Pseudobulbs* on a rather thick and short rhizome which is densely covered with brown remaining sheaths, closely spaced or nearly clustered, obovoid-oblong or subellipsoid, more or less narrowed toward the base, 1.3–1.8 × 0.7–0.9 cm, wrinkled or ridged when dried, sheathing bracts at base. *Leaves* 2, narrowly obovate-lanceolate to oblong-lanceolate, occasionally elliptic, acuminate, coriaceous, (4–) 7–13 × 1–2 (–3) cm with short petiole. *Inflorescence* synanthous, lower part of peduncle enclosed with sheaths, 8–12.5 cm long, rachis, 2- to 4-flowered, flowers opening ?simultaneously, floral bracts deciduous, pedicel and ovary 1.2–2.2 cm long. *Flowers* cream-coloured (according to the collector), side-lobes brown-veined, keels deep yellow. *Sepals* subelliptic, acute, 3.2–3.5 × 1.5 cm, lateral sepals slightly narrower than dorsal sepal. *Petals* subrhombic-elliptic, 3.3–3.5 × ca 1.7 cm. *Lip* 3-lobed, c. 2 cm long, side-lobes nearly semiorbicular, entire, mid-lobe ovate c. 0.8 × 0.6 cm, callus of 3 keels, extending from base of lip to middle of mid-lobe, median keel much shorter. *Column* ca 1.3 cm long, winged on both sides, wings ca 1.5 cm across in upper part.

DISTRIBUTION. China (Northwest Yunnan (Gong Shan County)).
HABITAT. Epiphytic in coniferous forests or on the branches of shrubs in thickets.
ALTITUDE. 2800–3200 m.
FLOWERING. May when surrounding snow has melted.
NOTES. The epiphet refers to Gong Shan County, a region of NW Yunnan, where the type specimen was found. A reference to this species was given by H. Li in *Flora Dulongjian Region*: 341 (1993) but there was no description of the species. The type specimen was collected in 1960. The Dulongjiang Expedition Team collected two herbarium specimens 5355 (KUN) and 6940 (KUN).

159. COELOGYNE HITENDRAE

Coelogyne hitendrae S. Das & S.K. Jain in *Orchid Rev.* 86(1020): 195 (1978). Type: Northeast India, Nagaland, Pulebadje, *Kataki* 60202 (holo. CAL).

DESCRIPTION. *Pseudobulbs* slightly oblique, closely packed, oblong-ellipsoid, irregularly ridged, green, 2.5–3.5 × 1.2–1.6 cm, enclosed with bracts at base. *Leaves* 2, narrowly elliptic-oblong, acute, entire, tapering towards the base, coriaceous, dark green, with 7 nerves, 8–10 ×

1.5–2.5 cm with 1–1.5 cm long, grooved petiole. *Inflorescence* proteranthous, peduncle erect, rachis up to 12 cm long, 1- to 2-flowered. *Flowers* white, mildly fragrant, side-lobes and mid-lobe of lip blotched with continuous golden yellow band bordered with reddish-brown on the throat, streaked with reddish-brown and with a patch on inner side of side-lobes, keels white. *Dorsal sepal* elliptic, acute, 3–3.2 × 1.5 cm. *Lateral sepals* oblong-lanceolate, acute, 3.3 × 1.3 cm. *Petals* oblong-elliptic, acute, entire, glabrous, subfleshy, 3 × 0.9 cm. *Lip* 3 × 2 cm, 3-lobed, subsaccate at base, sessile to column, side-lobes 1.5 × 0.9 cm, rounded at front, crenulate margins arched over the column, mid-lobe 1.3 × 1.3 cm, broadly ovate, acute, rounded, crenulate, callus of 3 keels extending from base of lip to base of mid-lobe, median keel extends to middle of mid-lobe, crenulate, undulate. *Column* arcuate, 2 cm long, hooded, winged, serrulate at tip.

DISTRIBUTION. Northeast India (Nagaland).

HABITAT. Not known.

ALTITUDE. 500–1000 m.

FLOWERING. April–May.

NOTES. This species dedicated to Professor Hitendra Kumar Borwah, an eminent Indian botanist.

160. COELOGYNE NITIDA

Coelogyne nitida (Wall. ex D. Don) Lindl., *Gen. Sp. Orch. Pl.* 40 (1830). Type: Nepal, *Wallich* s.n. (Wall. Cat. no. 1954) (holo. BM; iso. K, CAL) *pro parte*. **Plate 22C.**

Cymbidium nitidum Wall. ex D. Don, *Prod. Fl. Nepal* 3: 35 (1825), *non Cymbidium nitidum* Roxb., *Fl. Ind.* 3: 459 (1832). Type: Bangladesh, Sylhet, *M.R. Smith* s.n. (holo. BM; iso. LINN).
Coelogyne conferta Hort in *Gard. Chron.* n.s. 3: 314 (1875). Type: Burma, Moulmein, *Parish* 150 (holo. K-LINDL).
C. ochracea Lindl. in *Bot. Reg.* t. 69 (1846); Hook.f., *Fl. Brit. India* 5: 831 (1890). Type: India, Darjeeling, *Griffith* 24 (holo. K-LINDL).

DESCRIPTION. *Pseudobulbs* on a stout rhizome, somewhat clustered, 2–2.5 cm apart, erect, ovoid to conical, furrowed, 7.5–10 × 2–2.5 cm, enclosed with bracts at base. *Leaves* 1 or 2, narrowly elliptic-lanceolate, acute, with 8 nerves, 3 prominent beneath, 15–25 × 2.5–3 cm with 4–5.5 cm long, grooved petiole. *Inflorescence* proteranthous, peduncle erect or drooping, bare, 14–16 cm long, rachis slightly zig-zag, 8 cm long, 3- to 6-flowered, flowers opening simultaneously, flower bracts deciduous. *Flowers* 4 cm across, white with 4 yellow eye marks on side-lobes and 2 yellow disc bordered with red on mid-lobe. *Sepals* narrowly oblong, subacute to obtuse, 3.5 × 0.7 cm. *Petals* narrowly oblong-lanceolate, subacute, 2.5 × 0.5 cm. *Lip* 3-lobed, almost ovate, 1.9 × 1.6 cm, side-lobes oblong to rounded, mid-lobe rotund to cordate, tip rounded to somewhat acute, callus of 3 keels extending from base of lip, lateral keels extend onto base of mid-lobe. *Column* slender, slightly arcuate, 1.2 cm long, expanding into hood, margin entire to slightly erose, rounded.

DISTRIBUTION. Nepal, Bhutan, Upper Burma, China (Tibet, Yunnan (Pu Bia)), Northeast India (Himalayas, Sikkim, Khasia Hills, Lushai Hills, (including Bangladesh)), Laos, Thailand (Chiang Mai).

HABITAT. Epiphytic, sometimes lithophytic, on trees and mossy rocks in lower and upper montane forest.

ALTITUDE. 1300–2600 m.

FLOWERING. March–June.

NOTES. The epithet refers to the luminous or shining colour of the flowers. *C. conferta* Hort. could be described as *C. nitida* Lindl. var. *conferta*, see *Parish* 150 (K, LINDL) and C.S.P. Parish & Rchb.f in *Trans. Linn. Soc.* 30:146, t. 30, B6-8 (1874).

161. COELOGYNE OCCULATA

Coelogyne occulata Hook.f., *Fl. Brit. India* 5: 832 (1890). Type: India, Sikkim Himalayas, *Griffith* 5159 (holo. K).

var. **occulata**

DESCRIPTION. ***Pseudobulbs*** on a creeping rhizome, obliquely attached 1–2 cm apart, ovoid, acute at apex, narrowed towards base, 2.5–4 × 0.7–1.4 cm, green, glossy, faintly grooved. ***Leaves*** 2, elliptic, acute, with 3–5 nerves, 2.5–8 × 0.9–2.2 cm with 1 cm long, grooved petiole. ***Inflorescence*** hysteranthous, peduncle slender, erect, within pale green scales, 4 cm long, rachis not very evident with flowers half concealed by the scales on rachis, 3- to 4-flowered. ***Flowers*** 2.5 cm across, sepals and petals white, lip white with side-lobes streaked with dull red, column white with orange-yellow band in front of base. ***Sepals*** spreading, oblanceolate, acute, 2.2 × 0.6 cm. ***Petals*** narrow, acute, 2 × 0.4 cm. ***Lip*** 3-lobed, side-lobes large, erect, rounded at front with crenulate margins, mid-lobe ovate, obtuse, crenate, callus of 3 keels, a short median keel, 2 lateral keels, thick, crenate, extending from base of lip and terminating on mid-lobe. ***Column*** slender, 1.4 cm long, hood broadly winged, crenulate margins.

DISTRIBUTION. Bhutan, Northeast India (Himalayas, Sikkim), China.

HABITAT. Epiphytic on trees and lithophytic on rocks in upper montane forest.

ALTITUDE. 2000 m.

FLOWERING. April.

NOTES. The epithet refers to the eye-like markings on the side-lobes of the lip of the flowers.

var. **uniflora** N.P. Balakr. in *J. Bombay Nat. Hist. Soc.* 75(1): 159 (1978). Type: Bhutan, Nyath Forest, *Balakrishan* 43041 (holo. CAL; iso. ASSAM).

DESCRIPTION. ***Pseudobulbs*** on a creeping rhizome, obliquely attached 1–2 cm apart, oblanceolate, clavate, acute at apex, narrowed towards base, green, glossy, faintly grooved, 3–4 cm long. ***Leaves*** 2, elliptic, acute, with 3–5 nerves, 5–7.5 × 1.5–2.3 cm with 1–1.8 cm long, grooved petiole. ***Inflorescence*** synanthous or hysteranthous, peduncle sheathed within pale green scales, slender, erect, 4 cm long, rachis not very evident, 1- to 2-flowered, flowers opening simultaneously, floral bracts persistent. ***Flowers*** 2.5 cm across, sepals and petals white, lip white with reddish-brown veins, dark yellow spot bordered by reddish-brown band on each side-lobe near mid-lobe and 2 similar spots on mid-lobe, column white with orange-yellow band in front of base. ***Sepals*** spreading, elliptic-oblong, subacute, 2.2 × 0.6 cm. ***Petals*** linear-lanceolate, subacute, 2 × 0.4 cm. ***Lip*** 3-lobed, side-lobes large, erect, rounded at front crenulate margins, mid-lobe ovate, acute, reflexed, minute wave on margin, callus of 2 keels extending from base of lip and terminating abruptly at yellow spots, also a smaller, slender median keel exists. ***Column*** slender, 1.4 cm long, hood broadly winged, crenulate margins.

DISTRIBUTION. Bhutan (Nyath Forest), Burma (Mandalay), Northeast India (Darjeeling, Sikkim).

HABITAT. epiphytic on moss-covered trees and lithophytic on rocks in upper montane forest.

ALTITUDE. 1830–3500 m.

FLOWERING. April–July.

NOTES. The epithet for this variation refers to the mainly 1-flowered inflorescence.

162. COELOGYNE PUNCTULATA

Coelogyne punctulata Lindl., *Coll. Bot.* sub. t. 33 (1821). Type: Nepal, *Wallich* s.n. (holo. K). **Plate 22D.**

Cymbidium nitidum Roxb., *Hort. Bengal.* 63; *Fl. Ind.* 3: 459 (1832). Type: Bangladesh, Sylhet, *M.R. Smith* (holo. K-W).

Coelogyne ocellata Lindl., *Gen. Sp. Orch. Pl.* 40 (1830). Type: India, Khasia Hills, *Wallich* s.n. (Wall. Cat. no. 1953) (holo. K-LINDL).

C. brevifolia Lindl., *Fol. Orch.-Coelog.* 7 (1854). Type: India, Khasia Hills, *Hooker & Thomson* s.n. (holo. K, iso. CAL).

C. goweri Rchb.f. in *Gard. Chron.* 1869: 443 (1869). Type: Northeast India, Assam, *Gower* s.n. (holo. W).

C. ocellata Lindl. var. *maxima* Rchb.f. in *Gard. Chron.* ser. 2, 11: 524 (1879). Type: provenance unknown, cult. Williams (holo. W, not located).

C. ocellata Lindl. var. *boddaertiana* Rchb.f. in *Gard. Chron.* ser. 2, 18: 776 (1882). Type: provenance unknown, cult. Belgium, Ghent, B. van Cutjen s.n. (holo. W, not located).

C. nitida (Roxb.) Hook.f., *Fl. Brit. India* 5: 837 (1890), *non* (Wall. ex D. Don) Lindl., *Gen. Spec. Orch. Pl.* 40 (1830). Type: Northeast India, Khasia Hills, *Griffith* s.n. (holo. K; iso. CAL).

Pleione goweri (Rchb.f.) Kuntze, *Rev. Gen. Pl.*: 680 (1891).

Pleione nitida (Roxb.) Kuntze, *Rev. Gen. Pl.*: 680 (1891).

Coelogyne punctulata Lindl. var. *hysterantha* Tang & Wang in *Act. Phytotax. Sin.* 1(1): 79 (1951). Type: China, Yunnan, *Forrest* 26146 (holo. K).

DESCRIPTION. *Pseudobulbs* somewhat pyriform, angular, polished green, bracts persistent, 3.8–7.6 cm long. *Leaves* 2, narrowly lanceolate, acute, with 5–7 nerves, 18–25.5 cm long, narrowing towards base into 1.5–2 cm long, grooved petiole. *Inflorescence* hysteranthous, peduncle erect, bare, 7.5–9.5 cm long, rachis erect, slightly zig-zag, 3–4 cm long, 5- to more-flowered, floral bracts lanceolate, acute, reddish-brown, deciduous. *Flowers* about 2.5 cm across, sepals and petals pure white, lip white with 2 very bright orange-yellow spots on each side-lobe and 2 smaller spots of the same colour at the base of the mid-lobe, also some lateral streaks of brown on the mid-lobe, column bordered with bright orange-yellow. *Dorsal sepal* oblong-lanceolate, obtuse, 9-nerved, 2.7 × 1 cm. *Lateral sepals* oblique, oblong-lanceolate, acute, 7-nerved, 3 × 1 cm. *Petals* truncate at base, linear-oblong, obtuse, 5-nerved, 2.8 × 1 cm. *Lip* 3-lobed, side-lobes truncate, mid-lobe ovate, callus of 3 dentate keels which converge towards base of mid-lobe, then 2 lateral keels diverge over the arched part of the lip, then disappear, at this point, at each side of lateral keels are further keels which are curled and dentate. *Column* slightly arcuate, 1.9 cm long, expanding into a winged hood, tip mucronate.

DISTRIBUTION. Nepal, Bhutan, Northeast India (Sikkim, Khasia Hills), Burma, China (Yunnan).
HABITAT. Epiphytic on trees in upper montane forest.
ALTITUDE. 2200 m.
FLOWERING. March–April.
NOTES. The epithet refers to the markings on the lip of the flowers which is characteristic of all sect. *Ocellatae* Pfitzer species.

SECTION *LENTIGINOSAE*

Section 21. **Lentiginosae** Pfitzer & Kraenzlin in Engler, *Pflanzenr. Orch.-Mon.-Coelog.*: 49 (1907). Type: *Coelogyne lentiginosa* Lindl.

Pseudobulbs generally close together on the rhizome, ovoid, elliptic, fusiform or cylindric with 2 leaves, rarely 1 leaf at the apex. Inflorescence usually proteranthous to synanthous, rarely heteranthous. Emerging young leaves and peduncle enclosed in loosely imbricate or a few bracts at the base. Flowers open simultaneously, very rarely in succession, on an arcuate peduncle and rachis. The sepals and petals are of equal length and spread away from the column. The lip mid-lobe is large in relation to the size of the other flower parts. Margins of the lip are membraneous and generally undulate. The flowers are usually white with yellow or brown marking on the lip. Distribution map 22.

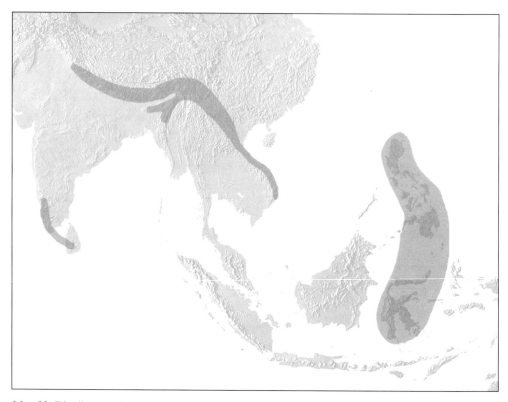

Map 22. Distribution of *Coelogyne* Sect. *Lentiginosae*.

Section Lentiginosae

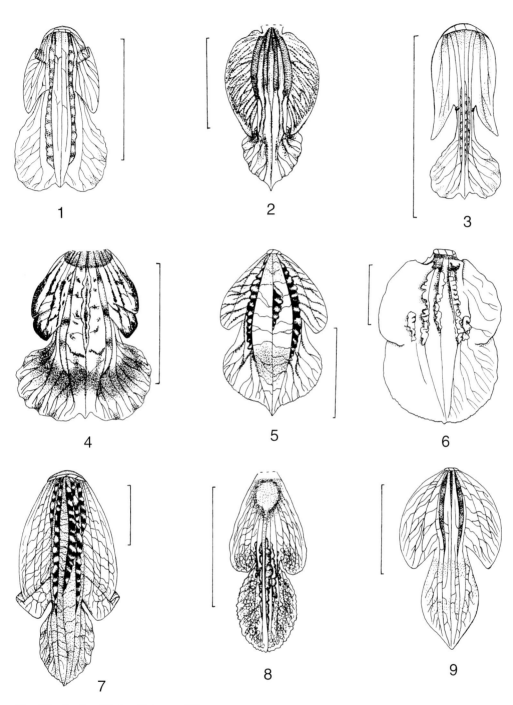

Fig. 25a. **Section 21. Lentiginosae** Pfitzer
1. Coelogyne breviscapa Lindl. [*O'Brien* s.n.]; **2. C. chloroptera** Rchb.f. [Cult. D.A. Clayton DAC 62]; **3. C. lacinulosa** J.J. Sm. [*M.J. Sands* 257]; **4. C. lentiginosa** Lindl. [Type]; **5. C. loheri** Rolfe [*A. Loher* 549]; **6. C. marmorata** Rchb.f. [Cult. Atagawa B.G.]; **7. C. merrillii** Ames [*E.D. Merrill* 6620]; **8. C. monticola** J.J. Sm. [*V. Balgooy* 3288]; **9. C. mossiae** Rolfe ex S.K. Jain & S. Das [Type]. Drawn by Linda Gurr. Scale bar = 1 cm.

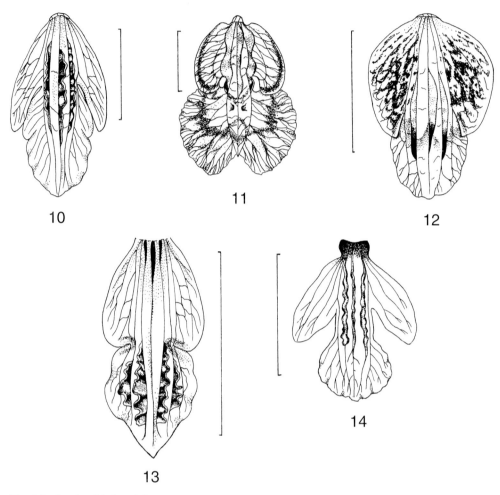

Fig. 25b. **Section 21. Lentiginosae** Pfitzer **contd.**
10. C. odoratissima Lindl. [*Wight* V41]; **11. C. schilleriana** Rchb.f. & C. Koch. [Type]; **12. C. sparsa** Rchb.f. [Cult. J. Veitch 2306]; **13. C. suaveolens** (Lindl.) Hook.f. [Type]; **14. C. zeylanica** Hook.f. [after Dassanayake & Forbery]. Drawn by Linda Gurr. Scale bar = 1 cm.

Key to species in section Lentiginosae

1. Lip with 3 or 5 keels 2
   Lip smooth and without keels. [Pseudobulbs 3–7 cm long. Inflorescence with peduncle bare, 2- to 6-flowered. Lip mid-lobe orbicular, margins undulate.] .... **C. chlorophaea**

2. Lip with 3 keels 3
   Lip with 5 keels 15

3. Inflorescence heteranthous 4
   Inflorescence proteranthous to synanthous 5

4. Pseudobulbs obovoid, 4 cm long. Mid-lobe of the lip transverse oblong, emarginate, with 3 undulate keels terminating at the base of the mid-lobe. [Sepals and petals green, lip white with brown spots on the side-lobes and mid-lobe.] .... *C. sparsa*
    Pseudobulbs ellipsoid, 5–6 cm long. Mid-lobe semi-elliptic, with 3 slightly undulate keels extending from the base of the lip to the middle of the mid-lobe. [Flowers white and yellow.] .... *C. confusa*

5. Pseudobulbs fusiform or cylindric 6
    Pseudobulbs ovoid or ellipsoid 8

6. Dorsal sepal more than 2 cm long 7
    Dorsal sepal less than 1 cm long. [Pseudobulbs more or less 3.5 cm long. Lip mid-lobe spathulate, with 3 prominent keels terminating at the tip of the mid-lobe, verrucose on the mid-lobe. Flowers small, white, lip yellow.] .... *C. lacinulosa*

7. Pseudobulbs more or less 5 cm long. Lip mid-lobe suborbicular, margins undulate, erose, with 3 keels, the lateral keels thicken and terminate two thirds onto the mid-lobe, median keel shorter. [Sepals and petals greenish-yellow, lip orange at the base, side-lobes streaked light brown.] .... *C. undatialata*
    Pseudobulbs 4 cm long. Lip mid-lobe clawed, elliptic, emarginate, with 3 keels terminating at the middle of the mid-lobe, lateral keels elevated near the base, median keel irregularly dentate .... *C. vanoverberghii*

8. Dorsal sepal more than 2 cm long 9
    Dorsal sepal less than 2 cm long 13

9. Pseudobulbs close together on the rhizome 10
    Pseudobulbs 2–3 cm apart on the rhizome. [Dorsal sepal at least 2.5 cm long. Lip mid-lobe clawed, ovate, with 3 low, undulate keels terminating at the base of the mid-lobe, median keel continues to the tip of the mid-lobe. Sepals and petals pale green, lip white marked red-brown or dark brown, orange area at centre of the mid-lobe.] ....
    .... *C. lentiginosa*

10. Lip with lateral keels extending from the base of the lip to the middle or tip of the mid-lobe, keels thickened 11
    Lip with lateral keels extending from the base of the lip to the base of the mid-lobe, keels entire. [Lip mid-lobe oblong, margins entire, with 2 keels. Flowers white, lip with yellow-brown blotches on the mid-lobe.] .... *C. mossiae*

11. Pseudobulbs ovoid, more than 2 cm long 12
    Pseudobulbs globose-ovoid, less than 2 cm long. [Mid-lobe of the lip orbicular-cordate, bi-lobed, margins undulate, with 3 smooth keels terminating at the middle of the mid-lobe, the lateral keels diverge, then parallel on the mid-lobe and become thicker. Sepals and petals greenish or ochre, lip yellow mottled brown.] .... *C. schilleriana*

12. Dorsal sepal 2 cm long. Lip mid-lobe orbicular, margins serrate, with 3 keels terminating at the tip of the mid-lobe, lateral keels form papillae on the mid-lobe ...... *C. chloroptera*

Dorsal sepal more than 3 cm long. Lip mid-lobe ovate, margins membraneous, with 3 keels terminating at the tip of the mid-lobe, median keel half as long. [Flowers yellow, lip spotted with brown 'holes'.] .... *C. taronensis*

13. Lip mid-lobe orbicular **14**

Lip mid-lobe spathulate. [Lip margins undulate, dentate, with 3 undulate keels terminating at the base of the mid-lobe, lateral keels continue to the middle of the mid-lobe. Flowers white with median yellow strip down the lip.] .... *C. odoratissima*

14. Lip mid-lobe orbicular-ovate, with 3 keels, lateral keels thickened, median keel slender. [Flowers white with yellow veins on the lip.] .... *C. breviscapa*

Lip mid-lobe orbicular, with 3 undulate keels. [Flowers white, lip with 2 ochre spots on the mid-lobe.] .... *C. zeylanica*

15. Pseudobulbs close together **16**

Pseudobulbs 3–6 cm apart. Lip mid-lobe ovate, with 5 keels; 3 crenate keels terminating at the base of the mid-lobe; 2 further keels continue to the tip of the mid-lobe. [Flowers white, keels on the lip yellow.] .... *C. suaveolens*

16. Dorsal sepal less than 2 cm long **17**
    Dorsal sepal more than 2 cm long **18**

17. Pseudobulbs more or less 5.3 cm long. Lip mid-lobe ovate, margins undulate, with 5 keels; 3 of the keels terminating at the tip of the mid-lobe, verrucose; 2 additional verrucose keels on the outside. [Flowers small, white, lip with yellow spots.] .... *C. monticola*

Pseudobulbs 4–6 cm long. Lip mid-lobe oval, margins undulate, emarginate, with 5 keels, crisped, undulate; with 3 keels extending from the base of the lip to a third way onto the mid-lobe; further 2 keels on outside but not always present. [Flowers white and yellowish-brown.] .... *C. loheri*

18. Pseudobulbs fusiform, to 10 cm long. Lip mid-lobe cordate, tip rounded, with 5 fleshy and undulate keels .... *C. marmorata*

Pseudobulbs pyriform, to 4 cm long. Lip mid-lobe oblong, tip rounded, with 5 lacerate keels .... *C. merrillii*

163. COELOGYNE BREVISCAPA

Coelogyne breviscapa Lindl., *Fol. Orch.-Coelog.* 4 (1854). Type: Sri Lanka, *Walker* s.n. (holo. K). **Plate 22E.**

Coelogyne angustifolia Wight, *Icon.* 5 (1), t. 1641 (1851) *non* A. Rich. (1841). Type: South India, Nilgiri Hills, *Wight* s.n. (holo. K).

DESCRIPTION. *Pseudobulbs* on a stout rhizome, 1 cm apart, narrowly ovoid, coriaceous, reddish-brown, 4–6.5 × 1.5 cm, enclosed with bracts at base. *Leaves* 1 or 2, narrowly lanceolate, acute, with 1 prominent dorsal nerve, 7.5–16.5 × 1.5 cm with 2 cm long, grooved petiole. *Inflorescence* proteranthous, peduncle slender, arching, rachis slender, arching, overall about 10 cm long, loosely 4- to 8-flowered, flowers opening simultaneously, floral bracts persistent. *Flowers* about 1.5 cm long, white with yellow veins on lip, hypochile with 2 deep golden-yellow lateral keels. *Dorsal sepal* oblong-lanceolate, acute, 1.5 cm long. *Lateral sepals* ovate-oblong, acute, 1.5 cm long. *Petals* linear-oblong, 1.5 cm long. *Lip* 3-lobed, side-lobes short, embracing the column, mid-lobe orbicular-ovate, shallowly notched or apiculate, callus of 3 keels extending from base of lip, lateral keels thickened, median keel slender. *Column* long, slender, winged and crenate at the apex.

Distribution. South India (Nilgiri Hills, Mysore), Sri Lanka (Central Province).

Habitat. Epiphytic on branches of trees in montane forest and lithophytic on rocks in grasslands.

Altitude. 1800–2000 m.

Flowering. March–April, December–January.

Notes. The epithet refers to the short form of the inflorescence.

164. COELOGYNE CHLOROPHAEA

Coelogyne chlorophaea Schltr. in *Repert. Sp. Nov. Regni Veg. Beih.* 10: 17 (1911). Type: Sulawesi, Minahassa, Gunung Masanang, *Schlechter* 20420 (holo. B†).

Description. *Pseudobulbs* on short, thick rhizome with thread-like roots, conical or cylindric-conical, 4-angled, smooth, 3–7 × 1–1.8 cm. *Leaves* 2, erect-spreading, elliptic-lanceolate, acuminate, smooth, 13–20 × 2–5 cm, gently narrowing into narrow petiole. *Inflorescence* synanthous, peduncle bare, 13–20 cm long, rachis erect, lax 2- to 6-flowered. *Flowers* possibly with greenish sepals and petals and similar lip with some additional colour. *Dorsal sepal* ovate, apiculate, dorsal nerve keeled, 2 cm long. *Lateral keels* oblique, lanceolate, apiculate, smooth, dorsal nerve keeled, 2 cm long. *Petals* linear, somewhat obtuse, slightly broad at base, 1.9 cm long. *Lip* broadly oval, 3-lobed, side-lobes round, minutely crenulate, mid-lobe much larger, somewhat orbicular, obtuse, margin undulate, lip totally smooth. *Column* semi-erect, smooth, slightly arcuate, 1 cm long, tip expanding into 2-lobed wings, dorsal section ending abruptly in a point.

Distribution. Sulawesi (Minahassa).

Habitat. Epiphytic in lower montane forest.

Altitude. 1200 m.

Flowering. October.

Notes. The epithet presumably refers to grey-green foliage.

165. COELOGYNE CHLOROPTERA

Coelogyne chloroptera Rchb.f. in *Gard. Chron.* ser. 2, 19: 466 (1883). Type: Philippines, Luzon, Mt. Mariveles, Lamao River, *Roebelin* s.n. (holo. W). **Plate 22F.**

Description. *Pseudobulbs* ovate-oblong or ellipsoid, somewhat compressed, 4-angled, rather glossy, 4–5 × 3 cm. *Leaves* 2, narrowly lanceolate to oblong-lanceolate, acute, with 7 nerves, coriaceous, 12–15 × 3.5–5 cm with no obvious petiole. *Inflorescence* peduncle enclosed with bracts at base, rachis arching, 15–20 cm long, loosely 4- to 8-flowered, flowers open in short sequence. *Flowers* about 2.5–2.8cm across, fragrant, waxy, sepals and petals pale yellowish-green, lip paler yellowish-green to whitish, red-brown veins on side-lobes and brownish-yellow keels on mid lobe. *Sepals* oblong, acute, dorsal keel prominent beneath, tip reflexed, 2 × 0.5 cm. *Petals* linear-elliptic, acute, tip reflexed, 1.5 × 0.1–0.2 cm. *Lip* oblong, about 1.5 cm long, 3-lobed, side-lobes erect, semi-oblong, obtuse, rounded in front, alongside column, mid-lobe nearly orbicular, acute to acuminate, margin serrate, callus of 3 keels extending from base of lip, lateral keels along the base of the side-lobes, keels dividing near the tip of mid-lobe to form papillose. *Column* short, thickened tip.

Distribution. Philippines (Luzon (Mount Mariveles, Lamao River)).
Habitat. Epiphytic on trees and lithophytic on boulders in upper montane pine forest.
Altitude. 1000–1700 m.
Flowering. November–February.
Notes. The epithet refers to the greenish flowers with a winged column.

166. COELOGYNE CONFUSA

Coelogyne confusa Ames, *Orch.* 5: 49 (1915). Type: Philippines, Camiguin de Mindanao, *Ramos* 14430 (holo. AMES).

Description. *Pseudobulbs* on a stout rhizome, angled, close together, erect, swollen in middle, 5.7 × 1 cm, no bracts at base except when developing. *Leaves* 2, elliptic-lanceolate, acuminate-acute, with 15 nerves, slightly coriaceous, 22 × 4.5 cm with rigid, 2.8 cm long, grooved petiole. *Inflorescence* heteranthous, peduncle erect 8–10 cm long, enclosed with loose imbricate bracts at base, rachis lax, 16.5 cm long, 8-flowered, flowers opening simultaneously, floral bracts persistent. *Flowers* white and yellow. *Dorsal sepal* lanceolate, acute, fleshy, 5-nerved, 2 × 0.7 cm. *Lateral sepals* lanceolate, acute, fleshy, 5-nerved, 2 × 0.7 cm. *Petals* linear-lanceolate, acute, 1.8 × 0.4 cm. *Lip* 3-lobed, side-lobes semi-oval, obtuse, mid-lobe semi-elliptic with rounded tip, margins undulate, entire, membraneous, callus of 3 keels extending from base of the lip to near centre of mid-lobe, elevated, simple, slightly undulate, hypochile minutely hairy. *Column* strongly arcuate, 1.1 cm long, margin erose.

Distribution. Philippines (Mindanao).
Habitat. Epiphytic.
Altitude. Not known.
Flowering. March–April.
Notes. The epithet refers to possible confusion with other Philippine species such as *C. sparsa* Rchb.f.

167. COELOGYNE LACINULOSA

Coelogyne lacinulosa J.J. Sm. in *Bot. Jahrb. Syst.* 65 :463 (1933). Type: Sulawesi, Central Teil, Poko Pinjan, *Kjellberg* 1497a (holo. L).

Description. *Pseudobulbs* on a stout, initially sheathed, vigorous rhizome, forming acute angle with rhizome, 1.3–1.7 cm apart, fusiform, apex attenuated, often curved, 3.5–3.8 cm long, when dried less fusiform, strongly wrinkled, dingy yellow-brown. *Leaves* 2, lanceolate, gently narrowing, acute, lightly plicate longitudinally, with 3 nerves prominent beneath, finely coriaceous, 12–13 × 0.9–1.55 cm with extremely short narrowing petiole.

Inflorescence synanthous, peduncle 4 cm long initially but thereafter 9.75–15 cm long, rachis slender, arcuate, zig-zag, 7 cm long, internodes, 0.7–0.9 cm long, 15-flowered, flowers opening, possibly simultaneously, floral bracts deciduous. *Flowers* small, white with yellow lip. *Dorsal sepal* narrowly oblong, tip shortly narrowed, obtuse, minutely apiculate, 5-nerved, 0.93 × 0.3 cm. *Lateral sepals* oblique, lanceolate, tip shortly conduplicate towards narrowing dorsal below tip, apiculate, 5-nerved, 0.9 × 0.33 cm. *Petals* oblique, lanceolate, lightly narrowed towards tip, obtuse, base shortly narrowed, 3-nerved, 0.8 × 0.2 cm. *Lip* at base shortly saccate-concave, with 5 nerves branching on outside, 3-lobed, side-lobes erect, tip distinctly free, porrect, triangular, about 0.1 cm long, mid-lobe much larger, spathulate, 0.4–0.45cm long, claw linear-oblong, 0.15 × 0.1 cm, in front rounded, with shallow notches to form 2 projecting triangular, undulate lobes which are slashed into narrow divisions, overall 2.8 × 0.36 cm, callus of 3 keels prominent at base of lip and extending from base and then becoming verrucose on mid-lobe and continuing to tip of mid-lobe. *Column* slight, base curved and twisted, 0.6 cm long, at tip broadens to form inwardly curving wings, tip 6-angled.

Distribution. Sulawesi (Central Teil, Poko Pinjan).
Habitat. Epiphytic on trees in rainforest.
Altitude. 2200–2300 m.
Flowering. May.
Notes. The epithet refers to the two projecting lobes of the mid-lobe, which are narrowly slashed. In habit, it is said to resemble *Pholidota globosa* Lindl.

168. COELOGYNE LENTIGINOSA

Coelogyne lentiginosa Lindl., *Fol. Orch.-Coelog.* 3 (1854). Type: Vietnam, Lanbian, *Sigalde* 262 (holo. K). **Plate 23A.**

Description. *Pseudobulbs* on a creeping, sheathed rhizome, 2–3 cm apart, ovoid, narrowly ellipsoid or narrowly cylindrical, glossy, light green when young, yellow with age, 3.5–9.5 × 1.2–2 cm, enclosed with bracts at base. *Leaves* 2, arcuate to suberect, very narrowly elliptic to oblanceolate, acute, with 5 nerves, 8–32 × 1.6–4 cm with 1–3.5 cm long, grooved petiole. *Inflorescence* proteranthous, peduncle erect, bare, sheathed to rachis within developing leaves, 7.5 cm long, rachis erect, 6–16 cm long, 4- to 5-flowered, flowers opening simultaneously, floral bracts persistent. *Flowers* erect, sepals and petals pale green; lip white, marked with red-brown or dark brown and with a broad orange area in centre of mid-lobe. *Dorsal sepal* lanceolate, acuminate, 5-nerved, 2.5 × 0.8 cm. *Lateral sepals* oblique, lanceolate, acuminate, 5-nerved, 2.3 × 0.7 cm. *Petals* narrowly lanceolate, acuminate, 3-nerved, 2.5 × 0.4 cm. *Lip* 3-lobed, slightly arcuate, 1.8 × 1.2 cm, side-lobes suberect, narrowly oblong, shortly rounded in front, mid lobe spreading, broadly clawed, broadly ovate below and subacute above, callus of 3 low, undulate keels extending from base of lip to base of mid-lobe with median keel continuing to tip of mid-lobe. *Column* straight, fairly broad at base, 1.5 cm long, gradually expanding into hood.

Distribution. ?Lower Burma, Vietnam, Thailand (Kanburi).
Habitat. Epiphytic on small trees and shrubs in open ground.
Altitude. 800–1000 m.
Flowering. February–March.
Notes. The epithet refers to the freckled nature of the lip to the flowers.

169. COELOGYNE LOHERI

Coelogyne loheri Rolfe in *Bull. Misc. Inf., Kew* 1908: 414 (1908). Types: Philippines, Luzon, Benguet, *Loher* 552 (syn. K) & 549 (syn. AMES).

DESCRIPTION. *Pseudobulbs* on a stout, sheathed rhizome, close together, ovoid-oblong, somewhat flattened, 4–5.5 × 1.5–2.8 cm, enclosed in fibrous, long lived bracts at base. *Leaves* 2, lanceolate-oblong, acute to acuminate, somewhat undulate, with 3–5 prominent nerves, coriaceous, 12–16.8 × 1.3–2.6 cm with 1.2–1.8 cm long, grooved petiole. *Inflorescence* synanthous peduncle enclosed in bracts and developing leaves, erect, slender, 5.5–12.3 cm long, rachis short, 3.5 cm long, 5- to 7-flowered, flowers opening simultaneously, floral bracts deciduous. *Flowers* white and yellowish-brown. *Dorsal sepal* lanceolate, acute, 7-nerved, 2.2 × 0.9 cm. *Lateral sepals* oblique, oblong-lanceolate, acuminate, 5-nerved, 2.1 × 0.6 cm. *Petals* linear, obtuse to acute, 2–2.3 × 0.25 cm. *Lip* 3-lobed, 1.8–2 × 1–1.2 cm, side-lobes oblong, obtuse, short, margins entire, mid-lobe broadly oval, 1.1 cm across, 1.1 cm long, margins undulate, tip shallowly notched, callus of 5 keels, 3 extending from base of lip to one third way onto mid-lobe, 2 further keels on outside of lateral keels extending from union of side-lobes and mid-lobe to one third way onto mid-lobe, all keels crisped and undulate. *Column* straight, stiff, 1.2 cm long, expanding into hood without prominent wings, tip crenulate.

DISTRIBUTION. Philippines (Luzon).
HABITAT. Not known.
ALTITUDE. 1700 m.
FLOWERING. March–June.

NOTES. This species named after the collector, A. Loher. On the Ames Herbarium drawing of the lip of specimen *Loher* 549 (U.S. National Herbarium), there is a note stating that there are 3 keels on the lip and the lip is not emarginate whereas the specimen *Loher* 552 (Kew), definitely has 5 keels on the lip and the margin is entire. Ames suggested that *Loher* 549 should be defined as the holotype. However, there is no obvious justification for this conclusion and it is the specimen *Loher* 552 used by Rolfe in his description of the species.

170. COELOGYNE MARMORATA

Coelogyne marmorata Rchb.f. in *Linnaea* 41: 116 (1877). Type: Philippines, Luzon, Benguet, Sablan, *W. Schultze* s.n. (holo. W; iso. PNH†). **Plate 23B.**

Coelogyne zahlbrucknerae Kraenzl. in *Repert. Spec. Nov. Regni Veg. Beih.* 17: 389 (1921). Type: provenance unknown, cult. Berlin B.G. (holo. B†).

DESCRIPTION. *Pseudobulbs* elongated, somewhat pyriform, close together, to 10 cm long. *Leaves* 2, oblong-lanceolate, acute, 40 × 4 cm, petiole concealed by closely imbricating bracts. *Inflorescence* shorter than leaves and slender, 3- to 5-flowered. *Flowers* sepals and petals apple green, lip white with brown markings, column green, edged with orange. *Dorsal sepal* oblong-lanceolate, acute, 3.3–3.7 × 1–1.6 cm. *Lateral sepals* oblong-lanceolate, acute, with a prominent dorsal nerve, 3.3 × 0.7 cm. *Petals* linear, 3-nerved, 3 × 0.17 cm. *Lip* 3-lobed, side-lobes semi-obcordate, rounded, 1.7 cm long, about 0.5 cm wide, net-like veins with numerous brown spots near veins, mid-lobe round cordate, emarginate-apiculate at the apex, margins slightly crenulate, 1.7 cm long, callus of 5 keels, 3 extending from base of lip to base of mid-lobe or a little further, 2 similar starting near side-lobes, all fleshy and undulated. ***Column*** clavate.

DISTRIBUTION. Philippines (Luzon, Camiguin Island).
HABITAT. Epiphytic, sometime lithophytic.
ALTITUDE. Not known.
FLOWERING. Not known.
NOTES. The epithet refers to the marbled, irregularly striped or veined markings on the side-lobes of the lip of the flowers.

171. COELOGYNE MERRILLII

Coelogyne merrillii Ames in *Philipp. J. Sci.* 6: 40 (1911). Type: Philippines, Luzon, Benguet, Panai, *Merrill* 6620 (holo. AMES; iso. K). **Plate 23C.**

DESCRIPTION. ***Pseudobulbs*** on a stout rhizome, close together, pyriform or ovoid-oblong, wrinkled, 4 × 2.4 cm, enclosed with bracts at base. ***Leaves*** 2, oblong-lanceolate, acuminate-acute, with 7 nerves, prominent beneath, slightly coriaceous, 8–15 × 1.6–3.6 cm with 2.2 cm long, grooved petiole. ***Inflorescence*** synanthous, peduncle bare, erect, straight, 7.7–14.5 cm long, rachis zig-zag, stout, 2.2 cm long, 2- to 4-flowered, flowers opening simultaneously, floral bracts deciduous. ***Flowers*** straw-coloured. ***Dorsal sepal*** lanceolate, acuminate, 9-nerved, 3.7 × 1.1 cm. ***Lateral sepals*** oblique, oblong, acuminate, 9-nerved, 3.4 × 1 cm. ***Petals*** linear, acute, 3-nerved, dorsal prominent, 3.5 × 0.2 cm. ***Lip*** 3-lobed, 3.7 × 2 cm, side-lobes semi-elliptic, front obtuse, 0.3 cm long, margin entire, undulate, mid-lobe oblong with rounded tip, 1.5 × 1 cm, callus of 5 keels with undulate margins, 3 keels extending from base of lip to base of mid-lobe, 2 further keels on outside of lateral keels which traverse boundary between side-lobes and mid-lobe, all keels lacerated, zig-zag margins on sections from middle of epichile to base of mid-lobe. ***Column*** stout, arcuate, 3 cm long, expanding into definitely winged hood, tip entire but 2 mucronate on each wing.

DISTRIBUTION. Philippines (Luzon).
HABITAT. Lithophytic on boulders in upper montane pine forest.
ALTITUDE. 1800 m.
FLOWERING. April–June.
NOTES. The species dedicated to the collector Elmer D. Merrill.

172. COELOGYNE MONTICOLA

Coelogyne monticola J.J. Sm. in *Bot. Jahrb. Syst.* 65: 462 (1933). Type: Sulawesi, Teil, Poko Pinjar, *Kjellberg* 1497 (holo. L).

DESCRIPTION. ***Pseudobulbs*** on a short rhizome, close together, cylindric, tip attenuate, when dry strongly longitudinally wrinkled, 5–5.5 cm long, 2 main bracts, veined, overlapping, acute, coriaceous, 5.5 cm long. ***Leaves*** 2, narrowly lanceolate, acute, tip margin curved inwards, grooved, with 5 nerves prominent beneath, papyraceous to coriaceous, 18–18.5 × 1–1.4 cm, narrowing into short, grooved petiole. ***Inflorescence*** synanthous, peduncle bare, 13.5 cm long, rachis 8.25 cm long, lax 7-flowered, flowers opening simultaneously, floral bracts deciduous. ***Flowers*** small, white, lip with yellow spots. ***Dorsal sepal*** lanceolate, narrowed in middle, tip narrowly truncate, tip subulate-apiculate, lightly curved inwards, 5-nerved, dorsal keel prominent beneath, 1.75 × 0.56 cm. *Lateral sepals* oblique, lanceolate, narrowing towards tip, shortly conduplicate, dorsal keel prominent, keel on each side of tip to form mucronate, laterally compressed tip, 5-nerved, 1.7 × 0.5 cm. *Petals* oblique, lanceolate, narrowing towards tip, somewhat acuminate, obtuse, apiculate, 3-nerved, 1.6 × 0.35 cm. *Lip* undulate in character, base hollowed, in middle thickened on both sides, folded transversally, 9 distinct nerves, column well above, 1.54 × 0.85 cm across side-lobes, 3-lobed, side-lobes small, erect, front shortened, free, extending forward, oblique, rounded, mid-lobe larger at base, narrows slightly, 0.85 × 0.625 cm, angular, ovate, obtuse, slightly undulate margin, tip minutely erose; callus of 5 keels, 3 keels extending from base of lip to tip of mid-lobe, simple, verrucose, 2 dwarf verrucose keels on outside of lateral keels. *Column* slender, arcuate, 1.125 cm long, above broadly winged, below concave, wings rounded.

DISTRIBUTION. Sulawesi.

HABITAT. Epiphytic in upper montane rain forest.

ALTITUDE. 2300 m.

FLOWERING. May.

NOTES. The epithet refers to the montane habitat of the type.

173. COELOGYNE MOSSIAE

Coelogyne mossiae Rolfe in *Bull. Misc. Inf., Kew* 1894: 156 & *Gard. Chron.* ser. 3, 15: 400 (1894); em. S.K. Jain & S. Das in *Orchid. Rev.* 86: 195–199 (1978). Type: South India, Nilgiri Hills, cult. Glasnevin B.G. (holo. K) *pro parte*. **Plate 23D**.

DESCRIPTION. *Pseudobulbs* on a creeping rhizome, clustered, narrowly ovoid to ovoid, ridged, 3–7.5 × 1.7–2 cm, enclosed with bracts at base. *Leaves* 2, linear-oblong, acute, slightly undulate, coriaceous, with 5–7 nerves, tapering towards base, 30–34 × 2.5–3.5 cm with short petiole. *Inflorescence* synanthous or proteranthous, peduncle and rachis arching, about 24 cm long, loosely 8- to 10-flowered, flowers opening simultaneously, floral bracts deciduous, ovate, acute. *Flowers* about 4–4.5 cm across, fragrant, white, with 2 yellow-brown blotches on mid-lobe. *Sepals* oblong-lanceolate, acute, entire, glabrous, 7-nerved, 2–2.5 × 0.8–1 cm. *Petals* elliptic, acute, entire, glabrous, 7-nerved, not keeled, 2–2.25 × 0.9–1.1 cm. *Lip* 1.8–2 × 1.2–1.4 cm, deeply 3-lobed, side-lobes, oblong, obtuse, entire, mid-lobe broadly elliptic, acute, entire, callus of 2 keels extending from base of lip to near base of mid-lobe, entire. *Column* slightly arcuate, about 1.6 cm long, broadly winged, slightly bi-lobed at apex.

DISTRIBUTION. South India (Nilgiri Hills, Palni (Pulney) Hills).

HABITAT. Epiphytic on trees and lithophytic on vertical limestone rock in subtropical upper montane forest.

ALTITUDE. 2300–2700 m.

FLOWERING. August.

NOTES. The description of this species was amended by Das & Jain in *Orchid Review* 86:198 (1978). They considered there had been confusion between this species and *C. glandulosa* Lindl., chiefly because of the faulty protologue and misinterpretation of the *C. mossiae* type. The characters causing the confusion were the size and shape of the leaves and the number and nature of the keels. The result was that two varieties of *C. glandulosa* were recognised (see sect. Coelogyne, *C. glandulosa* Lindl. var. *sathyanarayanae* Das & Jain). It has not been possible to confirm to my satisfaction which aspects of the *C. mossiae* type were wrongly interpreted. The type specimens of *C. mossiae* and *C. glandulosa* are at Kew and there is no difficulty in differentiating the two specimens. The type sheet for *C. mossiae* contains only a leaf and two inflorescences, which were obtained from cultivated material at RBG Glasnevin. Rolfe's *Kew Bulletin* description on the type sheet is attached to the Kew sheet together with a copy of a drawing from Fyson's *Flora of the Nilgiris*. In my judgement, the drawing does not represent *C. mossiae*.

174. COELOGYNE ODORATISSIMA

Coelogyne odoratissima Lindl., *Gen. Sp. Orch. Pl.* 41 (1830). Type: Sri Lanka, Nuera Ellia, *MacRae* s.n. (Wall. Cat. no. 1960) (holo. K; iso. CAL). **Plate 23E.**

Coelogyne trifida Rchb.f. in *Hamburger Garten-Blumenzeitung* 19: 546 (1863). Type: Burma, *Parish* s.n. (holo. W).
C. angustifolia A. Rich. in *Ann. Sci. Nat. Bot.* ser. 2, 15. 16, Pt. 6 (1841). Type: South India, Nilgiri Hills, *Perrotet* 522, 868 (syn. P).

DESCRIPTION. ***Pseudobulbs*** in clusters, ovoid to almost round, 1–2.8 × 0.5–1.5 cm, enclosed with bracts at base. ***Leaves*** 2, vary in shape, elliptic-lanceolate, acute, with 5 nerves, prominent dorsal nerve, membraneous, 2–16 × 0.6–2 cm with 0.6 cm long petiole. ***Inflorescence*** synanthous, peduncle bare, slender, erect, rachis slender, overall 5–10 cm long, loosely 2- to 5-flowered, flowers opening simultaneously, floral bracts persistent. ***Flowers*** fragrant, waxy, about 4 cm across, white with median yellow strip down the lip. ***Dorsal sepal*** oblong, acute, 5-nerved, median nerve thickened into a cusp dorsally, 1.7 × 0.7 cm. ***Lateral sepals*** oblong, acute, 5-nerved, median nerve prominent, ending in a subterminal cusp on the dorsal side, 1.8 × 0.6 cm. ***Petals*** linear-elliptic, oblique, obtuse, 3-nerved, about 1.8 cm long. ***Lip*** 1.6 × 1.4 cm, 3-lobed, side-lobes obliquely oblong, erect, obtuse, mid-lobe spathulate, undulate, dentate, callus of 3 undulate keels extending from base of lip to base of mid-lobe, lateral keels continue to nearly halfway onto mid-lobe. ***Column*** 0.8–1 cm long, winged.

DISTRIBUTION. South India (Nilgiri Hills), Sri Lanka (Nuera Ellia, Horton Plains).
HABITAT. Epiphytic in subtropical lower and upper montane forest. 1000–2700 m.
FLOWERING. March–May, December–January.
NOTES. The epithet refers to the very fragrant nature of the flowers.

175. COELOGYNE SCHILLERIANA

Coelogyne schilleriana Rchb.f. & C. Koch in *Allg. Gartenzeitung* 26: 189 (1858). Type: Burma, Moulmein, collected for Veitch (holo. W).

Pleione schilleriana (Rchb.f.) B. S. Williams, *Orch. Grow. Man.* ed. 6, 551 (1885).

DESCRIPTION. ***Pseudobulbs*** on a stout, sheathed rhizome, close together, globose-ovoid, 1.35–2 × 1.2 cm, enclosed with bracts at base. ***Leaves*** 2, elliptic-lanceolate, acute-acuminate, with 1 prominent nerve, coriaceous, 5–10 × 0.9 cm with no clearly defined petiole. ***Inflorescence*** synanthous, peduncle sheathed with new leaves and imbricating bracts at base, slender, 3 cm long, rachis short, suberect, 1.5 cm long, 1-flowered, floral bract persistent.

Flowers with green or ochreous sepals and petals, lip yellow mottled with brown. ***Dorsal sepal*** oblong-lanceolate, acute, 1-nerved, 2.5–4 × 0.8 cm. ***Lateral sepals*** lanceolate, acute, reflexed, 1-nerved, 2.5–4 × 0.8 cm. ***Petals*** narrowly linear, 2–3 cm long. ***Lip*** 3-lobed, 2.3 cm long, 1.7 cm across, side-lobes erect, oblong, short, rounded, mid-lobe large, orbicular-cordate, margin undulate, erose, distinctly bi-lobed, callus of 3 smooth keels extending from base of lip to middle of mid-lobe, lateral keels diverging then remaining roughly parallel on mid-lobe, keels thickened. ***Column*** (no detail).

DISTRIBUTION. Burma (Lower (Moulmein)), Thailand (Mek Lik).

HABITAT. Epiphytic on trees in valleys and on ridges.

ALTITUDE. 200 m.

FLOWERING. May.

NOTES. The species is dedicated to Consul Schiller who developed an important orchid collection in Hamburg. He sent many of his flowering plants to H.G. Reichenbach for identification. P.F. Hunt & Summerhayes suggested that this species be in sect. *Lentiginosae* Pfitzer because it has a considerable affinity with *C. lentiginosa* Lindl.

176. COELOGYNE SPARSA

Coelogyne sparsa Rchb.f. in *Gard. Chron.* ser. 2, 19: 306 (1883). Type: Philippines, Luzon, Banlog, cult. Sander 2867 (holo. W; iso. AMES, K).

Pleione sparsa (Rchb.f.) Kuntze in *Rev. Gen. Pl.* 2: 680 (1891).

DESCRIPTION. ***Pseudobulbs*** obovoid or narrowly obovoid to ovoid, frequently curved, 4 × 1.5 cm at base, becoming grooved with age. ***Leaves*** 2, oblong-lanceolate, acute, 8–10 × 4 cm with short petiole. ***Inflorescence*** heteranthous, peduncle sheathed from base to rachis, rachis arching, not zig-zag, about 5–7.5 cm long, loosely 3- to 5-flowered, flowers open simultaneously. ***Flowers*** about 3.5 cm across, fragrant; sepals and petals green, lip white with many irregular brown spots, brown spots on side-lobes, brown patch in front of each side-lobe, 2 brown patches at base of mid-lobe where keels terminate, median keel extends between patches, keels pale yellow, 2- to 3-intense yellow spots near base of lip, column white. ***Sepals*** oblong, acute, curved inwards, 1.5 × 0.5–0.7 cm. ***Petals*** linear-oblong, obtuse, reflexed, 1 × 0.2 cm. ***Lip*** broadly ovate, side-lobes semi-ovate, obtuse, mid-lobe transversally oblong, front emarginate or acute, callus of 3 keels extending from base of lip to base of mid-lobe, somewhat zig-zag. ***Column*** uneven tip, slightly sinuous margin, margin entire.

DISTRIBUTION. Philippines (Luzon (Lamao River, Mount Mariveles)).

HABITAT. Epiphytic on trees in lower montane forest.

ALTITUDE. 900–1100 m.

FLOWERING. June–August.

NOTES. The epiphet refers to the wide separation of the flowers on the rachis of the inflorescence.

177. COELOGYNE SUAVEOLENS

Coelogyne suaveolens (Lindl.) Hook.f., *Fl. Brit. India* 5: 832 (1890). Type: India, provenance unknown, cult. Bishop of Winchester s.n. (holo. K).

Pholidota suaveolens Lindl. in *Gard. Chron.* 1856: 372 (1856).

DESCRIPTION. ***Pseudobulbs*** on a stout, creeping, sheathed rhizome, 3–6 cm apart, ovoid, obtuse, 5–7.5 × 1.5–3 cm, enclosed with bracts at base. ***Leaves*** 2, elliptic-lanceolate, acute to acuminate, with 9 nerves, with 5 prominent below, particularly the dorsal keel, coriaceous, margins undulate, 15–23 × 5–6 cm with 1.5–3.5 cm long, grooved petiole. ***Inflorescence*** synanthous, peduncle sheathed to rachis within developing new leaves, 3–4 cm long, rachis stiff, zig-zag, 9–10 cm long, many-flowered, flowers opening in succession with 2 or 3 open together, floral bracts deciduous. ***Flowers*** with white sepals and petals, lip white with 4–6 crenate yellow keels on mid-lobe. ***Sepals*** curved inwards, oblong-lanceolate, acute, 1.4–1.8 × 0.3–0.7 cm. ***Petals*** curved inwards, oblong-ovate, somewhat acute, 1.4–1.8 × 0.3–0.7 cm. ***Lip*** 3-lobed, side-lobes rounded, short, mid-lobe broadly ovate or orbicular, callus of 5 keels, 3 crenate keels extending from base of lip to base of mid-lobe, further 2 keels on outside of initial keels, extending from base of mid-lobe to tip of mid-lobe. ***Column*** very short, 0.7 cm long, expanding into hood.

    DISTRIBUTION. Northeast India (Mungpoo, Manipur, Khasia Hills), China.
    HABITAT. Epiphytic in lower montane forest.
    ALTITUDE. 900–1500 m.
    FLOWERING. April–May.
    NOTES. The epithet refers to the fragrance of the flowers.

178. COELOGYNE TARONENSIS

Coelogyne taronensis Hand.-Mazz. in *Anz. Akad. Wiss. Wien, Math-Nat.* 59: 254 (1922). Type: China, Yunnan, between R. Salween & R. Irrawaddy, *Boeumen* 9163 (holo. WU; iso. W).

DESCRIPTION. ***Pseudobulbs*** on a short, creeping rhizome, close together, rounded at base, 2–4 cm long, longitudinally wrinkled, base enclosed with bracts which become fibrous but do not adhere. ***Leaves*** 2, linear-lanceolate, acute or obtuse, with 7 major nerves and a further 4–7 minor nerves, 7–18 cm × 1.2–3.3 cm with short, indistinct petiole. ***Inflorescence*** synanthous, peduncle enclosed in 3–4 straw-coloured bracts, peduncle to 7 cm long, rachis to 10 cm, 1- to 3-flowered, flowers opening in succession, floral bracts deciduous. ***Flowers*** mainly yellow but spotted in form of brown holes or slits. ***Dorsal sepal*** oblong-lanceolate, acute, 11-nerved, 3.5–4 × 1–1.3 cm. ***Lateral sepals*** oblique, oblong-lanceolate, acute, 11-nerved, 3.5–4 × 1–1.3 cm. ***Petals*** oblique, lanceolate, base saccate, 3.5 × 0.5–0.6 cm. ***Lip*** 3-lobed, side-lobes narrow, front rounded, mid-lobe ovate, tip acute, margins membraneous, veined, callus of 3 keels extending from base of lip, lateral keels extend to tip of mid-lobe, median keel half as long. ***Column*** slightly arcuate, middle section thickened, expanding to form winged hood, tip rounded, denticulate.

    DISTRIBUTION. China (Yunnan).
    HABITAT. Epiphytic in montane forest on pass.
    ALTITUDE. 2450–3450 m.
    FLOWERING. July.
    NOTES. The epithet refers to the Taron Gorge in Yunnan, China.

179. COELOGYNE UNDATIALATA

Coelogyne undatialata J.J. Sm. in *Bull. Jard. Bot. Buitenzorg* ser. 3, 9: 449 (1928). Type: Moluccas [Maluku], Pulau Buru, cult. Bogor, *Toxopeus* 1555 (holo. BO; iso. L).

DESCRIPTION. *Pseudobulbs* close together, elongated, 5.2 cm long. *Leaves* 2, lanceolate, gradually narrowing from base to apex, acute, with 5 nerves, firm paper-like texture, 21 × 2.3–2.5 cm with 2–2.2 cm long, grooved petiole. *Inflorescence* synanthous, peduncle bare, 4.5 cm long, rachis zig-zag, 9 cm long, 6-flowered. *Flowers* 2 cm across, sepals and petals greenish-yellow, lip orange at base, mid-lobe dark brown, side-lobes streaked light brown. *Dorsal sepal* base erect but then hooded over column, oblong, obtuse and minutely apiculate, 9-nerved, 2.4 × 0.95 cm. *Lateral sepals* spreading, angled acutely, oblique, oblong, obtuse and apiculate, falcate, strongly winged dorsal keel, 7-nerved, 2.2 × 0.675 cm. *Petals* rolled, linear, obtuse, falcate, 3-nerved, 2 × 0.175 cm. *Lip* porrect, column above, 3-lobed, side-lobes erect, 1.1 × 1.3 cm, trapezium-triangular, irregular margin, mid-lobe suborbicular, tip apiculate, undulate, erose, callus of 3 keels, winged in shape with slightly irregular margins, median keel ends at middle of mid-lobe, lateral keels gradually diverge and thicken, sulcate, terminating suddenly two thirds from base of lip. *Column* slender, curved, 1.2 cm long, hood narrowly winged, winged apex 3-lobed, side-lobes shorter, recurved, erose, mid-lobe obtuse, reflexed with 6 teeth.

DISTRIBUTION. Maluku (Buru).
HABITAT. Not known.
ALTITUDE. Not known.
FLOWERING. October.
NOTES. The epiphet refers to the undulate margins of the mid-lobe of the lip.

180. COELOGYNE VANOVERBERGHII

Coelogyne vanoverberghii Ames, *Orch.* 5: 53 (1915). Type: Philippines, Luzon, Bantog, Bauko, *Vanoverbergh* 2865 (holo. AMES).

DESCRIPTION. *Pseudobulbs* on a slender rhizome, close together, fusiform, parallel-sided, about 4 × 1 cm, base initially enclosed with bracts which become fibrous. *Leaf* 1, broadly elliptic, tapering to acute, with 7 nerves, 9–15 × 4.5–5.5 cm with a short 1.3 cm long, channelled petiole. *Inflorescence* synanthous, peduncle slender, erect with base enclosed with loosely imbricate bracts, 9.5 cm long, rachis slender, pendulous, slightly zig-zag, about 6-flowered, flowers opening simultaneously, floral bracts deciduous. *Flower* colour not known. *Dorsal sepal* ovate-lanceolate, acute, lightly keeled beneath, 2 × 0.8 cm. *Lateral sepals* oblong-lanceolate, acute, strongly keeled beneath, 2 × 0.6 cm. *Petals* linear-triangulate, acute, 3 nerves, 2 × 0.4 cm. *Lip* 3-lobed, 2 cm long, side-lobes very prominent, ovate-obtuse, 0.3 cm long from tip to sinus with mid-lobe which is short and clawed, elliptic, emarginate, callus of 3 keels, lateral keels elevated at base of lip and extending to middle of mid-lobe, median keel also elevated and irregularly dentate. *Column* slender, arcuate, 1.5 cm long, expanding into winged hood, tip slightly erose, rounded.

DISTRIBUTION. Philippines (Luzon).
HABITAT. Not known.

ALTITUDE. 1700 m.
FLOWERING. Not known.
NOTES. The species is dedicated to Father M. Vanoverbergh, the collector of the type specimen.

181. COELOGYNE ZEYLANICA

Coelogyne zeylanica Hook.f. in Trimen, *Fl. Ceylon* 4: 161 (1898). Type: Sri Lanka, Ambagamuwa, *Thwaites* 4003 (holo. K).

DESCRIPTION. ***Pseudobulbs*** pyriform or ovoid, green, somewhat wrinkled, fibrous and reddish-brown at the base with the decaying remnants of old sheaths. ***Leaves*** 1 or 2, narrowly linear-lanceolate, acute, narrowing at the base into a petiole, coriaceous, keeled, 7–12 × 0.7 cm, narrowing into a petiole. ***Inflorescence*** synanthous, peduncle with bracts at the base, slender, erect, rachis drooping, 2.5–3.7 cm long overall, 1- to 2-flowered. ***Flowers*** white with two ochre spots on the middle of the lip. ***Dorsal sepal*** oblong, subacute, 5-nerved, 1.2–1.8 × 0.7 cm. ***Lateral sepals*** ovate-oblong, 5-nerved, 1.2–1.8 × 0.45 cm. ***Petals*** narrowly linear, 3-nerved, 1.65 × 0.1 cm. ***Lip*** 1.8 cm long, 3-lobed, recurved, side-lobes oblong, obtuse, mid-lobe 0.8 × 0.7 cm, orbicular, apiculate, callus of 3 wavy keels extending from near the base of the lip to just onto the mid-lobe, median keel slightly shorter at both ends. ***Column*** 1 cm long, slightly bent forward.
DISTRIBUTION. ?South India (Nilgiri Hills), Sri Lanka (Ambagamuwa, Maskeliya).
HABITAT. Epiphytic on forest trees.
ALTITUDE. Not known.
FLOWERING. October–November.
NOTES. The epiphet refers to the island of Ceylon (Sri Lanka) where the type specimen was collected. In a revision of the genus *Panisea* by I. D. Lund (1987), *C. uniflora* Lindl. is treated as a synonym of *Panisea uniflora* Lindl. More significantly, Joseph (1982), in *Orchids of the Nilgiris* 115, Fig. 100a–100c, indicates that *C. uniflora* auct. non Lindl. should be considered an excluded or misinterpreted *Coelogyne* sp. In a later edition of *Orchids of the Nilgiris* in 1987 he repeats this. In my judgement the illustrations used by Joseph closely resemble *C. zeylanica* Hook. f. In addition, the species has been reported from the Nilgiri Hills in South India.

SECTION *FLACCIDAE*

Section 22. ***Flaccidae*** Lindl., *Fol. Orch.-Coelog.* 10 (1854). Type: *Coelogyne flaccida* Lindl. Section *Carinatae* Pfitzer & Kraenzlin in Engler, *Pflanzenr. Orch.-Mon.-Coelog.*: 43 (1907). Type not selected.

Pseudobulbs generally close together of the rhizome, ovoid, sometimes cylindric-oblong, grooved with age. Inflorescence heteranthous or proteranthous to synanthous. Scape with a few bracts at the base of the peduncle, up to about 10-flowered, flowers opening simultaneously, floral bracts generally deciduous. Lip with marks and blotches but no eye-shaped spots on side-lobes and/or mid-lobe, dorsal sepal, lateral sepals and petals almost of equal length. The keels are generally undulate. Distribution map 23.

THE GENUS COELOGYNE

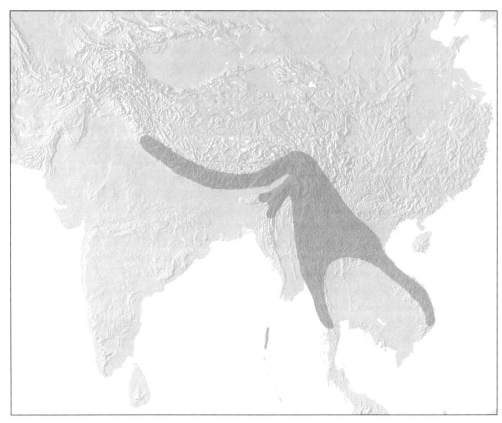

Map 23. Distribution of *Coelogyne* Sect. *Flaccidae*.

KEY TO SPECIES IN SECTION FLACCIDAE

1. Inflorescence heteranthous **2**
   Inflorescence proteranthous to synanthous **3**

2. Inflorescence with arching peduncle and rachis. Lip mid-lobe ovate-lanceolate, crenate at the sinuses, with 3 undulate keels terminating at the base of the mid-lobe. [Flowers (creamy)-white, side-lobes of the lip inside reddish-brown veins, mid-lobe yellow in the middle, red spot near base of the mid-lobe, keel white except reddish-brown near the tip of the mid-lobe.] .... ***C. flaccida***
   Inflorescence with arching peduncle and rachis, rachis slender. Lip mid-lobe ovate, tip sharply acute, not crenulate at the sinuses, with 3 undulate keels terminating at the base of the mid-lobe, median keel simple, straight and shorter. [Flowers white, side-lobes of the lip inside brownish veins, mid-lobe citron yellow and brown marks.] ....
   .... ***C. huettneriana***

3. Inflorescence with arching peduncle and rachis **4**
   Inflorescence with erect peduncle fully enclosed in bracts of young, emerging leaves **6**

4. Floral bracts deciduous **5**

SECTION FLACCIDAE

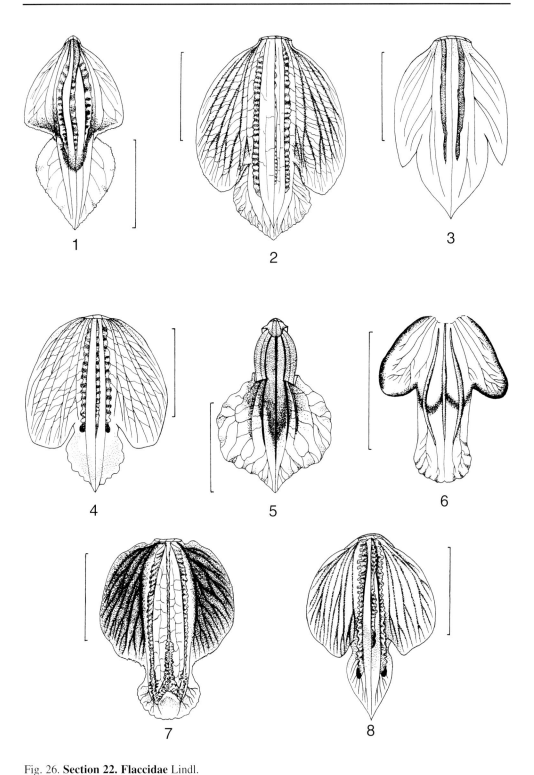

Fig. 26. **Section 22. Flaccidae** Lindl.
1. Coelogyne albolutea Rolfe [Type]; **2. C. flaccida** Lindl. [cult. D.A. Clayton DAC 21]; **3. C. hajrae** Phukan [after Orchid Review (1997)]; **4. C. huettneriana** Rchb f. [Cult. Glasnevin B.G. (1891)]; **5. C. integerrima** Ames [*A. Loher* 550A]; **6. C. quadratiloba** Gagnep. [after Seidenfaden]; **7. C. trinervis** Lindl. [Cult. D.A. Clayton DAC 41]; **8. C. viscosa** Rchb.f., Otto & Dietr. [Cult. D.A. Clayton DAC 45]. Drawn by Linda Gurr. Scale bar = 1 cm.

Floral bracts persistent. Lip mid-lobe ovate, margins undulate, with 3 strongly undulate keels terminating at the base of the mid-lobe. [Sepals and petals white, greater part of the side-lobe of the lip yellow and the mid-lobe a paler yellow, keels white except where they are on the yellow sections of the lip.] .... *C. albolutea*

5. Lip mid-lobe (sub) orbicular, tip obtuse and apiculate, with 3 slightly verrucose, parallel keels terminating at the base of the mid-lobe, median keel shorter .... *C. esquirolei*
Lip mid-lobe broadly ovate, tip acuminate, entire at the sinuses, with 2 keels, initially thick then narrower on the mid-lobe. [Flowers white, streaked dark brown, keels maroon coloured.] .... *C. hajrae*

6. Leaves lanceolate or lanceolate-linear, prominently nerved 7
Leaves linear, finely nerved. [Lip mid-lobe lip initially with parallel sides then expands, 3 crenate keels terminating at about one third onto the mid-lobe. Flowers white and orange-yellow at the front of the mid-lobe of the lip and the tips of the side-lobes.] .... *C. viscosa*

7. Leaves lanceolate-linear or lanceolate-elliptic, with 3 prominent nerves beneath. Lip distinctly 3-lobed 8
Leaves lanceolate, 3–5 nerved. Lip with no defined side-lobes. [Lip oblanceolate, tip acuminate, 3 keels terminating at the tip of the mid-lobe, lateral keels more prominent near the middle of the mid-lobe.] .... *C. integerrima*

8. Leaves lanceolate-linear, with 3 very prominent nerves. Lip distinctly 3-lobed, side-lobes semi-orbicular, mid-lobe oblong-quadrate, margins undulate, with 3 undulate keels terminating at the tip of the mid-lobe, lateral keels more prominent on the mid-lobe, 2 additional tuberculae keels on the mid-lobe .... *C. trinervis*
Leaves lanceolate-elliptic, with 3 prominent nerves. Lip mid-lobe square, with 3 smooth, elevated and slightly undulate keels, terminating well onto mid-lobe, further 2 keel-like elements on outside of lateral keels, near base of lip only .... *C. quadratiloba*

182. COELOGYNE ALBOLUTEA

Coelogyne albolutea Rolfe in *Bull. Misc. Inf., Kew* 1908: 414 (1908). Type: Northeast India, Himalayas, cult. Glasnevin B.G. (holo. K). **Plate 23F.**

DESCRIPTION. **Pseudobulbs** on a sheathed rhizome, ovoid-oblong, grooved, 8 cm long, enclosed with bracts. **Leaves** 2, lanceolate-elliptic, acute, coriaceous, 18–28 × 4–5 cm with short petiole. **Inflorescence** synanthous or proteranthous, peduncle arching, covered with overlapping bracts and emerging leaves, rachis arching, overall 15–18 cm long, 5- to 7-flowered, flowers opening simultaneously, floral bracts lanceolate-oblong, acute, 2.5–3.5 cm long, persistent. **Flowers** fragrant, pure white with greater part of side-lobes yellow, base of mid-lobe rather lighter yellow, keels white except at apex where they extend into yellow area. **Dorsal sepal** somewhat spreading, oblong-lanceolate, acute, prominent dorsal slightly fleshy, 2.5–3 × 0.8 cm. **Lateral sepals** oblong-lanceolate, acute, 1-nerved, slightly fleshy, 3 × 0.8 cm. **Petals** lanceolate, acute, with branching nerves, 2.7 × 0.5 cm. **Lip** 3-lobed, 2 cm long, side-lobes erect, oblong, obtuse, margin entire, mid-lobe ovate, acute, convex, margin undulate, callus of 3 keels extending from base of lip to base of mid-lobe, strongly undulate. **Column** straight, clavate, 1.3 cm long, tip bi-lobed.

DISTRIBUTION. Northeast India (Himalayas).
HABITAT. Not known.
ALTITUDE. Not known.
FLOWERING. May.
Notes. The epithet refers to the white and yellow colouring of the flowers. Described as a distinct, handsome species, introduced by Messrs. Sander & Sons and flowered in their nursery in May 1896 and subsequently at Glasnevin B.G. The species is only known from the type and it is yet to be rediscovered. If the species is known to be growing in its natural habitat, then some material should be compared with the type.

183. COELOGYNE ESQUIROLEI

Coelogyne esquirolei Schltr. in *Repert. Sp. Nov. Regni Veg. Beih.* 4: 60 (1919), (as *esquirolii*). Type: China, Guizhou, *Esquirol* s.n. (holo. B†).

DESCRIPTION. Pseudobulbs on a strong, compact rhizome, close together, cylindric-oblong, longitudinally grooved, 5–6 × 1.8 cm. **Leaves** 2, erect-spreading, lanceolate, acuminate, overall 20 × 2.5–3 cm with petiole. **Inflorescence** proteranthous, peduncle and rachis pendulous, to 15 cm long, lax 4- to 8-flowered, flowers open simultaneously, floral bracts deciduous, ovary glabrous. **Flowers** generally white. **Dorsal sepals** oblong, somewhat acute, glabrous, about 1.7 cm long. **Lateral sepals** oblique, somewhat acuminate, prominent keel beneath, about 1.7 cm long. **Petals** oblique, narrowly linear, obtuse, narrows gradually towards base, about 1.7 cm long. **Lip** 1.3 × 1.1 cm, 3-lobed, side-lobes rounded, strongly obtuse, mid-lobe somewhat orbicular, obtusely apiculate, callus of 3 slightly verrucose keels, parallel and extending from base of lip to middle of mid-lobe, median keel slightly shorter than lateral keels. **Column** slightly arcuate, slender, glabrous, about 1 cm long.

DISTRIBUTION. China (Guizhou).
HABITAT. Epiphytic.
ALTITUDE. Not known.
FLOWERING. March.
NOTES. Similar to *C. flaccida* Lindl. but differs in the shorter form of the lip.

184. COELOGYNE FLACCIDA

Coelogyne flaccida Lindl., *Gen. Sp. Orch. Pl.* 39 (1830). Type: Nepal, Toka, *Wallich* s.n. (Wall. Cat. no. 1961) (holo. K; iso. CAL). **Plate 24A.**

Coelogyne lactea Rchb.f. in *Gard. Chron.* ser. 2, 23: 692 (1885). Type: provenance unknown, cult. J. Day (holo. W 42299).

DESCRIPTION. Pseudobulbs on a stout, sheathed rhizome, close together, conical to ovoid-cylindric, lightly grooved, 3.5–12 × 1.3–4 cm, bracts at base. **Leaves** 2, oblong-elliptic to lanceolate, acute to acuminate, with 3 prominent nerves, 5–20 × 1–5 cm with 2.5–4 cm long, grooved petiole. *Inflorescence* heteranthous, peduncle sheathed at base, to 7 cm long, rachis somewhat zig-zag, pendulous, 20 cm long, 5- to 12-flowered, flowers opening simultaneously, floral bracts deciduous. *Flowers* 3–5 cm across, creamy-white with yellow on middle of lip, veins marked reddish-brown on inside of side-lobes and spotted red at base of mid-lobe, keels white away from tip, reddish-brown near tip. *Sepals* lanceolate, acute, 1.5–2.6 × 0.4 cm. *Petals* linear, acute, reflexed, 1.5–2.3 × 0.2 cm. *Lip* 3-lobed, side-lobes oblong, erect, rounded, mid-lobe reflexed, ovate-lanceolate, acute, callus of 3 undulate keels from base of lip to base of mid-lobe, flexuous. *Column* straight, 1.5 cm long, expands into winged hood, tip slightly bi-lobed.

DISTRIBUTION. Nepal, Northeast India (Himalayas, Sikkim, Khasia Hills), Burma, China (Yunnan (Chiengbwang, Muang Cha, Tengyueh)), Laos, Vietnam, Thailand.

HABITAT. epiphytic on trees in lower and upper montane forest and by streams, lithophytic on rocks in open spaces.

ALTITUDE. 900–2300 m.

FLOWERING. February–May, depending on altitude.

NOTES. The epithet refers to the pendulous nature of the inflorescence. Seidenfaden considered that *C. flaccida* may prove to be merely a variety of *C. lactea* differing only in having a thinner pendulous inflorescence with more flowers. Comparative examination of a number of examples of *C. flaccida* and *C. lactea* indicates they are conspecific and *C. flaccida* Lindl. takes precedence.

185. COELOGYNE HAJRAE

Coelogyne hajrae Phukan in *Orchid Rev.* 105 (1214): 94–95 (1997). Type: North-east India, Arunachal Pradesh, Tirap, Namdapha, *B.K. Shukla* 87943 (holo. CAL).

DESCRIPTION. *Pseudobulbs* on a stout rhizome, oblong ovoid, ridged, slightly curved, 6–9 × 0.8–1.3 cm, sheathed at base; sheaths purplish-brown, scarious, the uppermost about 6–9 cm long. *Leaves* 2, narrowly lanceolate, acuminate, with 7 nerves, coriaceous, 17–20 × 2.8–4.2 cm with 6–7 cm long petiole. *Inflorescence* proteranthous, peduncle and rachis 18–22 cm long, laxly 9- to 10-flowered, flowers opening ?simultaneously, floral bracts deciduous. *Flowers* 2.5–3 cm across, white, streaked dark brown, keels maroon-coloured. *Dorsal sepal* oblong-lanceolate, acuminate, 7-nerved, 2.5 × 0.7 cm. *Lateral sepals* oblique, oblong-lanceolate, acuminate, 7-nerved, 2.8 × 0.6 cm. *Petals* oblong-lanceolate, acuminate, 5-nerved, 2 × 0.5 cm. *Lip* 2.2 × 1.3 cm, 3-lobed, entire at sinuses; side-lobes acute or subacute, mid-lobe broadly ovate, acuminate, callus of 2 keels extending from base of lip to middle of mid-lobe, keels initially thick but become narrower on mid-lobe. *Column* 1 cm long, anther 0.3 cm, 2-chambered, pollinia 4 in two pairs.

DISTRIBUTION. Northeast India (Himalayas).

HABITAT. Epiphytic in lower montane forest.

ALTITUDE. 850 m.

FLOWERING. April.

NOTES. This species is dedicated to Dr. P.K. Hajra, Director, Botanical Survey of India. It is closely allied to *C. flaccida* Lindl. but differs in having acuminate sepals, petals and an acuminate lip, the side-lobes are acute or subacute, sinuses entire, and a callus of 2 thick keels unlike *C. flaccida* which has a callus of 3 keels.

186. COELOGYNE HUETTNERIANA

Coelogyne huettneriana Rchb f. in *Flora* 55: 277 (1872). Type: Burma, Tenasserim, *Parish* s.n. (holo. W; iso. K). **Plate 24B.**

DESCRIPTION. *Pseudobulbs* on rhizome, close together, oblong-ovate, 4–10 × 2–2.5 cm, enclosed with bracts at base. *Leaves* 2, elliptic, acuminate, with 5–7 nerves, 10.5–14 × 1.9–3.2 cm with 1.3–2 cm long, grooved petiole. *Inflorescence* heteranthous, peduncle loosely sheathed with bracts, arching, to 10 cm long, rachis slender, arcuate, 4 cm long, 4- to 8-flowered, flowers opening simultaneously, floral bracts deciduous. *Flowers* musk-scented, about 5 cm across, white, lip with citron-yellow and brown on mid-lobe, brownish striations on side-lobes. *Dorsal sepal* lanceolate, acute, with 1 prominent nerve, 2.2 × 0.2 cm. *Lateral sepals* lanceolate, acute, with 1 prominent nerve, 2.1 × 0.2 cm. *Petals* narrowly linear, 2 cm long. *Lip* 3-lobed, side-lobes elliptic, front acute, branched nerves, mid-lobe ovate, tip sharply acute, callus of 3 keels extending from base of lip, lateral keels undulate and continue to base of mid-lobe, median keel simple, straight, shorter than lateral keels. *Column* slender, slightly arcuate, 1.4 cm long, expands into slightly winged hood, tip bilobed.

DISTRIBUTION. Lower Burma, Thailand.

HABITAT. Epiphytic on trees and lithophytic on rocks in open spaces in evergreen forest.

ALTITUDE. 1100–1200 m.

FLOWERING. May–June.

NOTES. The name *C. huettneriana* Rchb.f. var. *lactea* (Rchb.f.) Pfitzer reflects the confusion that occurs between *C. flaccida* Lindl. and *C. huettneriana* Rchb.f. In my judgement they are two distinct species. See notes under *C. flaccida* above.

187. COELOGYNE INTEGERRIMA

Coelogyne integerrima Ames in *Philipp. J. Sci.* 4: 665 (1910). Type: Philippines, Luzon, Benguet, Mt. Anlog, *Merrill* 6350 (holo. AMES; iso. K).

DESCRIPTION. *Pseudobulbs* on a stout, sheathed rhizome, close together, ovoid, 4.6–7.4 × 1.5–3.2 cm, enclosed with fibrous bracts at base. *Leaves* 2, lanceolate, acuminate, with 3–7 main nerves, 17–20 × 3.5–6.4 cm with 1.8–3.2 cm long, grooved petiole. *Inflorescence* synanthous, peduncle enclosed at base with developing leaves and bracts, otherwise bare, slender, nearly erect, 10–18 cm long, rachis slender, erect to pendulous, not zig-zag, 15–16.2 cm long, 7- to 9-flowered, flowers opening simultaneously, floral bracts deciduous. *Flowers* greenish-yellow. *Dorsal sepal* elliptic-lanceolate, acute, 7-nerved, 2.9 × 1.5 cm. *Lateral sepals* slightly oblique, lanceolate, acute, 5-nerved, 2.4 × 0.95 cm. *Petals* linear-lanceolate, acute, 3-nerved, 2.5 × 0.5 cm. *Lip* without defined side-lobes, overall oblanceolate, 2.4 × 1 cm, acuminate tip, margins entire, callus of 3 keels extending from base of lip to tip of lip, smooth, elevated sections, lateral keels more prominent near middle of lip. *Column* slender, arcuate, 1.1 cm long, expanding into winged hood, margin erose.

DISTRIBUTION. Philippines (Luzon).

HABITAT. Epiphytic in mossy upper montane forest.

ALTITUDE. 2200 m.

FLOWERING. May–June.

NOTES. The epithet refers to the entire lip of the flowers.

188. COELOGYNE QUADRATILOBA

Coelogyne quadratiloba Gagnep. in *Bull. Mus. Hist. Nat. Paris* sér. 2, 22: 507 (1950). Type: Vietnam, without exact loc., cult. Hanoi, *Deschamps* s.n. (holo. P).

C. thailandica Seidenf. in *Dansk Bot. Ark.* 29(4): 46 (1975). Type: Thailand, Kanburi, *Kerr* 260 (holo. K).

DESCRIPTION. *Pseudobulbs* ovoid, 4-angled, yellowish-green, 4–5 × 2–2.5 cm. *Leaves* 2, lanceolate, acute-acuminate, with 7 major nerves, 12–13 × 3–3.5 cm with 2 cm long petiole. *Inflorescence* hysteranthous, peduncle slender, nearly enclosed with sheathing bracts, 3–4 cm long, rachis slender, 3–4 cm long, 3-flowered, flowers opening simultaneously, floral bracts deciduous. *Flowers* with transparent white sepals, lip white with brown margins on side-lobes and an inverted Y-shaped marking on mid-lobe, column yellow with orange stripe at base of lip. *Dorsal sepal* lanceolate, acute, 7-nerved, 2–2.2 × 0.7–0.8 cm. *Lateral sepals* oblique, lanceolate, acute, 7-nerved, 2 × 0.7–0.8 cm. *Petals* linear, acuminate, 3-nerved, 1.8–2 × 0.15 cm. *Lip* 1.7 cm long, 3-lobed, side-lobes broad, obtuse, rounded at front, mid-lobe rather square, 0.8 × 0.5 cm, callus of 5 keels, 3 extending from base of lip to half or two thirds the way onto mid-lobe, convergent, median keel terminates near base of mid-lobe, 2 further keel-like elements near base of lip on outside of lateral keels, keels slightly undulate and elevated on mid-lobe. *Column* slender, straight, 1.1 cm long, expanding into winged hood, tip erose.

DISTRIBUTION. ?Andaman Islands (*C. thailandica* Seidenf.), Vietnam, Thailand (Khao Yai, Haew Suwat, Ta Khanum, Kanburi).

HABITAT. Epiphytic on trees in evergreen, hill forest, near streams.

ALTITUDE. 400–700 m.

FLOWERING. January–February.

NOTES. The epithet refers to the shape of the lip of the flower. In a note concerning *C. thailandica* Seidenf., the author suggested the following—"So far as is known the species is endemic to Thailand. The collection from the Andaman Islands is of phytogeographic interest as it extends the distribution of this species to the Indian flora. If this is so, then the species probably exists in Burma and a search needs to be made. This is the only (first) record of the genus from the Andaman and Nicobar islands."

189. COELOGYNE TRINERVIS

Coelogyne trinervis Lindl., *Gen. Sp. Orch. Pl.* 41 (1830). Type: Burma, Tenasserim, Tavoy, *Wallich* s.n. (Wall. Cat. no. 1955) (holo. K-LINDL). **Plate 24C.**

Coelogyne cinnamomea Lindl. in *Gard. Chron.* 1858: 37 (1858); Teijsm & Binn. in *Natuurk. Tijdschr. Ned.-Indie* 24: 360 (1862).Type: provenance unknown (holo. K-LINDL).

C. rhodeana Rchb.f. in *Gard. Chron.* 1867: 901 (1867). Type: Maluku, location unknown, cult. Schiller s.n. (holo. W).

C. rossiana Rchb.f. in *Gard. Chron.* ser. 2, 22: 808 (1884)). Type: Burma, location unknown, *Ross* s.n. (holo. W).

C. angustifolia Ridl. in *J. Linn. Soc.* 32: 322 (1896). Type: Peninsular Malaysia, Pulau Langkawi, *Ridley* s.n. (holo. SING).

C. pachybulbon Ridl. in *J. Linn. Soc.* 32: 324 (1896). Type: Thailand, Punga, Lower Siam, *Curtis* s.n. (holo. SING).

C. wettsteiniana Schltr. in *Osterr. Bot. Z.* 69: 124 (1920). Type: Vietnam, cult. Vienna (holo. WU).

C. stenophylla Ridl. in *Fl. Mal. Penins.* 4: 132 (1924). Type: Peninsular Malaysia, Lankawi, *Curtis* s.n. (holo. K).

DESCRIPTION. *Pseudobulbs* on a prostrate, thick, densely sheathed rhizome, 1–3.5 cm apart, oblique, ovoid, longitudinally 4-grooved, 5–8 cm long, in middle 1.7–3 cm dia., enclosed with bracts at base. *Leaves* 2, erect-spreading to suberect, linear-lanceolate, acute, coriaceous, smooth, with 3 prominent nerves, 34–45 × 2.4–4.5 cm with 9–12 cm long, grooved petiole. *Inflorescence* proteranthous, maybe hysteranthous, peduncle erect, bare, in sections, sheathed at base, to 30 cm long, rachis slender, very slightly zig-zag, 9 cm long, 4- to 8-flowered, flowers open simultaneously, floral bracts deciduous. *Flowers* about 4 cm across, widely open, sepals and petals white, side-lobes with reddish-brown markings. *Dorsal sepal* lanceolate, acute-acuminate, 7-nerved, dorsal keel, base concave, 2 × 0.6 cm. *Lateral sepals* oblique, lanceolate, acuminate, 7-nerved, dorsal keel, 1.9 cm × 0.4 cm. *Petals* oblique, linear-lanceolate, obtuse, 5-nerved, 2 × 0.25 cm. *Lip* ovate, base concave, 1.8 cm long, 1.2 cm across, 3-lobed, side-lobes semi-oblong, front oblique, obtuse, not prominent, mid-lobe in front oblong-quadrate, obtuse, margins at sides slightly undulate, ovate, acute apex curled under, margins erose, callus of 3 undulate keels extending from base of lip to tip of mid-lobe, lateral keels more prominent than median keel, on mid-lobe there are 2 additional short tuberculate keels. *Column* slender, semi terete, lightly curved, 1.4–1.6 cm long, expanding into hood with wings, tip truncated, rounded with tiny mucronate at each end.

DISTRIBUTION. Burma, Cambodia, Laos, Vietnam (Da Lat), Thailand (Khao Yai Nat. Pk.), Peninsular Malaysia (Kelantan, Pahang, Pinang), Sumatra, Java (West & Central), Flores.

HABITAT. Epiphytic on tree trunks in lowland rain forest and sometimes lithophytic on rocks.

ALTITUDE. 100–1600 m.

FLOWERING. September–October.

NOTES. The epiphet refers to the three very prominent nerves on the leaves.

190. COELOGYNE VISCOSA

Coelogyne viscosa Rchb.f. in Otto & Dietr., *Allg. Gartenzeitung* 24, 28: 218 (1856). Type: Northeast India, Eastern Himalayas, cult. Flottbeck, near Altona, *Booth* s.n. (holo. W). **Plate 24D.**

Coelogyne graminifolia C.S.P. Parish & Rchb.f. in *Trans. Linn. Soc. London* 30: 146 (1874). Type: Burma, Moulmein, *Parish* 252 (holo. W; iso. K).

DESCRIPTION. *Pseudobulbs* on a short, sheathed rhizome 1.5 cm apart, narrowly ovoid, ribbed with age, 4–7 × 1–1.5 cm, enclosed with bracts at base. *Leaves* 2, linear, acute, finely nerved, 25–53 × 0.8–1.5 cm with 1–5 cm long, grooved petiole. *Inflorescence* proteranthous, peduncle erect or arching, clothed in overlapping scales, then bare, 6–15 cm long, rachis indistinct, slender, slightly zig-zag, 3- to 7-flowered, flowers opening simultaneously, floral bracts deciduous. *Flowers* fragrant, about 6 cm across, sepals and petals white, lip white except for orange-yellow on front of mid-lobe and tips of the side-lobes, some dark brown veins on side-lobes and 3 crisped dark brown keels. **Dorsal sepal** oblong-lanceolate, acute, nerves inconspicuous, 3 × 0.8 cm. **Lateral sepals** oblique, oblong-lanceolate, acute, nerves inconspicuous, 3 × 0.7 cm. **Petals** linear-lanceolate, acute, 2.5 × 0.4 cm. **Lip** 3-lobed, side-lobes erect, level with column, elliptic, 1.4 × 0.4 cm, heavily veined, front acute, mid-lobe 2 cm long, at base 0.3 cm broad, parallel sides then widens to 0.6 cm, callus of 3 crenate keels from base of lip to about one third of the way onto mid-lobe. **Column** slender, slightly arcuate, 1.5 cm long, widens to 0.5 cm hood, tip rounded and notched.

Distribution. Northeast India (Himalayas, Khasia Hills), Burma, China (Yunnan (Tengchong)), Laos, Vietnam (Da Lat), Thailand (Khao Yai National Park), Peninsular Malaysia.

Habitat. Epiphytic along streams in hill forest and lower montane forest.

Altitude. 700–1000 m.

Flowering. January–April, July in China.

Notes. The epithet seems to refer to glutinous or sticky elements but it is unclear why Reichenbach chose such an epithet for this species. The epithet for the synonym, *C. graminifolia* is a more apt description, as the leaves are grass-like in this species. The species fits naturally with sect. *Flaccidae* but some authors have taken the view it should be in sect. *Coelogyne*. I consider *C. viscosa* and *C. graminifolia* to be conspecific but there seems to be another *C. viscosa* Boxall ex Naves *nom. nud.* which may refer to another species.

INDEX OF SYNONYMS AND EXCLUDED SPECIES

Coelogyne advena (Hook.f.) C.S.P. Parish & Rchb.f., *Otia Bot. Hamb.* 1: 47 (1878)
  = **Pholidota advena Hook.f.**
alata A. Millar, *Orchids of Papau New Guinea* 75 (1978) *nom. nud.*
  = **Coelogyne carinata Rolfe**
alba (Lindl.) Rchb.f., *Walp. Ann.* 7: 236 (1861)
  = **Otochilus albus Lindl.**
amplissima Ames & C. Schweinf. in *Orch.* 6: 21 (1920)
  = **Chelonistele amplissima (Ames & C. Schweinf.) Carr**
amplissima Ames & C. Schweinf. var. *schweinfurthiana* J.J. Sm. in *Bull. Jard. Bot. Buitenzorg* ser. 3, 11: 101 (1931) = **Chelonistele amplissima (Ames & C. Schweinf.) Carr**
angustifolia A. Rich, *Ann. Sc. Nat.* ser. 2, 15. 16, t. 6 (1841)
  = **Coelogyne odoratissima Lindl.**
angustifolia Wight in *Icon.* 5 (1), t. 1641 (1851)
  = **Coelogyne breviscapa Lindl.**
angustifolia Ridl. in *J. Linn. Soc. Bot.* 32: 322 (1896)
  = **Coelogyne trinervis Lindl.**
annamensis (Lindl. & Rchb.f.) Rolfe in *Bull. Misc. Inf. Kew* 211 (1914)
  = **Coelogyne assamica Lindl. & Rchb.f.**
annamensis Ridl. in *J. Nat. Hist. Soc. Siam* 4: 117 (1921)
  = **Coelogyne sanderae Kraenzl.**
apiculata (Lindl.) Rchb.f., *Walp. Ann.* 6: 225 (1861)
  = **Panisea apiculata Lindl.**
arthuriana Rchb.f. in *Gard. Chron.* ser. 2, 15: 40 (1881)
  = **Pleione maculata (Lindl.) Lindl.**
articulata (Lindl.) Rchb.f., *Walp. Ann.* 6: 238 (1861)
  = **Pholidota articulata Lindl.**
balfouriana Sander, Cat. 6 (1896); in *Bull. Misc. Inf. Kew, App.* 2: 48 (1897)
  – Nomen nudum
beccarii Rchb.f. var. *micholitziana* Kraenzl. in *Gard. Chron.* ser. 3, 10: 300 (1891).
  = **Coelogyne beccarii Rchb.f.**
beccarii Rchb.f. var. *tropidophora* Schltr., *Orchidaceen von Deutsch-Neu-Guinea* 172 (1914)
  = **Coelogyne beccarii Rchb.f.**
bella Schltr. in *Bot. Jahrb. Syst.* 45: 104, 5 (1911)
  = **Coelogyne salmonicolor Rchb.f.**
beyrodtiana Schltr. in *Orchis* 9: 90 (1915)
  = **Chelonistele sulphurea (Blume) Pfitzer** var. **sulphurea**

biflora C.S.P. Parish ex Rchb.f. in *Gard. Chron.* 1035 (1865)
    = **Panisea uniflora (Lindl.) Lindl.**
bihamata J.J. Sm. in *Mitt. Inst. Bot. Hamb.*, 7: 29 (1927)
    = **Coelogyne tenuis Rolfe**
bimaculata Ridl. in *J. Linn. Soc. Bot.* 32: 327 (1896)
    = **Coelogyne flexuosa Rolfe**
birmanica Rchb.f. in *Gard. Chron.* ser. 2, 18: 840 (1882)
    = **Pleione praecox (J.E. Sm.) D. Don**
brevifolia Lindl., *Fol. Orch. Coelog.* 7 (1854)
    = **Coelogyne punctulata Lindl.**
brevilamellata J.J. Sm. in *Bull. Jard. Bot. Buitenzorg* ser. 3, 11: 103 (1931)
    = **Chelonistele brevilamellata (J.J. Sm.) Carr**
brunnea Lindl. in *Gard. Chron.* 71 (1848)
    = **Coelogyne fuscescens Lindl.** var. **brunnea Lindl.**
bulbocodioides Franch. in *Nouv. Arch. Mus. Paris*, sér. 2, 10: 84 (1888)
    = **Pleione bulbocodioides (Franch.) Rolfe**
calceata Rchb.f., *Walp. Ann* 6: 238 (1861)
    = **Pholidota pallida Lindl.**
camelostalix Rchb.f., *Walp. Ann* 6: 238 (1861)
    = **Pholidota camelostalix Rchb.f.**
carnea (Blume) Rchb.f., *Walp. Ann.* 6: 237 (1861)
    = **Pholidota carnea (Blume) Lindl.**
carnea Hook.f., *Fl. Brit. Ind.* 5: 838 (1890)
    = **Coelogyne radicosa Ridl.**
casta Ridl. in *J. Linn. Soc. Bot.* 32: 322 (1896)
    = **Coelogyne cumingii Lindl.**
caulescens Griff. in *Natul.* 3: 282 (1851)
    = **Bromheadia finlaysoniana (Lindl.) Miq.**
chinensis (Lindl.) Rchb.f., *Walp. Ann.* 6: 237 (1861)
    = **Pholidota chinensis Lindl.**
ciliata Boxall ex Naves in *Blanco Fl. Philipp.* ed. 3: Nov. App. 237 (1822)
    - **Nomen nudum**
cinnamomea Lindl. in *Gard. Chron.* 37 (1858)
    = **Coelogyne trinervis Lindl.**
clarkei Kraenzl. in *Gard. Chron.* ser. 3, 13: 741 (1893)
    = **Coelogyne micrantha Lindl.**
clypeata Rchb.f., *Walp. Ann.* 6: 237 (1861)
    = **Pholidota gibbosa (Blume) de Vriese**
conchoidea Rchb.f., *Walp. Ann.* 6: 237 (1861)
    = **Pholidota imbricata Hook.**
conferta Hort in *Gard. Chron.* ser. 2, 3: 314 (1875)
    = **Coelogyne nitida Lindl.**
convallariae C.S.P. Parish & Rchb.f. in *Flora* 55: 277 (1872)
    = **Pholidota convallariae (Rchb.f.) Hook.f.**
corniculata Rchb.f. in *Gard. Chron.* 746 (1865)
    = **Pholidota chinensis Lindl.**
coronaria Lindl. in *Bot. Reg. Misc.* 83 (1841)
    = **Eria coronaria (Lindl.) Rchb.f.**

corrugata Wight in *Bot. Reg.* 5 (1), t. 1639 (1851)
 = **Coelogyne nervosa A. Rich.**
crassifolia (Carr) Masamune, *Enum. Phan. Born.* 140 (1942)
 = **Chelonistele sulphurea (Blume) Pfitzer** var. **crassifolia (Carr) Masamune**
croockewittii Teysm. & Binn., *Nat. Tijd. Ned. Ind.* C. 5: 488 (1853)
 = **Chelonistele sulphurea (Blume) Pfitzer** var. **sulphurea**
crotalina Rchb.f., *Walp. Ann.* 6: 238 (1861)
 = **Pholidota imbricata Hook.**
cuneata J.J. Sm. in *Bull. Jard. Bot. Buitenzorg* ser. 3, 11: 97 (1931)
 = **Chelonistele sulphurea (Blume) Pfitzer** var. **sulphurea**
cymbidioides Ridl. in *J. Linn. Soc., Bot.* 38: 329 (1908)
 = **Coelogyne tomentosa Lindl.**
cymbidioides Rchb.f., *Walp. Ann.* 6: 239 (1861)
 = **Gynoglottis cymbidioides (Rchb.f.) J. J. Sm.**
cynoches C. S. P. Parish & Rchb.f. in *Trans. Linn. Soc.* 30: 147 (1874)
 = **Coelogyne fuscescens Lindl.** var. **brunnea Lindl.**
dalatensis Gagnep. in *Bull. Mus. Hist. Nat. Hist. Paris*, sér. 2, 2: 423 (1930)
 = **Coelogyne assamica Linden & Rchb.f.**
darlacensis Gagnep. in *Bull. Mus. Hist. Nat. Paris* sér. 2, 22: 505 (1950)
 = **Coelogyne sanderae Kraenzl.**
dayana Rchb.f. in *Gard. Chron.* ser. 2, 21: 826 (1884).
 = **Coelogyne pulverula Teijsm. & Binn.**
dayana Rchb.f. var. *massangeana* Ridl. in *Mat. Fl. Mal. Pen.* 1: 127 (1907)
 = **Coelogyne tomentosa Lindl.**
decipiens Sander, *Orch. Guide* 31: 30 (1901)
 = **Chelonistele sulphurea (Blume) Pfitzer** var. **sulphurea**
decora Wall. ex Voigt, *Hort. Suburb. Calc.* 621 (1845)
 = **Coelogyne ovalis Lindl.**
delavagi Rolfe in *Bull. Misc. Inf. Kew* 1896: 195 (1896)
 = **Pleione bulbocodioides (Franch.) Rolfe**
densiflora Ridl. in *J. Roy. Asiat. Soc. Straits Branch.* 39: 81 (1903)
 = **Coelogyne tomentosa Lindl.**
diphylla Lindl., Fol. Orch.-Coelog. 15 (1854)
 = **Pleione maculata (Lindl.) Lindl.**
edelfeldtii F. Muell. & Kraenzl. in *Oest Bot. Zeilschr* 44: 421 (1894)
 = **Coelogyne asperata Lindl.**
elata Hook. in *Bot. Mag.* 83: t. 5001 (1857): *non* Lindl.
 = **Coelogyne holochila P.F. Hunt & Summerh.**
elata Lindl., *Gen. Spec. Orch. Pl.* 40 (1830); Wall. Cat. no. 1959 (1829)
 = **Coelogyne stricta (D. Don) Schltr.**
elegans Rchb.f. in *Flora* 561 (1886), *Gard. Chron.* ser. 3, 1: 145 (1887).
 = **Cypripedium elegans**
elegantula Kraenzl. in *Feddes Repert. Spec. Nov. Regni. Veg.* 17: 111 (1921)
 = **Bletilla stricta (Thunb.) Rchb. f.**
falcata Anders. ex Hook.f. in *Ann. Roy. Bot. Gard.* (*Calcutta*) 5: 29, Pl. 43 (1895) in syn.
 = **Panisea uniflora Lindl.**
flavida Lindl., Fol. Orch.-Coelog. 10 (1854)
 = **Coelogyne prolifera Lindl.**

fleuryi Gagnep. in *Bull. Mus. Hist. Nat. Paris* sér. 2, 2: 424 (1930)
 = **Coelogyne lawrenceana Rolfe**
forstenebrarum P. O'Byrne in *Malayan Orchid Rev.* 34: 49-51 (2000)
 = excluded from synopsis; review of published description inconclusive.
fusca (Lindl.) Rchb.f., *Walp. Ann.* 6: 236 (1861)
 = **Otochilus fuscus Lindl.**
fusco-lutea Teijsm. & Binn. in *Cat. Hort. Bog.* 378 (1866)
 – **Nomen nudum**
gardneriana Lindl. in *Wall. Pl. As. Rar.* i. 33, t. 38; *Gen. Spec. Orch.* 41 (1830)
 = **Neogyna gardneriana (Lindl.) Rchb.f.**
gibbosa (Blume) Rchb.f., *Walp. Ann.* 6: 237 (1861)
 = **Pholidota gibbosa (Blume) de Vriese**
globosa (Blume) Rchb.f., *Walp. Ann.* 6: 236 (1861)
 = **Pholidota globosa (Blume) Lindl.**
goweri Rchb.f. in *Gard. Chron.* 443 (1869)
 = **Coelogyne punctulata Lindl.**
graminifolia C.S.P. Parish & Rchb.f. in *Trans. Linn. Soc. London* 30: 146 (1874)
 = **Coelogyne viscosa Rchb.f.**
grandiflora Rolfe in *J. Linn. Soc.* 34: 22 (1903)
 = **Pleione grandiflora (Rolfe) Rolfe**
henryi Rolfe in *Bull. Misc. Inf. Kew* 1896: 195 (1896)
 = **Pleione bulbocodioides (Franch.) Rolfe**
hookeriana Lindl., *Fol. Orch.-Coelog.* 14 (1854)
 = **Pleione hookeriana (Lindl.) B.S. Williams**
hookeriana Lindl. var. *brachyglossa* Rchb.f. in *Gard. Chron.*, ser. 3, 1: 833 (1887)
 = **Pleione hookeriana (Lindl.) B.S. Williams**
humilis (J.E. Sm.) Lindl. *Collect. Bot.*, sub. t. 37 (1821)
 = **Pleione humilis (J.E. Sm.) D. Don**
humilis (J.E. Sm.) Lindl. var. *albata* Rchb.f. in *Gard. Chron.*, ser. 3, 3: 392 (1888)
 = **Pleione humilis (J.E. Sm.) D. Don**
humilis (J.E. Sm.) Lindl. var. *tricolor* Rchb.f. in *Gard. Chron.*, ser. 2, 13: 394 (1880)
 = **Pleione humilis (J.E. Sm.) D. Don**
imbricata (Hook.) Rchb.f., *Walp. Ann.* 6: 238 (1861)
 = **Pholidota imbricata Hook.**
ingloria J.J. Sm. in *Bull. Jard. Bot. Buitenzorg* ser. 3, 11: 95 (1931)
 = **Chelonistele ingloria (J.J. Sm.) Carr**
integrilabia (Pfitzer) Schltr. in *Orchis* 9: 13 (1915)
 = **Coelogyne fuscescens Lindl.** var. **integrilabia Pfitzer**
interrupta Lindl. ex Heynh. in *Nom.* 2: 153 (1840)
 - **Nomen nudum**
javanica Lindl., *Fol. Orch.-Coelog.* 17 (1854)
 = **Nervila crispata (Blume) Schltr.**
khasiyana Rchb.f., *Walp. Ann.* 6: 238 (1861)
 = **Pholidota articulata Lindl.**
kingii Hook.f. in *Ann. Bot. Gard. Calc.* 6: 25, t. 38 (1895)
 = **Coelogyne foerstermannii Rchb.f.**
kutaiensis J.J. Sm. in *Bull. Jard. Bot. Buitenzorg* ser. 3, 11: 99 (1931)
 = **Chelonistele sulphurea (Blume) Pfitzer var. sulphurea**

lactea Rchb.f. in *Gard. Chron.* ser. 2, 23: 692 (1885)
　= **Coelogyne flaccida Lindl.**
lagenaria Lindl., *Fol. Orch.-Coelog.* 15 (1854)
　= **Pleione × lagenaria Lindl. & Paxt.**
lamellata Rolfe in *Bull. Misc. Inf., Kew* 36 (1895)
　= **Coelogyne macdonaldii F. Muell. & Kraenzl.**
lamellulifera (Carr) Masamune, *Enum. Phan. Born.* 142 (1942)
　= **Chelonistele lamellulifera Carr**
laotica Gagnep. in *Bull. Mus. Hist. Nat. Hist. Paris*, sér. 2, 2: 425 (1930)
　= **Coelogyne fimbriata Lindl.**
lauterbachiana Kraenzl. in *Notizbl. Bot. Gart. Berlin* 1: 113 (1896)
　= **Coelogyne miniata (Blume) Lindl.**
leungiana S.Y. Hu in *Quart. J. Taiwan Mus.*, 25(3–4): 223 (1972)
　= **Coelogyne fimbriata Lindl.**
ligulata Teijsm. & Binn. in *Cat. Hort. Bog.* 46 (1866)
　- **Nomen nudum**
limminghei Hort. ex Gentil. in *Pl. Cult. Serres. Jard. Bot. Brux.* 59 (1907)
　- **Nomen nudum**
longebracteata Hook.f. *Fl. Brit. India* 6: 194 (1890)
　= **Coelogyne cumingii Lindl.**
longipes Lindl. var. *verruculata* S.C. Chen in *Acta Phytotax. Sin.* 21(3): 346 (1983)
　= **Coelogyne schultesii S.K. Jain & S. Das**
longipes Hook.f., *Fl. Brit. India* 6: 195 (1890).
　= **Phaius longipes (Hook.f.) Holttum**
loricata (L. Lindl. & Cogn.) Rchb.f., *Walp. Ann* 6: 238 (1861)
　= **Pholidota imbricata Hook.**
lowii Paxt. in *Paxt. Mag. Bot.* 26: 225 (1849)
　= **Coelogyne asperata Lindl.**
lurida Ames & C. Schweinf. in *Orch.* 6: 35 (1920)
　= **Chelonistele lurida (L. Lind. & Cogn.) Pfitzer**
macrobulbon Hook.f., *Fl. Brit. India* 5: 830 (1890)
　= **Coelogyne rochussenii de Vries.**
macroloba J.J. Sm. in *Mitt. Inst. Allg. Bot. Hamburg* 7: 30 (1927).
　= **Coelogyne gibbifera J.J. Sm.**
macrophylla Teijsm. & Binn. in *Tijdschr. Ned.-Indie* 29: 241 (1867)
　= **?Coelogyne asperata Lindl.**
maculata Lindl., *Gen. Spec. Orch. Pl.* 43 (1830)
　= **Pleione maculata (Lindl.) Lindl.**
maingayi Hook.f., *Fl. Brit. India* 5: 831 (1890)
　= **Coelogyne foerstermannii Rchb.f.**
mandarinorum Kraenzl. ex Diels in *Engl. Jahrb.* 29: 269 (1896)
　= **Ischnogyne mandarinorum (Kraenzl.) Schltr.**
massangeana Rchb.f. in *Gard. Chron.* ser. 2, 10: 684 (1878)
　= **Coelogyne tomentosa Lindl.**
media Wall. ex F. Voigt in *Hort Suburb.* Calc. 621 (1845)
　- **Nomen nudum**
membranifolia Carr in *Gard. Bull. Straits Settlements*, 7: 2 (1932)
　= **Coelogyne septemcostata J.J. Sm.**

micholitziana Kraenzl. in *Gard. Chron.* ser. 3, 10: 300 (1891), Rchb.f. in *Xen. Orch.* ser. 3, 6: 100, t.256 (1892) = **Coelogyne beccarii Rchb.f.**
minutissima Kraenzl. in *Feddes Repert. Spec. Nov. Regni. Veg.* 17: 390 (1921)
= **Dendrochilum ?**
modesta J.J. Sm. in *Orchids Java* 141 (1905).
= **Coelogyne prasina Ridl.**
nervillosa Rchb.f., *Walp. Ann* 6: 236 (1861)
= **Pholidota nervosa (Blume) Rchb.f.**
nervosa Wight in *Icon.* 5 (1), t. 1638 (1851)
= excluded as synonym of '**Coelogyne glandulosa Lindl.**'; requires further evaluation.
nigrescens P. Don ex Loudon in *Hort. Brit. Suppl.* 3: 520 (1850)
- **Nomen nudum**
nigro-furfuracea Guillaumin in *Bull. Mus. Hist. Nat. Paris* sér. 2, 27: 143 (1955)
= **Tainia barbata (Lindl.) Schltr.**
nitida (Roxb.) Hook.f., *Fl. Brit. India* 5: 837 (1890); *non* (Wall. ex D. Don) Lindl., *Gen. Spec. Orch.* 40 (1830) = **Coelogyne punctulata Lindl.**
ocellata Lindl., *Gen. Spec. Orch. Pl.* 40 (1830)
= **Coelogyne punctulata Lindl.**
ocellata Lindl. var. *boddaertiana* Rchb.f. in *Gard. Chron.* ser. 2, 18: 776 (1882)
= **Coelogyne punctulata Lindl.**
ocellata Lindl. var. *maxima* Rchb.f. in *Gard. Chron.* ser. 2, 11: 524 (1879)
= **Coelogyne punctulata Lindl.**
ochracea Lindl. in *Bot. Reg.* t.49 (1846)
= **Coelogyne nitida Lindl.**
oligantha Schltr. in *Repert. Spec. Nov. Regni Veg. Beih.* 16: 44 (1919)
= **Coelogyne carinata Rolfe**
pachybulbon Ridl. in *J. Linn. Soc. Bot.* 32: 324 (1896)
= **Coelogyne trinervis Lindl.**
palaelabellatum A. Gilli in *Ann. Naturhist. Mus. Wien,* 84: 22 (1980 publ. 1983)
= **Gynoglottis palaelabellatum (A. Gilli) Garay & Kittridge**
palawensis Tuyama in *J. Jap. Bot.* 17: 506 (1941)
= **Coelogyne guamensis Ames**
pallida (Lindl.) Rchb.f., *Walp. Ann.* 6: 238 (1861)
= **Pholidota pallida Lindl.**
papagena Rchb.f. in *Bot. Zeit.* 20: 214 (1862), sphalm *C. paparina* .
= **Coelogyne micrantha Lindl.**
parviflora Lindl., *Gen. Spec. Orch. Pl.* 44 (1830)
= **Panisea demissa (D. Don) Pfitzer**
peltastes Rchb.f. var. *unguiculata* J.J. Sm. in *Mitt. Allg. Bot. Hamburg* 7: 33, t. 33, f. 23 (1927). = **Coelogyne pandurata Lindl.**
perakensis Rolfe in *Bot. Mag.* t. 8203 (1908)
= **Chelonistele perakensis (Rolfe) Ridl.**
phaiostele Ridl. in *J. Roy. Asiat. Soc. Straits Branch* 54: 51 (1910)
= **Geesinkorkis phaiostele (Ridl.) de Vogel**
pholas Rchb.f., *Walp. Ann* 6: 237 (1861)
= **Pholidota chinensis Lindl.**
pilosissima Planch. in *Hort. Dinat.* 144 (1858)
= **Coelogyne ovalis Lindl.**

pinniloba J.J. Sm. in *Bull. Jard. Bot. Buitenzorg* ser. 3, 11: 98 (1931)
    = **Chelonistele sulphurea (Blume) Pfitzer** var. **sulphurea**
plantaginea Lindl. in *Gard. Chron.* 20 (1855)
    = **Coelogyne rochussenii de Vries**
platyphylla Schltr. in *Feddes Repert.* 21: 129 (1925)
    = **Coelogyne celebensis J.J. Sm.**
pogonioides Rolfe in *Bull. Misc. Inf. Kew* 1896: 196 (1896)
    = **Pleione bulbocodioides (Franch.) Rolfe**
porrecta (Lindl.) Rchb.f., *Walp. Ann* 6: 236 (1861)
    = **Otochilus porrectus Lindl.**
praecox (J.E. Sm.) Lindl., *Collect. Bot.*, sub. t. 37 (1821)
    = **Pleione praecox (J.E. Sm.) D. Don**
praecox (J. E. Sm.) Lindl. var. *sanguinea* Lindl., *Fol. Orch.-Coelog.* 16 (1854)
    = **Pleione praecox (J. E. Sm.) D. Don**
praecox (J. E. Sm.) Lindl. var. *tenera* Rchb.f. in *Gard. Chron.* ser. 2, 20: 294 (1883)
    = **Pleione praecox (J.E. Sm.) D. Don**
praecox (J. E. Sm.) Lindl. var. *wallichiana* Lindl., *Fol. Orch.-Coelog.* 16 (1854)
    = **Pleione praecox (J.E. Sm.) D. Don**
primulina Barretto in *Orchid Rev.* 98 (1156): 37-43 (1990)
    = **Coelogyne fimbriata Lindl.**
psectrantha Gagnep. in *Bull. Mus. Hist. Nat. Paris* sér. 2, 2: 425 (1930)
    = **Coelogyne mooreana Rolfe**
psittacina Rchb.f. in *Xenia Orch.* 2: 141 (1874)
    = **Coelogyne rumphii Lindl.**
psittacina Rchb.f. var. *huttonii* Rchb.f., *Gard. Chron.* 32: 1053 (1870)
    = **Coelogyne rumphii Lindl.**
pulverula auct. non Teijsm. & Binn.: Lamb & C.L. Chan in *Luping, Summit of Borneo* 238, pl. 28 (1978) = **Coelogyne rhabdobulbon Schltr.**
pumila Rchb.f., *Walp. Ann.* 6: 236 (1861)
    = **Dendrochilum pumilum Rchb.f.** var. **pumilum**
punctulata Lindl. var. *hysterantha* Tang & Wang in *Act. Phytotax. Sin.* 1(1): 79 (1951)
    = **Coelogyne punctulata Lindl.**
purpurascens (Thw.) Hook.f., *Fl. Brit. India.* 5: 842 (1890)
    = **Adrorrhizon purpurascens (Thw.) Hook.f.**
pusilla Ridl. in *J. Linn. Soc. Bot.* 32: 327 (1896)
    = **Chelonistele sulphurea (Blume) Pfitzer.** var. **sulphurea**
pustulosa Ridl. in *J. Bot.* 24: 353 (1886)
    = **Coelogyne asperata Lindl.**
quadrangularis Ridl. in *J. Linn. Soc. Bot.* 32: 323 (1896)
    = **Coelogyne swaniana Rolfe**
radiosa J.J. Sm. in *Bull. Jard. Bot. Buitenzorg* ser. 3, 11: 105 (1931)
    = **Coelogyne hirtella J.J. Sm.**
radicosus Ridl. in *J. Fed. Malay States Mus.* 6: 57 (1915).
    = **Coelogyne radicosa Ridl.**
ramosii Ames, *Orch.* 7: 90 (1922)
    = **Chelonistele sulphurea (Blume) Pfitzer** var. **sulphurea**
recurva Rchb.f., *Walp. Ann.* 6: 237 (1861)
    = **Pholidota recurva Lindl.**

reflexa J.J. Wood & C.L Chan in *Lindleyana* 5(2): 87 (1990)
    = **Coelogyne genuflexa Ames**
reichenbachiana T. Moore & Veitch in *Gard. Chron.* 1868: 1210 (1868)
    = **Pleione praecox (J.E. Sm.) D. Don**
rhizomatosa J.J. Sm. in *Recueil Trav. Bot. Néerl.* 1: 146 (1904)
    = **Coelogyne prasina Ridl.**
rhizomatosa J.J. Sm. var. *quinquelobata* J.J. Sm. in *Bull. Jard. Bot. Buitenzorg* ser. 3, 9: 140 (1927) = **Coelogyne prasina Ridl.**
rhodeana Rchb.f. in *Gard. Chron.* 901 (1867)
    = **Coelogyne trinervis Lindl.**
rhombophora Rchb.f. in *Linnaea* 41: 116 (1877)
    = **Dendrochilum rhombaphorum (Rchb.f.) Ames**
richardsii (Carr) Masamune, *Enum. Phan. Born.* 144 (1942)
    = **Chelonistele richardsii Carr**
ridleyana Schltr. in *Feddes Repert. Spec. Nov. Regni Veg.* 8: 561 (1910)
    = **Geesinkorchis phaiostele (Ridl.) de Vogel**
ridleyi Gagnep. in Lecompte & Humbert, *Fl. Gen. Indo-Chine*, 6: 320 (1933)
    = **Coelogyne sanderae Kraenzl.**
rossiana Rchb.f. in *Gard. Chron.* ser. 2, 22: 308 (1884)
    = **Coelogyne trinervis Lindl.**
rubens Ridl. in *Gard. Chron.* ser. 3, 7: 576 (1890); Hook.f., *Fl. Brit. India* 6: 195 (1890).
    = **Calanthe rubens Ridl.**
rubra (Lindl.) Rchb.f., *Walp. Ann.* 6: 238 (1861)
    = **Pholidota rubra Lindl.**
saigonensis Gagnep. in *Bull. Mus. Hist. Nat. Paris* sér. 2, 10: 435 (1938)
    = **Coelogyne assamica Linden & Rchb.f.**
salmonicolor Rchb.f. var. *virescentibus* J.J. Sm. ex Dakkus in *Orch. Ned. Ind.* 3: 89 (1935)
    = **Coelogyne salmonicolor Rchb.f.**
sarrasinorum Kraenzl. in Engler, *Pflanzenr., Orch.-Mon.-Coelog.* 29 (1907)
    = **Coelogyne carinata Rolfe**
sarawakensis Schltr. in *Notizbl. Bot. Gart. Berlin* 8: 15 (1921)
    = **Chelonistele lurida L. Lind. & Cogn.** var. **lurida**
siamensis Rolfe in *Bull. Misc. Inf. Kew* 373 (1914)
    = **Coelogyne assamica Linden & Rchb.f.**
simplex Lindl., *Fol. Orch.-Coelog.* 13 (1854)
    = **Coelogyne miniata (Blume) Lindl.**
speciosa (Blume) Lindl. var. *alba* Hort. in *Gard. Chron.* ser. 3, 37: 205 (1905)
    = **Coelogyne speciosa (Blume) Lindl. subsp. speciosa**
speciosa (Blume) Lindl. var. *albicans* H.J. Veitch, *Man. Orchid Pl.* 50 *(1890)*
    = **Coelogyne speciosa (Blume) Lindl. subsp. speciosa**
speciosa (Blume) Lindl. var. *fimbriata* J.J. Sm. in *Teysmannia* 31: 254 (1920).
    = **Coelogyne speciosa (Blume) Lindl. subsp. fimbriata (J.J. Sm.) Gravendeel**
speciosa (Blume) Lindl. var. *rubiginosa* Hort. in *Orch. Rev.* 30: 37 (1922)
    = **Coelogyne speciosa (Blume) Lindl. subsp. speciosa**
steffensii Schltr. in *Feddes Repert.* 21: 130 (1925)
    = **Coelogyne rochussenii de Vries**
stellaris Rchb.f. in *Gard. Chron.* ser. 2, 26: 8 (1886)
    = **Coelogyne rochussenii de Vries**

stenophylla Ridl. in *Fl. Mal. Penins.* 4: 132 (1924)
   = **Coelogyne trinervis Lindl.**
stipitibulbum Holttum in *Gard. Bull. Straits Settlem.* 11: 278 (1947)
   = **Coelogyne radicosa Ridl.**
subintegra J.J. Sm. in *Bull. Dep. Agric. Indes Néerl.* 22: 12 (1909)
   = **Coelogyne exalta Ridl.**
sulphurea (Blume) Rchb.f., *Bonplandia* 5:43 (1857)
   = **Chelonistele sulphurea (Blume) Pfitzer** var. **sulphurea**
sumatrana J.J. Sm. in *Bull. Dep. Agric. Indes Néerl.* 15: 5 (1908)
   = **Coelogyne testacea Lindl.**
tenuiflora Ridl. in *J. Linn. Soc. Bot.* 32: 287 (1896)
   = **Pholidota ?gibbosa (Blume) de Vries**
thailandica Seidenf. in *Dansk Bot. Ark.* 29(4): 46 (1975)
   = **Coelogyne quadratiloba Gagnep.**
thuniana Rchb.f., Otto & Dietr. in *Allg. Gartens* 23 (1855)
   = **Panisea uniflora (Lindl.) Lindl.**
tomentosa Lindl. var. *? penangensis* Hook.f., *Fl. Brit. India* 5: 830 (1885)
   = **Coelogyne velutina de Vogel**
tomentosa Lindl. var. *massangeana* (Rchb.f.) Ridl. in *Mat. Fl. Mal. Pen.* 4: 130 (1907)
   = **Coelogyne tomentosa Lindl.**
tomiensis O' Byrne in *Malayan Orchid Rev.* 29: 33-35 (1995), *nom. invalid.*
   = **Coelogyne tommii Gravendeel & P. O'Byrne**
treutleri Hook.f., *Fl. Brit. India* 5: 194 (1890)
   = **Epigeneium treutleri (Hook.f.) Ormd.**
tricarinata Ridl. in *J. Fed. Mal. States Mus.* 5: 156 (1915)
   = **Coelogyne rigida C.S.P. Parish & Rchb.f.**
trifida Rchb.f. in *Hamb. Gartenz.* 19: 246 (1863)
   = **Coelogyne odoratissima Lindl.**
triotos Rchb.f., *Walp. Ann* 6: 238 (1861)
   = **Pholidota imbricata Hook.**
triptera Brongn. in *Duperr. Voy. Cog. Bot.* 201 (1865)
   = **Epidendrum caespitosum Paepp & Endl.**
trisaccata Griff. in *Itin. Not.* 72 (1848)
   = **Neogyna gardneriana (Lindl.) Rchb.f.**
trunicola Schltr. in *Feddes Repert.* 1: 104 (1911)
   = **Coelogyne carinata Rolfe**
undulata Rchb.f., *Walp. Ann.* 6: 238 (1861)
   = **Pholidota rubra Lindl.**
undulata Wall. ex F. Voigt. in *Hort. Suburb. Calc.* 621 (1845)
   - **Nomen nudum**
undulata Wall ex Pfitzer & Kraenzl. in Engler, *Pflanzenr. Orch.-Mon.-Coelog.* 55 (1907)
   = ***Pholidota* suaveolens Lindl.**
unguiculata (Carr) Masamune, *Enum. Phan. Born.* 146 (1942)
   = **Chelonistele unguiculata Carr**
uniflora Lindl., *Fol. Orch. Panis.* 2 (1854)
   = **Panisea uniflora (Lindl.) Lindl.**
vagans Schltr. in *Bot. Jahrb. Syst.* 45(104): 5 (1911).
   = **Coelogyne prasina Ridl.**

ventricosa Rchb.f., *Walp. Ann.* 6: 237 (1861)
    = **Pholidota ventricosa (Blume) Rchb.f.**
viscosa Boxall ex Vidal in *Naves Novis App.* 237 (1882)
    - **Nomen nudum**
wallichii Hook. in *Bot. Mag.* 76: t. 4496 (1850), sphalm. *C. wallichiana*
    = **Pleione praecox (J.E. Sm.) D. Don**
wallichiana Lindl., *Gen. Spec. Orch. Pl.* 43 (1830)
    = **Pleione praecox (J.E. Sm.) D. Don**
wettsteiniana Schltr. in *Oesterr. Bot. Z.* 69: 124 (1920)
    = **Coelogyne trinervis Lindl.**
whitmeei Schltr. in *Feddes Repert.* 11: 41 (1912)
    = **Coelogyne lycastoides F. Muell. & Kraenzl.**
xanthoglossa Ridl. in *J. Fed. Malay States Mus.* 6, 3: 180 (1915).
    = **Coelogyne xyrekes Ridl.**
xerophyta Hand-Mazz. in *Symb. Sin.* 7: 1346 (1936).
    = **Coelogyne fimbriata Lindl.**
xylobiodes Kraenzl. in *Bot. Jahrb. Syst.* 27: 483 (1893)
    = **Gynoglottis cymbidioides (Rchb.f.) J.J. Sm.**
yunnanensis Rolfe in *J. Linn. Soc.* 36: 23 (1903)
    = **Pleione yunnanensis (Rolfe) Rolfe**
zahlbrucknerae Kraenzl. in *Feddes Repert.* 17: 389 (1921)
    = **Coelogyne marmorata Rchb.f.**

MOLECULAR PHYLOGENY OF COELOGYNE AND ALLIED GENERA: AN URGE TO REORGANIZE[1]

Barbara Gravendeel and Ed F. de Vogel *

Introduction

Subtribe Coelogyninae, with approximately 550 species, contains 16 genera including *Coelogyne*. Shared characters are the sympodial growth, pseudobulbs of one internode, terminal inflorescences, a winged column and 4 pollinia with massive caudicles (Butzin, 1992). *Coelogyne* Lindl. is defined by a free, never-saccate lip with high lateral lobes over the entire length of the hypochile and smooth, papillose, toothed or warty keels (Seidenfaden & Wood, 1992).

Several studies of sections of the genus *Coelogyne* were published in the last decade, but a comprehensive taxonomic treatment of all species is still lacking. Pfitzer and Kraenzlin (1907) grouped the species of *Coelogyne* into 14 sections. In contrast, Holttum (1964) proposed only four and de Vogel (1994) 23 subdivisions. These large differences in opinion are due not only to the rather large number of species in the genus, but also the relative lack of morphological characters available to define groups of species.

The aims of this chapter are to use phylogenetic analyses of molecular data to:
1. Investigate the evolutionary relationships of *Coelogyne* with its allies in subtribe Coelogyninae.
2. Address the generic circumscription and sectional and subgeneric relationships within *Coelogyne*. To accomplish these goals, PCR RFLPs of the plastid genome and sequence data from both the plastid *trnK* intron (mostly *matK*) and the ribosomal ITS1-5.8S-ITS2 regions were collected.

Material and Methods

In total, 44 taxa have been analysed; including 18 of the 23 sections/subgenera currently recognised within *Coelogyne* and ten of the 16 genera of Coelogyninae. Morphologically uniform sections/(sub)genera are represented by a single taxon only, whereas larger, more variable groups are represented by several species. Outgroups were sampled from tribe Arethuseae, subtribes Bletiinae and Thuniinae, because earlier studies using morphological

---

* Nationaal Herbarium Nederland, Universiteit Leiden, P.O. Box 9514, 2300 RA Leiden, The Netherlands. e-mail: Gravendeel@nhn.leidenuniv.nl

(Burns-Balogh & Funk, 1986) and molecular data (Neyland and Urbatsch, 1996; Cameron *et al.*, 1999; Chase *et al.*, unpubl.) indicate a close relationship between these taxa and *Coelogyne*. Information about voucher specimens, DNA extraction, sequencing and analysis methods can be found in Gravendeel *et al.* (in prep.) but see Gravendeel (2000), Chaps. 2 & 3.

Results and Discussion

Content of subtribe Coelogyninae: Analysis of the combined molecular data sets indicates that Coelogyninae are monophyletic (=derived from one common ancestor) and diverged early into three major clades (see Fig. 27).

Clade I comprises species of *Coelogyne* sect. *Coelogyne, Cyathogyne, Rigidiformes, Tomentosae, Veitchiae* and *Verrucosae*, together with *Bracisepalum, Chelonistele, Dendrochilum, Entomophobia, Geesinkorchis* and *Nabaluia*. Shared characters for this group of species are the simultaneously opening flowers (with the exception of *Geesinkorchis*) and inflorescences with relatively many flowers. Many other characters, such as small flower size, persistent floral bracts, hairy ovaries, ovate-oblong petals and a hypochilium with inconspicuous lateral lobes are present in the majority of taxa of clade I. The presence of many different generic names in clade I can be explained by the high number of unique characters present, such as the presence of a stipes in *Geesinkorchis* and a transverse callus on the lip of *Entomophobia*. *Bracisepalum selebicum* and both *Dendrochilum* species sampled form a well-supported subset of taxa in clade I. These three species have unifoliate pseudobulbs, pendulous inflorescences with sterile bracts on the base of the rhachis and small flowers with a hypochile with inconspicuous lateral lobes. Another well-supported subclade in clade I comprises *Chelonistele sulphurea* and *Entomophobia kinabaluensis*. Both species have erect inflorescences with sterile bracts on the base of the rhachis and small flowers with a relatively short column (de Vogel, 1986). Generic boundaries within clade I are not clear yet, as most internal nodes of this clade are only poorly supported. Additional taxon sampling is needed to improve resolution.

Clade II (*Coelogyne s.s.*) subsequently diverged into species of *Neogyna* and *Pholidota* nested within species of *Coelogyne* sect. *Bicellae, Brachypterae, Elatae, Flaccidae, Fuliginosae, Hologyne, Lawrenceanae, Lentiginosae, Longifoliae, Moniliformes, Ptychogyne* and *Speciosae*. Shared characters for this group of species are the glabrous ovaries, linear petals (with the exception of *Pholidota*) and broad, erect lateral lobes of the hypochile. Many other characters, such as a low flower number, deciduous floral bracts and large flower size are present in the majority of taxa of clade II. *Neogyna gardneriana* and *Pholidota imbricata* form a strongly supported subset of taxa in clade II. Both species have an epichile with semi-orbicular, widely retuse lateral lobes.

Clade III consists of species of *Pleione*. The relatively isolated position of *Pleione* is consistent with the purplish-pink colour of the flowers, and short-lived pseudobulbs with annually deciduous leaves of many species in this genus, which do not occur in any of the other Coelogyninae (Cribb *et al.*, 1983, Cribb & Butterfield, 2000).

Sectional and subgeneric relationships within *Coelogyne*

Monophyletic sections. From the 18 different sections of *Coelogyne* considered in this study, just two (with only two sampled species each) form strongly supported monophyletic

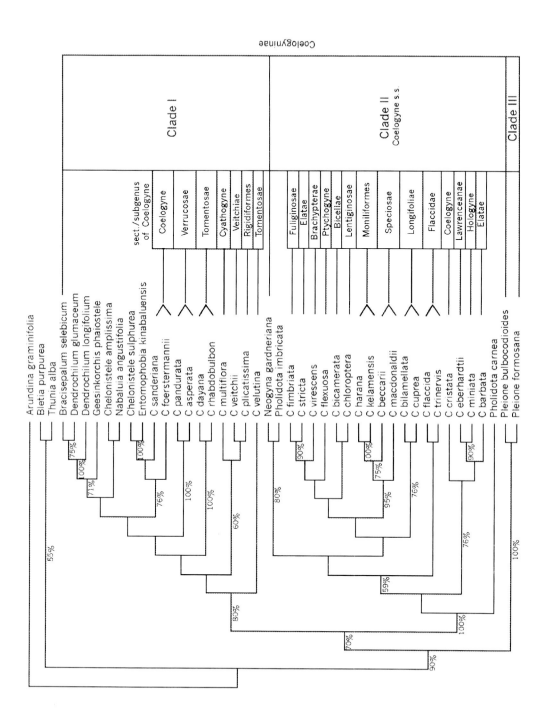

Fig. 27. One of the four cladograms produced by maximum parsimony analysis of the combined molecular datasets. Bootstrap percentages are indicated below the branches (only percentages > 50% are given). see Gravendeel *et al.* (in prep.) for more details.

groups in the combined analysis: sect. *Moniliformes* and *Verrucosae* (both 100% supported). This is consistent with the clear morphological characters that define both sections. Both species of sect. *Moniliformes* analysed have elongated, unifoliate pseudobulbs, a rhachis with distinctly swollen, short internodes and many flowers that open in succession (Carr, 1935). All species of sect. *Verrucosae* have rounded to strongly flattened bifoliate pseudobulbs, inflorescences with many, simultaneously opening flowers and a rachis with a few sterile bracts at the base and scattered minute scale-like hairs, which are also present on the pedicel, ovary and the abaxial side of the sepals and petals (Sierra, Gravendeel and de Vogel, 2000). *Coelogyne* sect. *Longifoliae* is also monophyletic, although support for this clade is weak (76%). The species of sect. *Longifoliae* all have bifoliate pseudobulbs, long and stiff inflorescences, a rachis with long internodes, and intermediate-sized flowers that open in succession. *Coelogyne* sect. *Flaccidae* is monophyletic in all shortest trees, but bootstrap support for this clade is low (<50%). This is in accordance with the few and not unique characters that define this section. *Coelogyne* sect. *Flaccidae* is characterised by a low number of simultaneously opening flowers with deciduous floral bracts and undulating keels on the lip, a combination of characters that is also found in sect. *Ocellatae*. *Coelogyne* sect. *Tomentosae* is not monophyletic, but none of the branches separating its two parts receive even low internal support. Unique characters for this section are lacking (de Vogel, 1992).

Polyphyletic sections. *Coelogyne* sect. *Coelogyne* and sect. *Elatae* are clearly polyphyletic (=not derived from one common ancestor). This is in accordance with the high variation in pseudobulb shape, inflorescence type, flower size and morphology of the keels on the lip in both sections. The only character that is present in all species currently assigned to sect. *Coelogyne* is the colour of the flowers: white, with yellow/brown spots on the lip. The species currently assigned to sect. *Elatae* only share the sterile bracts at the base of the rachis and the simultaneously opening flowers, a combination of characters present in many other Coelogyninae species.

Separate sections. *Coelogyne* sect. *Lawrenceanae* and sect. *Speciosae* are well separated, which is not in accordance with Seidenfaden (1975), who suggested they should be combined. Molecular data support our view that they should be considered different sections because of their clear morphological differences. All species of sect. *Lawrenceanae* have shiny green, smooth pseudobulbs, hysteranthous inflorescences and flowers with deeply incised, glabrous keels on the lip. All species of sect. *Speciosae* are characterised by angular, dull green pseudobulbs, synanthous or proteranthous inflorescences and flowers with hairy or warty keels on the lip (Gravendeel & de Vogel, 1999).

Subgenera. The species sampled of subgenera *Bicellae*, *Hologyne* and *Ptychogyne* seem well nested within clade II. *Coelogyne multiflora* of subgenus *Cyathogyne* is closely related to the species of several sections within *Coelogyne*. It seems therefore, that these taxa do not deserve the status of subgenera.

New sections. A well-supported subset of species is formed by *C. multiflora* (subgenus *Cyathogyne*), *C. plicatissima* (sect. *Rigidiformes*) and *C. veitchii* (sect. *Veitchiae*). Shared characters for this clade are the hairy ovaries, persistent floral bracts and small flowers. Another clade with high support consists of *C. fimbriata* (sect. *Fuliginosae*) and *C. stricta* (sect. *Elatae*). Both species have scales on the base of the scape and intermediate-sized flowers, but a close relationship has not been suggested so far (Pelser *et al.*, 2000). A third group of taxa supported by high bootstrap percentages is made up of *C. eberhardtii* (sect. *Lawrenceanae*) and *C. miniata* (subgenus *Hologyne*). Both species have bifoliate pseudobulbs and deciduous floral bracts. However, in other characters, such as plant size, leaf texture, inflorescence type and keel morphology, they show considerable differences. To investigate whether these three clades deserve the status of new sections, a much larger sampling within *Coelogyne* is needed.

Reorganisation of *Coelogyne*. Traditionally, the genus *Coelogyne* comprises the large-flowered members of the subtribe, lacking the special characters that define the other genera, such as a cup-shaped lip base (all other genera), a lip with sideways spreading lateral lobes (*Chelonistele*), a stipe (*Geesinkorchis*), or stelidia (*Dendrochilum*). In other words, the present circumscription of *Coelogyne* is mainly negative. Absence of characters often indicates artificial groups that did not evolve from one common ancestor. The polyphyletic nature of *Coelogyne* and several of its sections according to molecular data clearly shows how convergent floral morphology has confounded traditional taxonomy. Traditionally used classifications of *Coelogyne* and Coelogyninae are not supported by the molecular data presented here and should be abandoned.

There are two possible taxonomic solutions for a new phylogenetic classification of *Coelogyne* in which only monophyletic groups are recognised. The first would be to include all sampled species of Coelogyninae (excluding *Pleione*) within a single genus. According to the rules of priority, this genus should be called *Coelogyne*. However, given the large differences in floral morphology, the creation of many synonyms in widespread horticultural use, and the high number of species that *Coelogyne* would then encompass (approximately 530), this option is unsatisfactory. A second possibility would be to reduce *Coelogyne* to one of the three main clades found. The type species of *Coelogyne* (*C. cristata*) belongs to clade II (*Coelogyne s.s.*). The best solution seems therefore to take the following steps:

1. Inclusion of *Neogyna* and *Pholidota* in *Coelogyne*. The sample species of these two genera fit perfectly in *Coelogyne* when a new definition of the genus consists of glabrous ovaries only, a lip with a saccate or flat base, and a hypochile with broad, erect lateral lobes. Several taxonomists already considered these genera to be sections of *Coelogyne* (see de Vogel, 1988).
2. Removal of the species of *Coelogyne* sect. *Coelogyne* (in part), *Cyathogyne*, *Tomentosae*, *Rigidiformes*, *Veitchiae* and *Verrucosae*. These species fit better in clade I, because they share several characters with other genera in this clade, such as a relatively high number of simultaneously opening flowers with persistent floral bracts and hairy ovaries. Our phylogenetic analyses indicate that approximately 160 species would be left in *Coelogyne*.

In contrast with the *Coelogyne s.s.* clade, a good morphological delimitation of clade I is still difficult. Generic boundaries within clade I are not yet clear, as most internal nodes have only low support. Additional sampling in clade I is needed to find the limits of new monophyletic groups, and to justify any formal proposals for nomenclatural changes.

Acknowledgements

The authors thank Dudley Clayton (U.K.), Mark Clements (CSIRO, Australia), Mark Chase (Royal Botanic Gardens, Kew), Anton Sieder (Botanical Gardens, Vienna), Art Vogel (Hortus Botanicus, Leiden), Sofie Mursidawati and Dwi Murti Puspitaningtyas (Kebun Raya Bogor, Indonesia), Julaihi Lai (Semengoh Botanical Gardens, Sarawak) and Peter O'Byrne (Singapore) for plant specimens and valuable advice. Bertie Joan van Heuven, Jeffrey Joseph, Martyn Powell and Peter Kuperus assisted with sequencing. Fieldwork and a training visit to the Jodrell laboratories, Kew were supported by grants from the Alberte Mennega Stichting, Dutch Scientific Fund for Research in the Tropics (Treub Maatschappij), Stichting Fonds Dr. Christine Buisman, and the Netherlands Foundation for the Advancement of Tropical Research (WOTRO).

Literature cited

Burns-Balogh, P. & Funk, V.A. (1986). A phylogenetic analysis of the Orchidaceae. *Smithsonian Contributions to Botany* 61: 1–79.

Butzin, F. (1992). Subtribus Coelogyninae. pp. 914–958 in F.G. Brieger, R. Maatsch and K. Senghas (eds.). *Die Orchideeen* 1A. Verlag Paul Parey, Berlin, Germany.

Cameron, K.M., Chase, M.W., Whitten, W.M., Kores, P.J., Jarrell, D.C., Albert, V.A., Yukawa, T., Hills, H.G. & Goldman, D.H. (1999). A phylogenetic analysis of the Orchidaceae: evidence from *rbcL* nucleotide sequences. *American Journal of Botany* 86(2): 208–224.

Carr, C.E. (1935). Two collections of orchids from British North Borneo, part 1. *Gard. Bull. Str. Settlem.* 8: 207.

Cribb, P., Butterfield, I. & Tang, C.Z. (1983). The Genus *Pleione*. Royal Botanic Gardens, Kew.

Cribb, P.J. & Butterfield, I. (2000). The Genus *Pleione*. Second Edition. Natural History Publications (Borneo), Kota Kinabalu in association with The Royal Botanic Gardens, Kew.

de Vogel, E.F. (1986). Revisions in Coelogyninae (Orchidaceae) II. The genera *Bracisepalum*, *Chelonistele*, *Entomophobia*, *Geesinkorchis* and *Nabaluia*. *Orchid Monographs* 1: 17–86. Leiden.

―――― (1988). Revisions in Coelogyninae (Orchidaceae) III. The genus *Pholidota*. *Orchid Monographs* 3: 1–118. Leiden.

―――― (1992). Revisions in Coelogyninae (Orchidaceae) IV. *Coelogyne* section *Tomentosae*. *Orchid Monographs* 6: 4–42. Leiden.

―――― (1994). Character assessment for a subdivision of *Coelogyne* Lindley. pp. 203–205 in A.M. Pridgeon (ed.), *Proceedings of the 14th World Orchid Conference*, HMSO publications, Glasgow.

Gravendeel, B. & de Vogel, E.F. (1999). Revision of *Coelogyne* section *Speciosae* (Orchidaceae). *Blumea* 44: 253–320. Leiden

Gravendeel, B., Chase, M.W., de Vogel, E.F., Roos, M.C., Bachmann, K. & Mes, T.H.M. (in prep.). Molecular phylogeny of subtribe Coelogyninae (Epidendroideae; Orchidaceae) based on plastid RFLPs, *matK* and ITS sequences: evidence for polyphyly of *Coelogyne*.

Holttum, R.E. (1964). *Coelogyne* in H.M. Burkill (ed.). *A revised flora of Malaya 1: Orchids of Malaya*. Pp. 241–251.

Neyland, R. & Urbatsch, L.E. (1996). Phylogeny of subfamily Epidendroideae (Orchidaceae) inferred from ndhF chloroplast gene sequences. *American Journal of Botany* 83: 1195–1206.

Pelser, P.B., Gravendeel, B. & de Vogel, E.F. (2000). Revision of *Coelogyne* sect. *Fuliginosae*. *Blumea* 45: 253–273. Leiden.

Pfitzer, E. & Kränzlin, F. (1907). *Coelogyne* in H.G.A. Engler (ed.), *Das Pflanzenreich* 32: 20–82. Akademie-Verlag, Berlin, Germany.

Seidenfaden, G. (1975). Orchid Genera in Thailand III—*Coelogyne*. *Dansk Botanisk Arkiv* 29 (4): 7–94.
Seidenfaden, G. & Wood, J.J. (1992). *The Orchids of Peninsular Malaysia and Singapore*. Royal Botanic Gardens, Kew, Botanic Gardens, Singapore and Olsen and Olsen, Fredensborg, Denmark.
Sierra, S.E.C., Gravendeel, B. & de Vogel, E.F. (2000). Revision of *Coelogyne* section *Verrucosae* (Orchidaceae): a new sectional delimitation based on morphological and molecular evidence. *Blumea* 45: 275–318. Leiden.

HYBRIDISATION

Natural Hybrids

Natural hybridisation between *Coelogyne* species probably occurs but there does not seem to be much in the way of formal views in the literature. From *James Veitch & Sons* 1887–1894, *Manual of Orchidaceous Plants* we find the following quotation:

'The fact that most orchid flowers if fertilised at all must, in the wild state, be fertilised by insect agency being once recognised, it follows that where two allied species grow together, or in close proximity to each other, the pollen of the one is liable to be deposited on the stigma of the other, and crosses may thence be effected. This hypothesis and the structural evidence offered by intermediate forms that have appeared among importations of geographically combined species have suggested that such forms are of hybrid origin. Direct proof of the existence of natural hybrids has now been afforded by identical forms artificially raised from the same pair of species as those from the same pair of species as those from which the supposed wild hybrids were derived.'

In the case of coelogynes, it has been evident for some time that imports from northern India have contained plants which appear to be natural hybrids. There would seem to be no evidence that natural hybrids of *Coelogyne* species have been found in imports from Southeast Asia. In the *Orchid Review*, 106(1219) (1998) and 106(1227) (1999), I discussed the data I had collected regarding *Coelogyne* Intermedia (hybrid of *C. cristata* × *C. massangeana*), *C. intermedia* (sold as a species from India) and *C. granulosa* (sold as a species of unknown origin). The name, *C.* Intermedia was used in 1913 to register the artificial hybrid (*C. cristata* × *C. massangeana*) by Messrs. J. Cypher & Sons of Cheltenham in *Sander's List of Orchid Hybrids*. I doubt whether the 'Cypher' hybrid, *C.* Intermedia, is to be found in modern collections. In addition, the plants sold under names *Coelogyne intermedia* and *Coelogyne granulosa*, neither of which has been validly published, could be natural hybrids. Summarised below is some information that relates to the possibility that some natural hybrids exist.

Coelogyne cristata is an influential species and there are references to 13 named varieties and consequent ambiguity in descriptions to be found in the literature. Two such varieties, *C. cristata* var. *intermedia* and *C. cristata* var. *lemoniana* were described in Veitch's *Manual of Orchidaceous Plants* (1887–94). The former seems to have no pedigree while the latter first appeared in the collection of Sir Charles Lemon at Carclew, Cornwall, probably in the late 1870s. The morphology of the types of *C. cristata* and *C. cristata* var. *lemoniana* are similar but the lip of the former is orange whereas the variety has citron yellow on the disc and lamellae. *C. cristata* var. *intermedia* has an orange spot at the base of the lip and was deemed to be intermediate between the type and var. *lemoniana*. To add more confusion, Williams (1884) refers to *C. cristata* var. *lemoniana* as a synonym of *C. cristata* var. *citrina*, which was first described by Williams in Godfrey's *L'Orchidophile* in 1888. The name *Coelogyne albina* crops up from time to time and it is possible that this name is interchangeable with *C. cristata* var. *lemoniana*. The recognition of several varieties of *C. cristata* has been particularly unhelpful and only serves to illustrate the variability of the species. Today, it remains likely that plants with the name *C. intermedia* may prove to be variations of *C. cristata*, although they could be natural hybrids.

In the 1995 Sander's list of new orchid hybrids, David Banks of New South Wales, Australia, registered a *Coelogyne* with the grex name Unchained Melody (*C. cristata* × *C. flaccida*). Banks gave an interesting summary of his findings in relation to *C.* Intermedia, which led to the making of *Coelogyne* Unchained Melody (*Australian Orchid Review*, April 1995).

Banks believes that all plants with the grex name *C.* Intermedia were derived from India some decades ago. He has seen numerous catalogues from Indian nurseries listing either *C. cristata* var. *intermedia* or just plain *C. intermedia*. He believes that the plants are a natural hybrid (*C. cristata* × *C. flaccida*) as the two species commonly grow together. He has also seen plants simply listed as *C. cristata* × *C. flaccida*; these are identical to the commonly seen *C.* Intermedia. The distributions of the two species involved in the registered hybrid *C.* Intermedia (*C. cristata* × *C. massangeana* (now *C. tomentosa*)) do not overlap, so there can be no question of natural hybridisation. Banks has not seen any plants fitting the description of the true, registered hybrid and I have been unable to find a painting of the registered hybrid *C.* Intermedia at Kew or in the Lindley Library of the Royal Horticultural Society. There is, however, a specimen in the Kew Herbarium which came from Messrs. J. Cypher & Sons with the name *C.* Intermedia (*C. cristata* × *C. massangeana*). A study of this specimen suggests that it is *C. cristata* × *C. flaccida*. Unfortunately, even if the parents of this registered hybrid were wrongly named, the name 'Intermedia' has been registered and the terms '*intermedia*', 'x intermedia' and 'Intermedia' cannot be applied to any other *Coelogyne* species, natural hybrid or artificial hybrid.

Whilst the natural hybridisation of *C. cristata* and *C. flaccida* can be expected to occur, many plants judged to be this natural hybrid, have been in cultivation for some considerable time and they have no provenance. Even if bona-fide material could be found, there does not seem to be much sense in formally describing these natural hybrids and running the risk of introducing even more confusion with a new grex.

Plants named *Coelogyne granulosa* are grown in private collections. I have examined a number of specimens in flower and they all seem to represent the same taxon. They have been compared with various examples labelled '*C.* Intermedia' and the hybrid *C. cristata* × *C. flaccida*. The morphology of these shows that the influence of *C. cristata* is predominant and that they are virtually identical although *C. granulosa* is marginally smaller. As to colour, *C. granulosa* is mainly white with some pale yellow patches, whereas the other taxa are generally whitish and have deeper yellow markings on the lip and brownish-yellow veining on the inside of the side lobes. Vegetatively the pseudobulb, enclosing papery bracts and leaves are similar. In *C. granulosa* however, the growth of the inflorescence is more upright with the peduncle enclosed in loose, imbricating bracts which remain quite green whereas the inflorescences found in the other plants are slightly pendulous with the peduncle enclosed in tighter greenish-brown imbricating bracts. In all examples, the inflorescence is heteranthous; i.e. it is produced on a separate shoot, which does not develop to produce a pseudobulb and leaves. On balance, the plants have slightly different colouring but in all other respects they are virtually identical. My judgement is that it could be a natural hybrid of *C. cristata* × *C. albolutea*. The type specimen of *C. albolutea* which is at Kew, is a plant introduced by Sander & Sons in 1896 which flowered in their nursery in St. Albans and was subsequently cultivated at the Glasnevin

Botanic Gardens, Dublin. From the type description, the sepals and petals are a clear pure white and there are no brown markings on the lip, just white and yellow. The lip is not dissimilar to that of *C. flaccida* but the keels are more undulate and there are no crenulations at the sinuses. *C. albolutea* is no longer in cultivation at Glasnevin and, apart from the type collection, it has not been collected again. The form of *C. cristata* × *C. albolutea* must remain conjecture unless *C. albolutea* can be found and used for breeding. David Banks has re-made *C.* Unchained Melody using *C. cristata* var. *lemoniana* as the pod parent. The resulting flower is smaller than the original 'Unchained Melody' but it appears to be identical to the flower of *Coelogyne granulosa* (pers. comm.). Banks thinks that growers with *C. granulosa* should rename their plants as *C.* Unchained Melody and I am content with this interpretation.

The examples discussed above, would seem to be the only natural hybrids which have found their way into orchid collections. It is conceivable that there are some other natural hybrids possibly, between *Coelogyne asperata* and *C. pandurata* to produce what we know as *Coelogyne* Burfordiense (*C. asperata* × *C. pandurata*); also between sect. *Speciosae* species. From Vietnam, there may be a natural hybrid of *C. lawrenceana* and *C. mooreana* which as an artificial hybrid, we know as *Coelogyne* Memoria W. Micholitz (*C. lawrenceana* × *C. mooreana*). If there is any evidence of such events, then I have been unable to find any references in the literature.

Artificial Hybrids

The *Coelogyne* hybrids registered in the Sander's *Complete List of Orchid Hybrids*[1] and its supplements[2], *Royal Horticultural Society Orchid Hybrids* Addendum[3] and the summary, in *The Orchid Review*[4], of new orchid hybrids which have been registered are summarised below:

Brymeriana—*asperata* × *dayana* 1906 [1]
Col. W.E. Brymer, Islington House, Dorchester
Colmanii—*cristata* var. *alba* × *speciosa* var. *major* 1907 [1]
Sir Jeremiah Colman, Gatton Park, Reigate
Burfordiense (Stanny)—*asperata* × *pandurata* 1911 [1]
Sir Trevor Lawrence, Burford Lodge, Dorking
Intermedia—*cristata* × *massangeana* 1913 [1]
Messrs. James Cypher & Sons, Cheltenham
Albanense—*pandurata* × *sanderiana* 1913 [1]
Sander & Sons, St Albans
Gattonense—*speciosa* × *sanderae* 1915 [1]
Sir Jeremiah Colman, Gatton Park, Reigate
speciosa-Colmanii—*Colmanii* × *speciosa* var. *major* 1918 [1]
Sir Jeremiah Colman, Gatton Park, Reigate
Shibata—*flaccida* × *speciosa* 1923 [1]
Shibata?
Memoria W. Micholitz—*lawrenceana* × *mooreana* 'Brockhurst'. Plate 24F. 1950 [2]
Sanders & Sons, St Albans
Memoria Soedjana Kassan—*speciosa* × *asperata* 1976 [3] (1971–75)
A.S. Parnata, Bandung, Indonesia
Neroli Cannon—*speciosa* × *fragrans* 1981 [3] (1981–85)
D.M. Cannon, Victoria, Australia

Noel Wilson—*mossiae* × *cristata*	1984 [3] (1981–85)
Geyserland Orchids, Rotorue, N. Zealand	
Green Magic—*parishii* × *speciosa*	1986 [3] (1986–90)
T. Stevenson, Washington, USA	
Linda Buckley—*mooreana* × *cristata*	1989 [3] (1986–90)
R.C. Hull, Oregon, USA	
Green Dragon—*pandurata* × *massangeana*	1992 [3] (1991–95)
Burnham Nurseries (Geyserland, Rotorua, N.Z.)	
Memoria Louis Forget—*speciosa* × *mooreana* 'Brockhurst'	1994 [3] (1991–95)
Dr. P.N. Sander, Wimborne, Dorset	
Edward Pearce—*fragrans* × *mooreana*	1995 [3] (1991–95)
R. Pearce, Victoria, Australia	
Jannine Banks—*mooreana* × *flaccida*	1995 [3] (1991–95)
David Banks (Phillip Spence), NSW, Australia	
Unchained Melody—*cristata* × *flaccida*. Plate 24E.	1995 [3] (1991–95)
David Banks, NSW, Australia	
Andree Millar—*beccarii* × *speciosa*	1996 [4] (104:1209)
Phillip Spence, NSW, Australia	
Cosmo-Crista—**Intermedia** × *cristata*	1996 [4] (104:1210)
Kokusai	
South Carolina—**Burfordiense** × *pandurata*	1996 [4] (105:1213)
Carter & Holmes, USA	
Amber—*speciosa* × *ovalis*	1998 [4] (106:1223)
David Banks (Phillip Spence), NSW, Australia	
Danielle De Prins—*speciosa* × *fimbriata*	2001 [4] (109:1240)
A.De Prins	
Shinjuku—*speciosa* × *cumingii*	2001 [4] (109:1241)
Suwada Orch. (Shiniku)	
Memoria Fukuba—**Shinjuku** × *cristata*	2001 [4] (109:1241)
Suwada Orch. (Shiniku)	
Memoria Okami—**Shinjuku** × **Intermedia**	2001 [4] (109:1241)
Suwada Orch. (Shiniku)	
Memoria Sadako—*speciosa* × **Intermedia**	2001 [4] (109:1241)
Suwada Orch. (Shiniku)	
Memoria Tokiko—**Shinjuku** × *lawrenceana*	2001 [4] (109:1241)
Suwada Orch. (Shiniku)	

Depending on your point of view, the lack of *Coelogyne* hybrids may be a cause of disappointment or it may be of no consequence because your primary interest is the *Coelogyne* species. There are very few *Coelogyne* enthusiasts around the world when compared with the number of amateur and professional growers whose interests lie elsewhere. It is the commercial opportunities that dictate the level of interest in a particular genus but it is still surprising to discover that only 29 *Coelogyne* hybrids have been registered since 1906. Over the same period, tens of thousands of commercially produced hybrids of *Dendrobium, Paphiopedilum, Cymbidium, Odontoglossum, Oncidium, Vanda, Miltonia, Cattleya, Laelia, Epidendrum* and *Phalaenopsis* have been registered.

Hybridisation

The first artificial hybrid, *Coelogyne* Brymeriana flowered in the greenhouses of Colonel W.E. Brymer of Islington House, Dorchester, England in 1906. The hybrid, that is still to be found in amateur collections, is *C.* Burfordiense (Stanny) which was made by Sir Trevor Lawrence of Burford Lodge, Dorking, England in 1911. Sir Jeremiah Colman of Gatton Park, Reigate, England produced three hybrids whilst Sander & Sons of St Albans produced another two. In 1913, Messrs. James Cypher & Sons gave us *C.* Intermedia (*C. cristata* × *C. massangeana*) (see comments in Natural Hybrids above). By 1923 there were just nine registered hybrids after which there was no activity until 1950 when *C.* Memoria William Micholitz (*C. lawrenceana* × *C. mooreana* 'Brockhurst') appeared on the scene and it proved to be a very popular hybrid which is to be found in many modern collections. During the 1990s there has been a flurry of activity by some Australian enthusiasts, notably David Banks and Phillip Spence, which is sure to continue. Banks has indicated that he has had difficulty in hybridising within the genus but he has found that hybrids are more likely to take if the end flowers of an inflorescence are used. It should be emphasised, however, that there is no obvious reason why this should be. He has also found that the fertilisation has more chance of success if the maternal or pod parent is the smaller flowered species. It is interesting to note that in his *Orchids of Malaya*, Professor R.E. Holttum indicated that attempts in Singapore to cross *C. asperata* × *C. foerstermannii* and *C. asperata* × *C. rochussenii* both failed. My attempts to remake *C. cristata* × *C. massangeana*, using various sources of pollen have proved unsuccessful. *Coelogyne* enthusiasts should not be discouraged by the lack of success in hybridising the genus, as there are many species which are classified into a large number of sections which illustrate the differing growing habit, form of the inflorescence and character of the flowers. These differences provide enough variety to interest the enthusiast.

---

[1] Sander's *Complete List of Orchid Hybrids,* Vols. 1 & 2, p. 19 (1921).

[2] Sander's *List of Orchid Hybrids* Addendum.

[3] Royal Horticultural Society *Orchid Hybrids List* Addendum.

[4] *The Orchid Review.*

CULTIVATION

The cultivation of coelogynes seems to be outside the province of the majority of orchid growers although a few species, such as *Coelogyne cristata, C. nitida (ochracea)* and *C. flaccida*, are to be found in many collections. In recent years there would appear to be a greater interest in the genus and although, as mentioned in the previous chapter on Hybridisation, 29 hybrids have been created, it is the species which seem to interest the enthusiast. The number and character of the species reflect the range of geographical locations, topography and the climatic conditions in which *Coelogyne* species thrive. In a collection of the species there are some examples which flower at a particular time of the year, there are others which will flower twice a year, and others which produce flowers in sequence virtually throughout the year. The commonly available species, with some variations, are *C. cristata, C. nitida, C. fimbriata, C. stricta (elata), C. tomentosa (massangeana), C. speciosa, C. mooreana, C. flaccida* and *C. lawrenceana*. There remains a further 181 species which are not so easily found. Until the commercial growers of orchids realise there is a market and start propagating the plants, then the supply will be limited.

General Requirements and Temperature

In the temperate parts of the world a heated greenhouse, conservatory or an indoor space with artificial lighting is essential for the growing of a wide range of *Coelogyne* species and hybrids. Coelogynes seem to be reasonably adaptable and with careful observation of a newly acquired plant, the appropriate microclimate in a greenhouse can be found. There is a spectrum of climates appropriate to the various species. Firstly, there are those which grow in the cool night temperatures, at higher altitudes, in valleys and on ridges found in the mountainous areas of Nepal, India, Burma, China and Malaya. Secondly, there are those species, which prefer the higher temperatures and high humidity associated with the lowlands of New Guinea, Borneo and other parts of the Far East. Thirdly, at the intermediate altitudes within the regions of the tropical rain forests, there are conditions, enjoyed by the vast majority of the taxa, which lie between the cool and the warm environments. Coelogynes seem to fit in with the three culturally accepted temperature regimes, Cool, Intermediate and Warm, into which orchids are generally grouped, but as with many orchids, the boundaries of these groups are flexible. The **Cool-growing** species, such as *C. cristata, C. nitida* and *C. flaccida*, need a winter minimum temperature of 10°–13°C (50°–55°F), but prefer the lower temperature. An occasional cooler night at 7°C (45°F) will do no harm. For the cool-growing species, it is essential to avoid higher night temperatures as the plants will produce plenty of vegetative growth but they are unlikely to flower. A daytime temperature rise of 5°–10°C (9°–18°F) seems to be an ideal objective. In the summer, temperatures in excess of 27°C (81°F) can easily occur in a small greenhouse; these high temperatures need to be avoided if the growth of the plants is to be maintained. **Intermediate** conditions dictate a night temperature of 13°–15°C (55°–59°F) with a daytime rise of 10°C (18°F) but to no more than 27°C (81°F). There are a number of species found at altitudes between 900–1800 m in the tropical rain forests of Lower Burma, Thailand, Peninsular Malaya, Sumatra and Java which will thrive well in a greenhouse with intermediate conditions. The **Warm House** coelogynes need a minimum temperature of 18°C (65°F) all the year around. If the summer weather conditions prove to be unseasonably cool, then artificial heating is needed but a lower temperature of 15.5°C (60°F) can be tolerated. Species such as *C. pandurata, C. mayeriana* and many of the species from Borneo, found at the lower altitudes are suited to the Warm House.

The emphasis on minimum winter temperatures is important but it is also important to avoid excessive temperatures, over 27°C in the summer, as probably all the *Coelogyne* species rarely experience temperatures in excess of this level in their natural habitat. To ensure that the heating and ventilation system is maintaining the required temperature regime, it is essential to monitor conditions with thermometers or remote temperature reading sensors located at plant level in different parts of the greenhouse.

Lighting and Shading

Coelogynes enjoy the sunlight and it is essential in temperate regions, to make sure that the plants receive plenty of sunlight in the spring, summer and autumn. The problem is that with increased intensity of sunlight, temperatures in a greenhouse increase dramatically and the new growth on plants can be damaged and some types of leaves can be badly scorched. From about mid-March, some means of controlling the rise in temperature has to be applied and there are commercially available glasshouse paints that provide a temporary screen and keep the temperatures in the greenhouse down. By the beginning of May the temporary screen has to be supplemented by extra shading using laths, blinds or proprietary netting to give about 50 percent shade. Such material needs to be mounted 30 cm (12 in) away from the glass. This form of shading has a dual purpose: it reduces the amount of light entering the greenhouse from above; and the layer of air trapped between the glass and the shading material provides excellent insulation. It also prevents temperatures in the greenhouse from rising to damaging levels. In the autumn, the shading needs to be reduced by removing the shading material in about mid-October and washing off the shading paints at the end of October. None of the coelogynes can be defined as shade-loving, although the light they receive should be dappled. There are some species, such as those in sect. *Tomentosae,* sect. *Verrucosae* and sect. *Lawrenceanae,* which enjoy more sunlight than other taxa and they should be placed as near the glass at the top of the greenhouse or near the glass on the south-facing side of the greenhouse.

Ventilation and Humidity

All the *Coelogyne* species need fresh air and generally speaking this can be obtained by using ventilators at all times of the year. Common sense dictates the need to avoid drawing very cold air into the greenhouse during the winter. However, some means of bringing in fresh air, preferably below bench level is desirable. A good air circulation can be induced in the greenhouse by using low speed fans but better still, a ducted air system. The warm air at the top of the greenhouse can be circulated to ground level, thus providing a buoyant, even temperature environment in the different zones of the greenhouse. Such a system is suitable for use in both winter and summer.

The introduction of fresh air by opening ventilators causes a loss of humidity. The humidity needs to be maintained between 65 and 75 percent and damping down the floors and staging is the time-honoured method of achieving the requirement. It is a matter of personal preference whether damping down by hand or the automatic misting system is chosen. The use of under stage spray lines seems to offer the better solution and if it is operated on a time and humidistat controlled basis, then the desired humidity range can be maintained for virtually all times of the year. On those occasional summer days when the spray lines do not provide the optimum conditions, hand spraying of the plants on a summer evening will be beneficial. In the winter when the heating system is in operation, humidity levels can fall because of the drying out effects of the heating system. Again the under stage spray lines can be of service as the water in the vicinity of the heating pipes quickly raises the humidity.

Containers, Compost and Repotting

There is a diversity of habit to be found amongst *Coelogyne* species. There are the dense clumps formed by the sect. *Ocellatae* and sect. *Speciosae* species, which are generally grown in standard flower pots. Plants with a climbing, branching habit such as sect. *Fuliginosae* species are best grown in pans whilst those with a robust, extended rhizome, such as sect. *Elatae* and sect. *Verrucosae* species, need to be mounted on large blocks of tree fern. A hanging basket should be used for those species with a pendulous inflorescence, such as some of those in sect. *Tomentosae* and sect. *Flaccidae*.

The composition of the compost is very much a matter of personal choice and orchid growers with experience of growing other types of orchid will have their favourite mixture. There will be a number of devotees who use one of the commercially available rockwool compositions but on balance the use of conifer bark, in various grades, has proved to be a sound standard of compost which is dependable. The composition usually consists of the bark with added peat or chopped up Sphagnum moss to retain moisture, perlite, perlag or polystyrene granules to improve drainage and aeration, and granules of charcoal. A standard mix is 3 parts coniferous bark; 2 parts perlite or perlag; 2 parts coarse peat or chopped Sphagnum moss; 1 part charcoal. This mixture is suitable for all the taxa whether they are epiphytic, lithophytic or terrestrial. Depending on the size of the plant, the grade of the bark used should be altered from small (1–3 mm) to medium (4–6 mm) or large (7–12 mm). Whilst the bark compost will degrade with time, it is essential to note that coelogynes dislike root disturbance and should therefore only be repotted when absolutely necessary, usually when the plant has outgrown its container or the compost has deteriorated to an unacceptable condition. Because of the deterioration of the bark composts, there are many growers who simply put plants in conifer bark, polystyrene pieces having first been placed at the base of the container to allow for rapid drainage, thus giving a longer life to the compost and avoiding waterlogged mixtures.

Watering and Feeding

Overwatering is the way in which many orchids, including coelogynes, are destroyed. The standard rule is to allow the compost to dry out before further water is given. In the growing season the rate of growth and the amount of water required is the greatest and water should be given liberally to the plants but care must be taken not to allow water to lodge on the new growths or they may be damaged. As the plants mature, the rate of growth slows and the amount of water should be decreased. Most coelogynes benefit from a winter rest of some eight to ten weeks at a lower temperatures, 10°C (50°F), when only enough water should be given to prevent shrivelling of the pseudobulbs. The dormancy associated with this period helps to induce flowering. Watering can be resumed in the spring when species such as *C. cristata* and *C. flaccida* are well into their flowering period. Thereafter, through the spring, summer and autumn, the watering is maintained.

Coelogynes enjoy being fed and again the fertiliser a grower will use is a matter of personal choice, usually based on experience with growing other types of orchids. The preference for a weak fertiliser, which is readily taken up by the plant, leaving little or no residue is to be preferred. This avoids the large concentrations of unused fertiliser which accelerate the breakdown and decay of the compost. The use of a fertiliser every second or third watering seems to maintain a balance between plant growth and the well-being of the plant and the condition of the compost.

It cannot be overstressed, in the summer; all the coelogynes benefit from a quick hand misting in the evening on a warm day. However, the amount of misting will be a matter of judgement, taking account of the ambient conditions in the greenhouse.

Propagation

There is no well of experience with the raising of *Coelogyne* species from seed, presumably because of the slight commercial interest shown in the genus by commercial growers. The main form of propagation is by division when plants are being repotted. However, for the *Coelogyne* enthusiast, the aim is likely to be the development of large specimens rather than the production of a number of plants. The techniques for dividing plants apply to all orchids and there is sufficient information on this aspect in the available books on general orchid culture.

Pests and Diseases

Coelogynes would not appear to be susceptible, as many other orchids, to pests and diseases. Maintaining a clean greenhouse with conditions favoured by the plants seems to minimise the influenced of pests and certainly avoids diseases. Ants are troublesome as they bring aphids, mealybugs and scale insects of various kinds into the greenhouse. Ants are easy to control and thus the influence of the common pests can be diminished. Nevertheless, by a regular inspection of the plants, prevention of an infestation by the common pests that seem to affect other orchids is the best method of control. Slugs and snails will attack the new growths but they are easily countered using one of the proprietary brands of slug killer. Last but not least, there is nothing better than a friendly toad or frog in the greenhouse!

BIBLIOGRAPHY

Abraham, A. & Vatsala, P. (1981). *Introduction to Orchids.* St Joseph's Press, Trivandrum, India.

Ames, 0. (1915). 38. *Coelogyne* Lindl. *Orchidaceae* V: 48–54. Merrymount Press, Boston.

Ames, 0 & Schweinfurth, C. (1920). 14. *Coelogyne* Lindl. *Orchidaceae* VI: 21–43. Merrymount Press, Boston.

Banks, D. (1995). Ending the confusion surrounding "*Coelogyne intermedia*". *Australian Orchid Review* 60, 2: 47.

Bechtel, H., Cribb, P.J., Launert, E. (1992). *Manual of Cultivated Orchid Species.* 3rd edition. Blandford Press, London.

Bentham, G. & Hooker J.D. (1883). *Orchideae. Genera Plantarum* 3. L. Reeve & Co., London.

Blume, C.L. (1858). *Collection des Orchidies.* Sulpke, Amsterdam.

Bose, T.K. & Bhattacharjee, S.K. (1980). *Orchids of India.* Naya Prakash, Calcutta, India.

Brummitt, R.K. & Powell, C.E. (1992). *Authors of Plant Names.* Royal Botanic Gardens, Kew.

Butzin, F. (1974). Bestimmungsschlüssel für die in Kultur genommenen Arten der Coelogyninae (Orchidaceae). *Willdenowia* 7: 245–260.

—— (1992). *Coelogyne* Lindl. In F.G.Brieger, R.Maatsch & K. Segas (eds). *Die Orchideen* ed. 3, I A: 919–940. Paul Parey, Berlin.

Chan, C.L., Lamb, A., Shim, P.S. & Wood, J.J. (1994). *Orchids of Borneo.* Vol. 1. Royal Botanic Gardens, Kew.

Chen, S.C., Tsi, Z.H., Lang, K.Y. & Zhu, G.H. (1999). 101. *Coelogyne* Lindl. *Flora Reipublicae Popularis Sinicae* 18: Orchidaceae 2: 338–364. Science Press, Beijing.

Chen, S.C., Tsi, T.H. & Luo, Y.B. (1999). *The Native Orchids of China in Colour.* Science Press, Beijing & New York.

Clayton. D. (1998). *Coelogyne intermedia.* What do we know about it? *The Orchid Review,* 106: (1219).

—— (1999). *Coelogyne intermedia.* Further Thoughts. The *Orchid Review,* 107: (1227).

Comber. J.B. (1990). *Orchids of Java.* Bentham-Moxon Press, Royal Botanic Gardens, Kew.

Cribb, P. & Butterfield, I. (1988). *The Genus Pleione.* Royal Botanic Gardens, Kew in association with Christopher Helm, London & Timber Press, Portland, Oregon.

Cribb, P. & Butterfield, I. (1999). *The Genus Pleione.* 2nd edition, Natural History Publications (Borneo), Kota Kinabalu in association with Royal Botanic Gardens, Kew.

Das, S.J. & Jain, S.K. (1980). Orchidaceae. Genus *Coelogyne. Fascicles of Flora of India 5.* BSI, Howrah.

Deva, S. & Naithani, H.B. (1986). *The Orchids and Flora of Northwest Himalaya.* Print & Media Associates, New Delhi, India.

De Vogel, E.F. (1993). Character Assessment for a Subdivision of *Coelogyne* Lindley. *Proceedings of the]4th World Orchid Conference,* Glasgow: 203–205. HMSO, London.

―――― (1986). Revisions in Coelogyninae (Orchidaceae) II. The genera *Bracisepalum, Chelonistele, Entomophobia, Geesinkorkis* and *Nabaluia. Orchid Monographs 1:* 17–82. E.J. Brill, Leiden.

―――― (1988). Revisions in Coelogyninae (Orchidaceae) III. The genus *Pholidota. Orchid Monographs* 3: 1–118. E.J. Brill, Leiden.

―――― (1992). Revisions in Coelogyninae (Orchidaceae) IV. *Coelogyne,* Section *Tomentosae. Orchid Monographs* 6: 1–42. Hortus Botanicus, Leiden.

De Vogel, E. & Vermeulen, J. (1983). Revisions in Coelogyninae (Orchidaceae) 1. The genus *Bracisepalum. Blumea* 28: 413–418.

Dressler, R.L. (1981). *The Orchids, Natural History* and *Classification.* Harvard University Press, Cambridge, Mass.

Duthie, J.F. (1906). *The Orchids of the North Western Himalaya.* Royal Botanic Gardens, Calcutta.

Flenley, J.R. (1979). *The Equatorial Rain Forest: A Geological History.* Butterworths.

Fyson, P.F. (1915–1920). *Flora of the Nilgiri and Pulney Hill Tops.* Madras.

Gagnepain, F. & Guillaumin, A. (1932). 16. *Coelogyne* Lindl. in Gagnepain, F., ed., *Flore Generale de L'Indo-Chine: 6:* 307–321. Masson, Paris.

Grant, B. (1895). *The Orchids of Burma (including the Andaman Islands).* Hanthawaddy Press, Rangoon.

Gravendeel, B. & de Vogel, E.F. (1999). Revision of *Coelogyne* Section *Speciosae* (Orchidaceae). *Blumea* 44, 2: 253320.

Hawkes, A.D. (1965). *The Encyclopaedia of Orchids.* Faber & Faber, London & Boston.

Holttum, R.E. (1964). *Flora of Malaya. Orchids of Malaya,* 3rd edition. Government Printer, Singapore.

Hooker, J.D. (1890–1891). *Orchideae. Flora of British India* 5: 667–910; 6: 1–198. L. Reeve & Co., London.

―――― (1895). *A Century of Indian Orchids.* Royal Botanic Gardens, Calcutta.

Jackson, B.D. (1971). *A Glossary of Botanical Terms,* 4th edition. Duckworth, London.

Jayaweera, D.M.A. (1981). Orchidaceae. In Dassanayake, M.D. & Fosberg, F. R., eds. *Flora of Ceylon* 2: 4–386. A.A. Balkema, Rotterdam.

Joseph, J. (1987). *Orchids of the Nilgiris.* Botanical Survey India. Calcutta.

Kraenzlin, F. (1878–1900). *Xenia Orchidacea* 3. A. Brockhaus, Leipzig.

Lewis, B.A. & Cribb, P.J. (1991). *Orchids of the Solomon Islands.* Royal Botanic Gardens, Kew.

―――― (1991). *Orchids of Vanuatu.* Royal Botanic Gardens, Kew.

Linden, J. & Reichenbach, H.G. (1857). *Coelogyne assamica* Lind. & Rchb. fil. *Berliner Allegemeine Gartenzeitung* 25: 403.

Lindley, J. (1830–40). *Genera and Species of Orchidaceous Plants.* J. Ridgeway & Sons, London.

―――― (1854). *Folia Orchidacea Coelogyne.* J. Matthews, London.

Lund, I.D. (1987). The genus *Panisea* (Orchidaceae), a taxonomic revision. *Nordic Journal of Botany* 7: 511–527.

Matthew, K.M. (1999). *The Flora of the Palni Hills,* Part 3. St. Joseph's College, Tiruchinapalli, India.

Mueller-Dombois, D. & Ellenberg, H. (1974). *Aims and Methods of Vegetation Ecology.* John Wiley & Sons, New York.

O'Byrne, P. (1994). *Lowland Orchids of Papau New Guinea.* Singapore Botanic Gardens.

Ormerod, P. (1997). A Review of *Coelogyne* Lindl., Section *Proliferae* (Lindl.) Pfitzer. *The Australian Orchid Review* 62, 1: 19–25.

Pederson, H.Æ., Wood, J.J. & Comber, J.B. (1997). A revised subdivision and bibliographical survey of *Dendrochilum* (Orchidaceae). *Opera Botanica. 130:* 5–85.

Pelser, P., Gravendeel, B. & de Vogel, E.F. (2000). Revision of *Coelogyne* section *Fuliginosae* (Orchidaceae). *Blumea* 45: 253–273.

Pfitzer, E. & Kraenzlin, F. (1907). Orchidaceae – Monandrae - Coelogyninae. In: Engler, A., *Das Pflanzenreich. Regni vegetabilis conspectus.* IV. 50. 11. B.7: 1–169. W. Engelmann, Leipzig.

Pradhan, U.C. (1976–79). *Indian Orchids: Guide to Identijication and Culture.* Vols. 1 & 2. Thomson Press (India) Ltd.

Reichenbach, H.G. (1854, 1874). *Xenia Orchidacea.* 1 & 2. A. Brockhaus, Leipzig.

Ruangpanit, N. (1995). Tropical seasonal forests in monsoon Asia: with emphasis on continental southeast Asia. *Vegetatio* 121: 31–40.

Rumphius, G.E. (1750). *Herbarium Amboinense.* Vol. 6. Amsterdam.

Sander, F. (1861–64). *Reichenbachia.* Vols. 1–4 (1861–64). Sanders, St Albans.

Sierra, S.E.C., Gravendeel, B. & de Vogel, E.F. (2000). Revision *of Coelogyne* section *Verrucosae* (Orchidaceae). *Blumea* 45: 275–318.

Schlechter, R. (1982). *The Orchidaceae of German New Guinea.* (1911). Translated into English and edited by Rogers, R.S., Katz, H.J., Simmons, J.T. & Blaxell, D. Australian Orchid Foundation, Melbourne.

Seidenfaden, G. (1975). *Orchid Genera in Thailand* III. *Dansk Botanisk Arkiv* 29(4): 4–49.

—— (1975). Contribution to a revision of the orchid flora of Cambodia, Laos, Vietnam. Pp. 1–117. K. Olsen, Fredensborg.

—— (1982). Contribution to the Orchid Flora of Thailand 10. *Opera Botanica* 62

—— (1986). *Neogyna, Panisea, Otochilus. Opera Botanica* 89.

—— (1992). The Orchids of Indochina. *Opera Botanica* 114.

Seidenfaden, G. & Wood, J.J. (1992). *The Orchids of Peninsular Malaysia & Singapore.* Olsen & Olsen, Fredensborg, Denmark.

Smith, J.J. (1905). *Die Orchideen von Java. Flora von Buitenzorg* VI. E.J. Brill, Leiden.

—— (1908–14). *Die Orchideen von Java.* Figuren—Atlas. E.J. Brill, Leiden.

—— (1930–34). *Icones Orchidcearium Malayensium.* Oegstgust, Holland.

—— (1933). Enumeration of the Orchidaceae of Sumatra and Neighbouring Islands. *Fedde, Repertorium Beihefte* 32: 129–386.

Stearn, W.T. (1992). *Botanical Latin.* 4th edition. David & Charles, Newton Abbot, Devon.

Suriwang Book Centre. (1992). *Thai Orchid Species.* PO Box 44, Chiang Mai 5000, Thailand.

Veitch, H.J. (1887–94). *A Manual of Orchidaceous Plants.* Vols.1 & 2. J. Veitch & Sons, London.

Whitmore, T.C. (1984). *Tropical Rain Forests of the Far East.* 2nd edition. Oxford University Press.

Wood, J.J. & Cribb, P.J. (1994). *A Checklist of the Orchids of Borneo.* Royal Botanic Gardens, Kew.

Wood, J.J., Beaman, R.S. & Beaman, J.H. (1993). *The Plants of Mount Kinabalu 2. Orchids.* Royal Botanic Gardens, Kew.

Wood, J.J., Clayton, D.A. & Chan, C.L. (1998). *Coelogyne motleyi. Sandakania* 11: 35–42.

Wood, J.J. & Barkman, T.J. (1998). Notes on the orchid flora of Mount Kinabalu, Borneo. *Sandakania* 12: 22.

GLOSSARY OF TERMS

Abaxial: the side away from the stem, normally the lower surface.
Abrupt: suddenly ending as though broken off.
Acuminate: having a gradually tapering point.
Acute: distinctly and sharply pointed, but not drawn out.
Adnate: attached to the whole length.
Adventitious: applied to roots arising from a node on the stem.
Alternate: placed on opposite sides of the stem on a different line.
Anatomy: plant structure, or the study thereof, with an emphasis on tissues and their component cells in the interior of the plant.
Anther: that part of the stamen in which the pollen is produced.
Apical: applied to an inflorescence borne at the top of the pseudobulb.
Appressed: lying flat for the whole length of the organ.
Arching: curved like a bow.
Arcuate: curved like a bow.
Articulate: jointed; said of leaves or other parts which have an abscission layer or a joint at the base.
Auriculate: with small lobes or ear-like appendages.
Basal: applied to the inflorescence at the base of an organ or part such as the pseudobulb.
Bract: a much reduced leaf-like organ bearing a flower, inflorescence or partial inflorescence in its axil.
Callus: a thickened area of the lip.
Calyx: the outermost of the floral envelope.
Carnosus: fleshy, firm.
Cataphylls: the early leaf forms of a plant or shoot.
Caudicle: a slender, mealy. or elastic extension of the pollinium, or a mealy portion at the end of the pollinium; the structure is part of the pollen mass, and is produced within the anther.
Caulescent: becoming stalked, where the stalk is clearly apparent.
Channelled: hollowed out like a gutter, as in a leaf-stalk.
Clavate: club shaped, thickened towards the apex.
Claw: the conspicuously narrowed and attenuated base of an organ.
Column: an advanced structure composed of a continuation of the flower-stalk, together with the upper part of the female reproductive organ (pistil) and the male reproductive organ (stamen).
Concave: hollow, as the inside of a saucer.
Conduplicate: of leaf-like organs, with a single median fold, with each half being flat.
Convex: having a more or less rounded surface.
Convolute: when one part is wholly rolled up in another.

Cordate: heart-shaped.
Coriaceous: leathery.
Crenate: scalloped, toothed with crenatures.
Crenulate: crenate, but the teeth small.
Crisp: curled.
Crispate: curled.
Cristate: crested.
Crowded: closely pressed together or thickly set.
Decurrent: running down, as when leaves are prolonged beyond their insertion, an thus run down the stem.
Deflexed: bent outwards, the opposite of inflexed.
Deflexion: turned downwards.
Dentate: toothed.
Denticulate: minutely toothed.
Dilated: expanding into a blade.
Distichous: having leaves or other organs in two opposite rows, as opposed to spiral or whorled.
Dorsal: referring to the back, or attached thereto.
Duplicate: refers to the folding of the leaves during development; folded once with each half flat.
Ellipsoid: an elliptic solid.
Elliptic: ellipse-shaped, oblong with regularly rounded ends.
Emarginate: notched, usually at the apex.
Embracing: clasping at the base.
Entire: without tooth or divisions.
Epichile: the terminal part of the lip when it is distant from the basal portion.
Epiphyte: a plant growing on another plant but not parasitic.
Erect: upright, perpendicular to the ground or its attachment.
Erose: bitten or gnawed.
Falcate: sickle-shaped.
Filiform: thread-like.
Fimbriate: fringed.
Flaccid: withered and limp, flabby.
Flexuose: bent alternatively in opposite directions, zig-zag.
Free: not adhering.
Fusiform: spindle-shaped.
Genuflexed: bent.
Gibbous: more convex in one place than another
Globose: nearly spherical.
Globular: spheroidal.
Heteranthous: an apical inflorescence produced on a separate shoot which does not develop to produce a pseudobulb and leaves.
Hooded: a plane body, the apex or sides of which are curved inwards to resemble a hood.
Hypochile: the basal portion of the lip.
Hysteranthous: of an apical inflorescence produced after the pseudobulbs and leaves have developed.
Imbricate: overlapping as the tiles of a roof.
Incised: cutting sharply into the margin.

Glossary of Terms

Incurved: bending from without inwards.
Inflected: bent or flexed.
Inflexed: turned abruptly or bent outward, incurved, opposite of deflexed.
Inflorescence: the disposition of the flowers on the floral axis or the flowers, bracts and floral axis in toto.
Isthmus: the narrowed connection between parts of a flower.
Karst: limestone regions with underground drainage and many cavities and passages caused by the dissolution of the rock.
Keel: a ridge like the keel of a boat.
Knee: an abrupt bend in the stem or part of a flower.
Labellum: the enlarged, often highly modified third petal of the orchid flower.
Lacerate: torn, an irregular cleft.
Lacinate: slashed, cut into narrow lobes.
Lanceolate: narrow, tapering at each end, lance-shaped.
Lateral: fixed on or near side of the organ.
Lax: loose, distant.
Linear: at least 12 times longer than broad, with the sides about parallel.
Lip: the labellum of an orchid.
Lithophyte: growing upon stones and rocks.
Lobe: any division of an organ or especially rounded division.
Margin: the edge or boundary line of a body.
Median: belonging to the middle.
Membranaceous: thin or semi-transparent, like a thin membrane.
Mid-rib: the principle nerve in leaf.
Monilform: necklace shaped, like a string of beads.
Montane: pertaining to mountains, as a plant which grows on them.
Morphology: the study of form, particularly external structure.
Mucro: a sharp terminal point.
Mucronate: possessing a mucro.
Oblanceolate: tapering towards the base more than towards the apex.
Oblique: slanting or of unequal sides.
Oblong: much longer than broad, with nearly parallel sides.
Obovate: reversed ovate.
Obtuse: blunt or rounded at the end.
Obverse: the side facing, as opposed to reverse.
Operculate: lid like.
Ovate: egg-shaped, broader at the base.
Ovoid: egg-shaped solid.
Pandurate: fiddle-shaped.
Papillose: covered with soft superficial glands or protuberances, i.e. papillae.
Papyraceous: papery.
Pectinate: combed.
Pedicel: the stalk of a single flower in an inflorescence.
Peduncle: the stalk bearing an inflorescence or solitary flower.
Perloric: an abnormality in which the lip has the form of a petal; a mutant form.
Petal: a flower leaf.
Petiole: the foot stalk of a leaf.
Plicate: leaves having several or many longitudinal veins and usually folded at each one:

pleated.
Pollinia: pollen masses.
Porrect: directed outwards and forward.
Proteranthous: of an apical inflorescence produced before the pseudobulbs and leaves on the same shoot.
Pseudobulb: a thickened and a bulb-like internode in orchids.
Pyriform: pear-shaped.
Rachis: the axis of an inflorescence to which the pedicels are attached, above the peduncle.
Recurved: curved backwards and downwards.
Reflexed: abruptly bent downwards or backwards.
Resupinate: the flower having the lip on the lower side.
Retuse: with a shallow notch at a rounded apex.
Rhizome: rootstock bearing scale leaves and adventitious roots.
Rostellum: that portion of the stigma which aids in gluing the pollinia to the pollinating agent; tissue which separates the anther from the fertile stigma; sometimes beak-like
Saccate: a conspicuous hollow swelling.
Scale: any thin scarious body, usually a degenerate leaf, sometimes of epidermal origin.
Scape: the peduncle and the rachis of the inflorescence.
Sepal: term in universal use for each segment composing a calyx.
Serrate: sharp saw-like teeth.
Serrulate: finely serrate with a series of small notches.
Sessile: without a stalk.
Simultaneous: at the same time.
Spathulate: oblong and attenuated at the base, like a spatula.
Stelidia: column wings or teeth.
Stigma: the sticky receptive part of the pistil, produces a viscid, sugary material which receives the pollinia and permits the pollen grains to germinate.
Stipe: the stalk-like support of an organ, e.g. pollinium.
Subtend: to extend under, or be opposite to.
Sulcate: grooved or furrowed.
Synanthous: of an apical inflorescence produced at the same time as the pseudobulb and leaves.
Taxon (pl. Taxa): a taxonomic group of any rank.
Terete: circular in transverse section, cylindric and usually tapering.
Terrestrial: growing in the earth.
Undulate: wavy.
Unguiculate: contracted at the base into a claw.
Ventral: on the lower side; belly.
Verrucose: warty.
Wing: a membraneous expansion attached to an organ.
Zig-zag: having short bends or angles from side to side.

ACKNOWLEDGEMENTS

My interest in the Genus *Coelogyne* and the evolution of this book has been a most thought provoking experience. It has put into sharp relief the naivety of my initial approach, when it was necessary to retrace my steps on numerous occasions and the more confident approach I can now adopt when I am asked to name a specimen. I am grateful to Dr. Phillip Cribb, Jeffrey Wood and Sarah Thomas, of the Royal Botanic Gardens, Kew, who have offered me guidance and who have given answers to my questions over the past few years. I would also like to thank Sandra Bell, former Supervisor of the Kew orchid collection, for allowing me regular access to the Kew collection of *Coelogyne* plants. This enabled me to expand my knowledge about the growing habits of the various species. During my regular visits to Kew, I had the opportunity to meet Jim Comber, author of the *Orchids of Java* and *Orchids of Sumatra*. We exchanged information and he very kindly gave me copies of some of his *Coelogyne* slides, which have been invaluable during my studies. I have thoroughly enjoyed the exchange of ideas I have had with Barbara Gravendeel during her PhD research at the National Herbarium, the Netherlands in Leiden. Barbara, in association with Dr. Ed de Vogel has kindly provided the chapter containing ideas on the molecular phylogeny of *Coelogyne* and allied genera. The *Coelogyne* lip shapes come from a miscellany of sources and whilst the sources have been identified in the text, I specifically acknowledge the permission given by E.F. de Vogel and B. Gravendeel to use illustrations from the Monographs on sect. *Tomentosae* and sect. *Speciosae*.

I am grateful to the curators of the following herbaria for access to their collections and in some cases the loan of material: AMES, BM, E, K, L and P.

The interest shown by Dr. Ed de Vogel, of the National Herbarium, the Netherlands, in the results of my investigations was a great source of encouragement and I am grateful to him and to Phillip Cribb and Jeffrey Wood for their advice on the contents of the manuscript at various stages. I am indebted to Linda Gurr and Judi Stone, botanical artists associated with Kew, for the fine black and white line drawings produced for this book; Jeff Eden, also of Kew, for the conversion of my manuscript maps into their present form. I very much appreciated the opportunity to use the fine photographic skills of R.S. Beaman, C.L. Chan, C. Clarke, J.B. Comber, J. Cootes, P.J. Cribb, K. Jayaram, S.K. Jacobson, P. Jongejan, B. Kieft, C.G. Koops, S. Kumar, A. Lamb, J. Meijvogel, M. Perry, W.M. Poon, A. Schuiteman, C. Sussendran, D. Titmus, A. Vogel and I acknowledge their kind permission to use their work in this publication.

I have appreciated the regular contact with Dr. Satish Kumar of the Tropical Botanic Garden and Research Institute in Kerala, South India and his considerable assistance in obtaining photographs of a range of Indian species and other material. I thank David Banks of New South Wales, Australia, who is Editor of *Australian Orchid Review* and *The Orchadian* for reviewing and giving me advice on my chapters on Hybridisation and Cultivation. I welcomed the contact with Paul Ormerod of Cairns, Queensland, Australia with whom I have discussed many aspects of *Coelogyne* species and who sent me the article which was published in *Australian Orchid Review* based on the information he had gathered on the sect. *Proliferae* species and on some of the sect. *Elatae* species.

I am indebted to Dr. Kevin Davies, a regular contributor to *The Orchid Review*, for the stimulating exchanges of views we have had on a subject of mutual interest. I would like to single out a fellow *Coelogyne* enthusiast, Colin Howe, who has been in close contact with me during my studies and who gave me specimens to evaluate and plants for my collection.

I am grateful to the Royal Botanic Gardens, Kew and Natural History Publications (Borneo) for having the courage to publish this work.

My wife, Jill, has been a great source of strength during the evolution of this book and I am sure that without her constant encouragement, the work would still be in the notebooks, and I dedicate this volume to her.

Colour plates

Plate 1

A. *Coelogyne barbata* Griff. Cult. D.A. Clayton. (Photo: D.A. Clayton) (Page 32)

B. *Coelogyne calcicola* Kerr. China, Yunnan, Mong Pang. (Photo: P.J. Cribb) (Page 33)

C. *Coelogyne holochila* P.F. Hunt & Summerh. Cult. D.A. Clayton. (Photo: D.A. Clayton) (Page 35)

D. *Coelogyne stricta* (D. Don) Schltr. Cult. D.A. Clayton. (Photo: D.A. Clayton) (Page 40)

E. *Coelogyne tenasserimensis* Seidenf. Cult. R.B.G. Kew ex Thailand. (Photo: D.A. Clayton) (Page 40)

F. *Coelogyne ecarinata* C. Schweinf. China, Yunnan, Maguan. f. (Photo: P.J. Cribb) (Page 43)

Plate 2

A. *Coelogyne longipes* Lindl. Bhutan, Pele La. (Photo: P.J. Cribb) (Page 44)

B. *Coelogyne longipes* Lindl. Bhutan, Pele La. (Photo: P.J. Cribb) (Page 44)

C. *Coelogyne prolifera* Lindl. Cult. D.A. Clayton. (Photo: D.A. Clayton) (Page 44)

D. *Coelogyne raizadae* S.K. Jain & S. Das. Bhutan, Trashiyangtse. (Photo: P.J. Cribb) (Page 45)

E. *Coelogyne schultesii* S.K. Jain & S. Das. Bhutan, Trashiyangtse. (Photo: P.J. Cribb) (Page 46)

Plate 3

A. *Coelogyne fimbriata* Lindl. Cult. D.A. Clayton. (Photo: D.A. Clayton) (Page 51)

B. *Coelogyne fuliginosa* Lodd. ex Hook. Cult. OSGB Show, Syon Park. (Photo: D.A. Clayton) (Page 52)

C. *Coelogyne ovalis* Lindl. Cult. D.A. Clayton. (Photo: D.A. Clayton) (Page 53)

D. *Coelogyne padangensis* J.J. Sm. & Schltr. North Sumatra, Kg. Susuk. (Photo: J.B. Comber) (Page 54)

E. *Coelogyne pallens* Ridl. Cult. D.A. Clayton. (Photo: D.A. Clayton) (Page 54)

F. *Coelogyne triplicatula* Rchb.f. Cult. D.A. Clayton. (Photo: D.A. Clayton) (Page 55)

Plate 4

A. *Coelogyne brachyptera* Rchb.f. Cult. Rome EOC. (Photo: P.J. Cribb) (Page 59)

B. *Coelogyne parishii* Hook. Cult. Rome EOC. (Photo: P.J. Cribb) (Page 60)

C. *Coelogyne virescens* Rolfe. Cult. D.A. Clayton. (Photo: D.A. Clayton) (Page 60)

D. *Coelogyne beccarii* Rchb.f. Cult. Leiden 32230. (Photo: J. Meijvogel) (Page 67)

E. *Coelogyne carinata* Rolfe. Papua New Guinea, Laiagam. (Photo: P.J. Cribb) (Page 68)

F. *Coelogyne celebensis* J.J. Sm. Cult. D.A. Clayton. (Photo: D.A. Clayton) (Page 69)

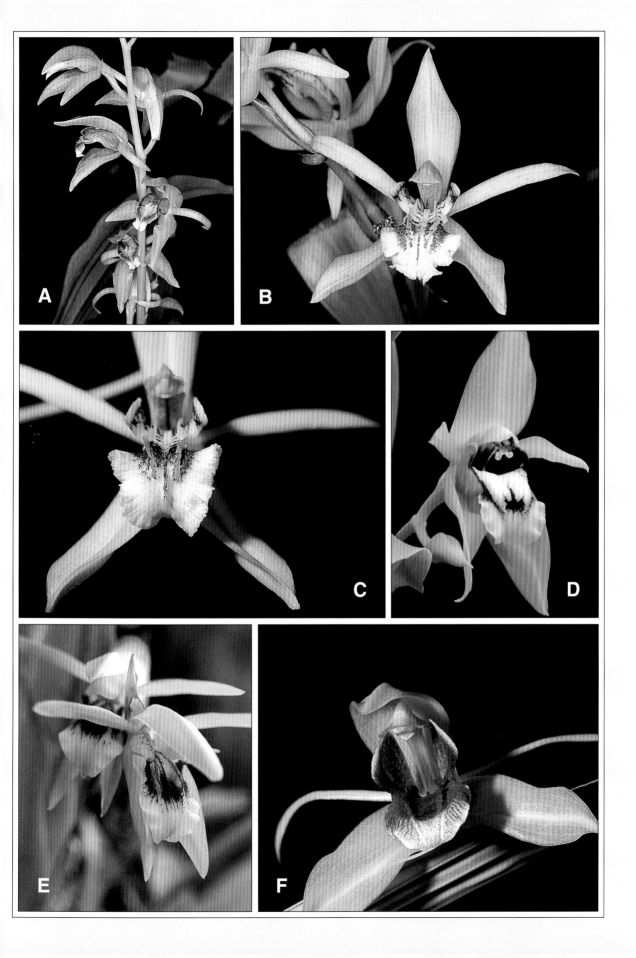

Plate 5

A. *Coelogyne fragrans* Schltr. Cult. Leiden. (Photo: D.A. Clayton) (Page 70)

B. *Coelogyne fuerstenbergiana* Schltr. Cult. Rizal, Bandung. (Photo: P.J. Cribb) (Page 71)

C. *Coelogyne lycastoides* F. Muell. & Kraenzl. Cult. M. Perry. (Photo: M. Perry) (Page 72)

D. *Coelogyne macdonaldii* F. Muell. & Kraenzl. Cult. R.B.G. Edinburgh. (Photo: J.B. Comber) (Page 73)

E. *Coelogyne rumphii* Lindl. Cult. D.A. Clayton. (Photo: D.A. Clayton) (Page 74)

F. *Coelogyne salmonicolor* Rchb.f. Cult. Bogor B.G. ex Sumatra, Jambi. (Photo: P.J. Cribb) (Page 75)

Plate 6

A. *Coelogyne septemcostata* J.J. Sm. Cult. R.B.G. Kew, ex Sarawak, Borneo. (Photo: J.B. Comber) (Page 76)

B. *Coelogyne speciosa* (Blume) Lindl. subsp. *speciosa*. Cult. D.A. Clayton. (Photo: D.A. Clayton) (Page 76)

C. *Coelogyne speciosa* (Blume) Lindl. subsp. *incarnata* Gravendeel. Cult. Leiden. 21412. (Photo: J. Meijvogel) (Page 76)

D. *Coelogyne susanae* P.J. Cribb & B.A. Lewis. Solomon Islands. (Photo: P.J. Cribb) (Page 79)

E. *Coelogyne tiomanensis* M.R. Henderson. Cult. D.A. Clayton. (Photo: D.A. Clayton) (Page 80)

F. *Coelogyne tommii* Gravendeel & P. O'Byrne. Cult. Leiden 21526. (Photo: J. Meijvogel) (Page 80)

Plate 7

A. *Coelogyne usitana* Röth & Gruss. Cult. M. Perry. (Photo: M. Perry) (Page 81)

B. *Coelogyne xyrekes* Ridl. Cult. Leiden 960160. (Photo: A. Vogel) (Page 82)

C. *Coelogyne bicamerata* J.J. Sm. Cult. Leiden 931067. (Photo: A. Vogel) (Page 84)

D. *Coelogyne gibbifera* J.J. Sm. Cult. Leiden 22074. (Photo: B. Kieft) (Page 89)

E. *Coelogyne harana* J.J. Sm. Cult. Leiden 27532. (Photo: P. Jongejan) (Page 89)

Plate 8

A. *Coelogyne incrassata* (Blume) Lindl. var. *incrassata*. Cult. D.A. Clayton. (Photo: D.A. Clayton) (Page 90)

B. *Coelogyne incrassata* (Blume) Lindl. var. *valida* J.J. Sm. Borneo, Kalimantan, Apokayan, Gunung Sungai, Pendan. (Photo: P.J. Cribb) (Page 91)

C. *Coelogyne kelamensis* J.J. Sm. Cult. Leiden 930568. (Photo: J. Meijvogel) (Page 91)

D. *Coelogyne longpasiaensis* J.J. Wood & C.L. Chan. Borneo, Kalimantan, Apokayan, Long Sungai, Barang. (Photo: P.J. Cribb) (Page 92)

E. *Coelogyne monilirachis* Carr. Borneo, Kalimantan, Apokayan, Gunung Sungai, Pendan. (Photo: P.J. Cribb) (Page 92)

F. *Coelogyne naja* J.J. Sm. Cult. Leiden 913470. (Photo: J. Meijvogel) (Page 93)

Plate 9

A. *Coelogyne vermicularis* J.J. Sm. Cult. Leiden 914400. (Photo. J. Meijvogel) (Page 94)

B. *Coelogyne bilamellata* Lindl. Borneo. (Photo: P.J. Cribb) (Page 101)

C. *Coelogyne brachygyne* J.J. Sm. North Sumatra, near Merek-Sidikalang road. (Photo: J.B. Comber) (Page 102)

D. *Coelogyne compressicaulis* Ames & C. Schweinf. Borneo, Sabah, Mount Kinabalu. (Photo: C.L. Chan) (Page 103)

E. *Coelogyne cuprea* H. Wendl. & Kraenzl. var. *cuprea.* Cult. R.B.G. Kew ex Sumatra, Lake Kawar. (Photo: J.B. Comber) (Page 104)

F. *Coelogyne cuprea* H. Wendl. & Kraenzl. var. *planiscapa* J. J. Wood & C. L. Chan Borneo, Sabah, Mount Kinabalu. (Photo: P.J. Cribb) (Page 105)

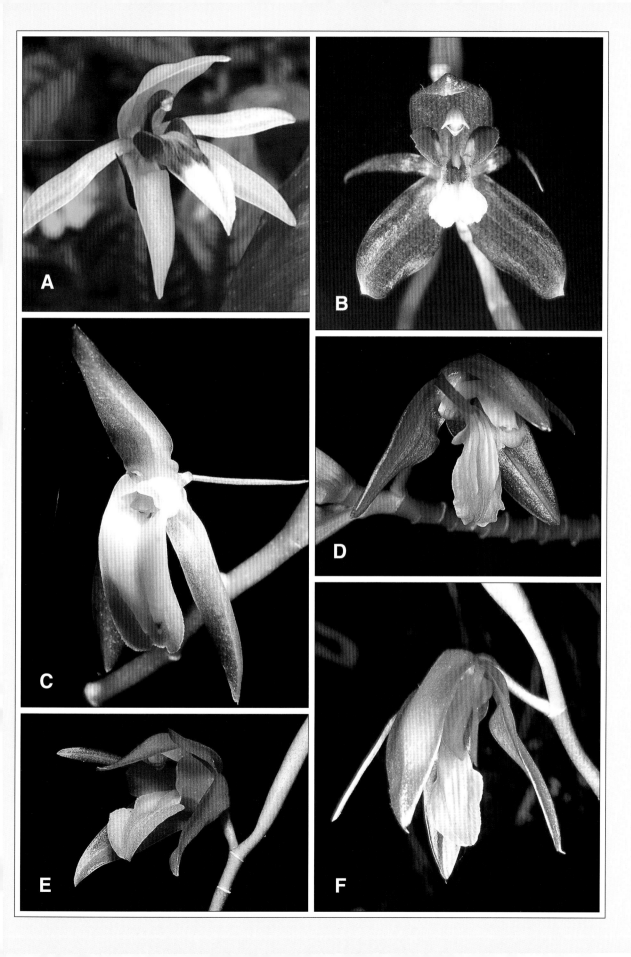

Plate 10

A. *Coelogyne endertii* J.J. Sm. Borneo, Long Sungai, Barang. (Photo: P.J. Cribb) (Page 107)

B. *Coelogyne kinabaluensis* Ames & C. Schweinf. Cult. R.B.G. Edinburgh. (Photo: J.B. Comber) (Page 108)

C. *Coelogyne longifolia* (Blume) Lindl. West Java, Puncak. (Photo: J.B. Comber) (Page 108)

D. *Coelogyne motleyi* Rolfe ex J.J. Wood, D.A. Clayton & C.L. Chan. Borneo, Sabah. Cult. Poring Orchid Centre. (Photo: C.L. Chan) (Page 109)

E. *Coelogyne planiscapa* Carr var. *grandis* Carr. Borneo, Sabah. (Photo: R.S. Beaman) (Page 110)

F. *Coelogyne prasina* Ridl. Cult. D.A. Clayton. (Photo: D.A. Clayton) (Page 112)

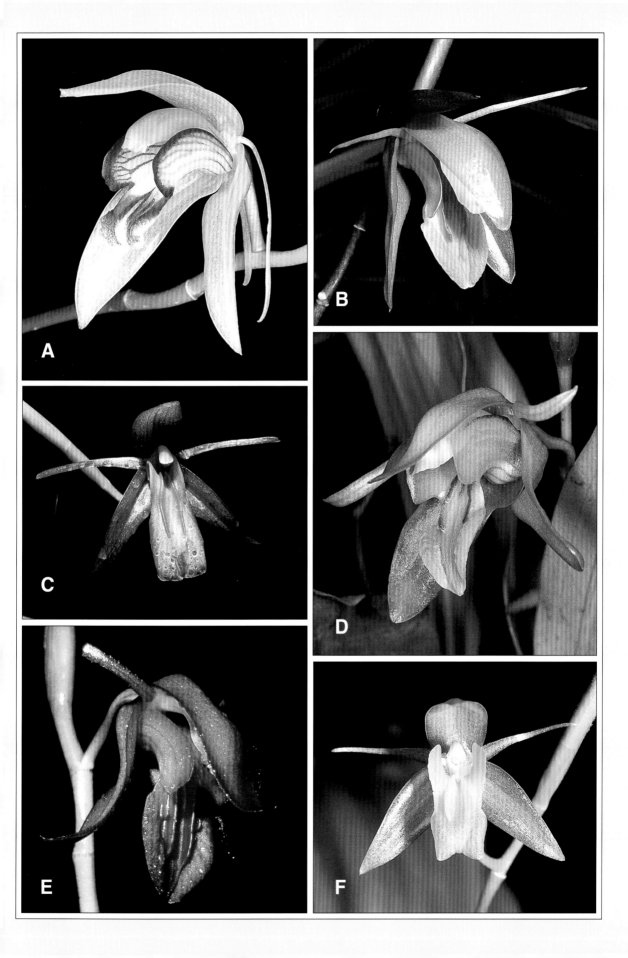

Plate 11

A. *Coelogyne quinquelamellata* Ames. Cult. J. Cootes. (Photo: D. Titmuss) (Page 112)

B. *Coelogyne quinquelamellata* Ames. Cult. J. Cootes. (Photo: D. Titmuss) (Page 112)

C. *Coelogyne radicosa* Ridl. Peninsular Malaysia, Genting highlands. (Photo: P.J. Cribb) (Page 113)

D. *Coelogyne stenochila* Hook.f. Cult. Leiden 960221. (Photo: A. Vogel) (Page 115)

E. *Coelogyne tenompokensis* Carr. Borneo, Sabah, Mount Kinabalu. (Photo: C. L. Chan) (Page 116)

F. *Coelogyne multiflora* Schltr. Cult. Leiden 20200. (Photo: B. Kieft) (Page 118)

Plate 12

A. *Coelogyne asperata* Lindl. Cult. D.A. Clayton. (Photo: D.A. Clayton) (Page 122)

B. *Coelogyne marthae* S.E.C. Sierra. Cult. Semonggoh Orchid Centre, Sarawak. (Photo: C.L. Chan) (Page 123)

C. *Coelogyne mayeriana* Rchb.f. (Photo: C. Suseendran) (Page 124)

D. *Coelogyne pandurata* Lindl. Cult. Tenom Orchid Centre, Sabah. (Photo: C.L. Chan) (Page 125)

E. *Coelogyne papillosa* Ridl. ex Stapf. Borneo, Sabah, Kinabalu. (Photo: P.J. Cribb) (Page 125)

F. *Coelogyne peltastes* Rchb.f. Cult. Leiden. (Photo: P.J. Cribb) (Page 126)

Plate 13

Coelogyne verrucosa S.E.C. Sierra. Cult. Tenom Orchid Centre, ex Nabawan kerangas forest. (Photo: C.L. Chan) (Page 126)

Plate 14

A. *Coelogyne zurowetzii* Carr. Borneo, Sabah. (Photo: P.J. Cribb) (Page 127)

B. *Coelogyne bruneiensis* de Vogel. Cult. Leiden 27697. (Photo: A. Vogel) (Page 136)

C. *Coelogyne echinolabium* de Vogel. Borneo, Sarawak. (Photo: C.G. Koops) (Page 137)

D. *Coelogyne genuflexa* Ames & C. Schweinf. Borneo, Sabah, Sinsuran Road. (Photo: C. L. Chan) (Page 138)

E. *Coelogyne hirtella* J.J. Sm. Borneo, Sabah, Mount Kinabalu. (Photo: W.M. Poon) (Page 139)

F. *Coelogyne judithiae* P. Taylor. Peninsular Malaysia, Taiping Hills. (Page 139)

Plate 15

A. *Coelogyne kaliana* P.J. Cribb. Peninsular Malaysia, Gunung Ulu Kali. (Photo: C. Clarke) (Page 140)

B. *Coelogyne longibulbosa* Ames & C. Schweinf. Cult. Leiden 914069. (Photo: J. Meijvogel) (Page 141)

C. *Coelogyne moultonii* J.J. Sm. Borneo, Sabah, near Mesilau in foothills of Mount Kinabalu. (Photo: J.B. Comber) (Page 142)

D. *Coelogyne muluensis* J.J. Wood. Borneo, Sarawak. (Photo: A. Lamb) (Page 143)

E. *Coelogyne odoardi* Schltr. Borneo, Sarawak. (Photo: P.J. Cribb) (Page 143)

Plate 16

A. *Coelogyne pholidotoides* J.J. Sm. Cult. Leiden 26848. (Photo: P. Jongejan) (Page 145)

B. *Coelogyne pulverula* Teijsm. & Binn. Cult. Poring Orchid Centre, Kinabalu Park, Sabah). (Photo: C. L. Chan) (Page 145)

C. *Coelogyne radioferens* Ames & C. Schweinf. Borneo, Sabah, Kinabalu, summit trail, 7000 feet asl. (Photo: C.L. Chan) (Page 146)

D. *Coelogyne rhabdobulbon* Schltr. Borneo, Sabah, Kinabalu. (Photo: S.K. Jacobson) (Page 147)

Plate 17

A. *Coelogyne rochussenii* de Vriese. Cult. R.B.G., Kew. (Photo: J.B. Comber) (Page 148)

B. *Coelogyne rupicola* Carr. Borneo, Sabah, Kinabalu. (Photo: S.K. Jacobson) (Page 149)

C. *Coelogyne squamulosa* J.J. Sm. Cult. Leiden. (Photo: André Schuiteman) (Page 149)

D. *Coelogyne swaniana* Rolfe. Borneo, Poring Orchid Centre. (Photo: C.L. Chan) (Page 150)

E. *Coelogyne testacea* Lindl. Cult. Tenom Orchid Centre, Sabah. (Photo: C.L. Chan) (Page 151)

Plate 18

A. *Coelogyne tomentosa* Lindl. Cult. D.A. Clayton. (Photo: D.A. Clayton) (Page 152)

B. *Coelogyne velutina* de Vogel. Cult. Leiden 25835. (Photo: A. Vogel) (Page 153)

C. *Coelogyne venusta* Rolfe. Borneo, Sabah, Mount Kinabalu. (Photo: P.J. Cribb) (Page 154)

D. *Coelogyne malipoensis* Z.H. Tsi. North Vietnam, Meo Vac. (Photo: P.J. Cribb) (Page 156)

E. *Coelogyne malipoensis* Z.H. Tsi. North Vietnam, Meo Vac. (Photo: P.J. Cribb) (Page 156)

Plate 19

A. *Coelogyne miniata* (Blume) Lindl. (Photo: J.B. Comber) (Page 156)

B. *Coelogyne clemensii* Ames & C. Schweinf. var. *clemensii*. Borneo, Sabah, Mount Kinabalu. (Photo: P.J. Cribb) (Page 160)

C. *Coelogyne craticulaelabris* Carr. Borneo, Sabah, Mount Kinabalu. (Photo: C.L. Chan) (Page 162)

D. *Coelogyne exalata* Ridl. Cult. Leiden 914394. (Photo: A. Vogel) (Page 162)

E. *Coelogyne plicatissima* Ames & C. Schweinf. Borneo, Sabah, Mount Kinabalu. (Photo: P.J. Cribb) (Page 163)

F. *Coelogyne veitchii* Rolfe. Cult. R.B.G Kew, ex Western New Guinea. (Photo: P.J. Cribb) (Page 165)

Plate 20

A. *Coelogyne flexuosa* Rolfe. Cult. D.A. Clayton. (Photo: D.A. Clayton) (Page 167)

B. *Coelogyne eberhardtii* Gagnep. Cult. Leiden 970797. (Photo: André Schuiteman) (Page 169)

C. *Coelogyne lawrenceana* Rolfe. Cult. D.A. Clayton. (Photo: D.A. Clayton) (Page 169)

D. *Coelogyne cristata* Lindl. Cult. D.A. Clayton. (Photo: D.A. Clayton) (Page 173)

E. *Coelogyne cumingii* Lindl. Cult. D.A. Clayton. (Photo: D.A. Clayton) (Page 174)

F. *Coelogyne foerstermannii* Rchb.f. Cult. Bogor Botanic Gardens, Indonesia. (Photo: J.B. Comber) (Page 175)

Plate 21

A. *Coelogyne kemiriensis* J.J. Sm. Sumatra, Mount Kemiri. (Photo: J.B. Comber) (Page 177)

B. *Coelogyne mooreana* Sand. ex Rolfe. Cult. D.A. Clayton. (Photo: D.A. Clayton) (Page 178)

C. *Coelogyne nervosa* A. Rich. South India. (Photo: C. Suseendran) (Page 178)

D. *Coelogyne sanderiana* Rchb.f. North Sumatra, kampung Susuk. (Photo: J.B. Comber) (Page 179)

E. *Coelogyne assamica* Linden & Rchb.f. Cult. D.A. Clayton. (Photo: D.A. Clayton) (Page 181)

F. *Coelogyne fuscescens* Lindl. var. *fuscescens*. Cult. ex Northeast India. (Photo: K. Jayaram) (Page 183)

Plate 22

A. *Coelogyne fuscescens* Lindl. var. *brunnea* Lindl. Cult. C. Howe. (Photo: D.A. Clayton) (Page 183)

B. *Coelogyne corymbosa* Lindl. Bhutan, East Trongsa. (Photo: P.J. Cribb) (Page 187)

C. *Coelogyne nitida* (Wall. ex D. Don) Lindl. OSGB Show. (Photo: D.A. Clayton) (Page 189)

D. *Coelogyne punctulata* Lindl. Cult. J. Cootes. (Photo: J. Cootes) (Page 191)

E. *Coelogyne breviscapa* Lindl. Cult. Trivandrum B.G. (Photo: Satish Kumar) (Page 196)

F. *Coelogyne chloroptera* Rchb.f. Philippines. (Photo: J. Cootes) (Page 197)

Plate 23

A. *Coelogyne lentiginosa* Lindl. Cult. D.A. Clayton. (Photo: D.A. Clayton) (Page 199)

B. *Coelogyne marmorata* Rchb.f. Cult. J. Cootes. (Photo: D. Titmuss) (Page 200)

C. *Coelogyne merrillii* Ames. Philippines. (Photo: D. Titmuss) (Page 201)

D. *Coelogyne mossiae* Rolfe em. S.K. Jain & S. Das. South India, Munnar. (Photo: Satish Kumar) (Page 202)

E. *Coelogyne odoratissima* Lindl. South India. (Photo: K. Jayaram) (Page 203)

F. *Coelogyne albolutea* Rolfe. Northeast India. (Photo: C. Suseendran) (Page 210)

Plate 24

A. *Coelogyne flaccida* Lindl. Cult. D.A. Clayton. (Photo: D.A. Clayton) (Page 211)

B. *Coelogyne huettneriana* Rchb f. O.S.G.B. Show. (Photo: D.A. Clayton) (Page 213)

C. *Coelogyne trinervis* Lindl. Cult. D.A. Clayton. (Photo: D.A. Clayton) (Page 214)

D. *Coelogyne viscosa* Rchb.f., Otto & Dietr. Cult. D.A. Clayton. (Photo: D.A. Clayton) (Page 215)

E. *Coelogyne* Unchained Melody. Cult. D.A. Clayton. (Photo: D.A. Clayton) (Page 238)

F. *Coelogyne* Mem. W. Micholitz. Cult. Burnham Nurseries (Photo: D.A. Clayton) (Page 237)

INDEX OF SCIENTIFIC NAMES

(Accepted names appear in roman type **bold**. Synonyms appear in *italics*. Numbers in **bold** type indicated page with detailed treatment. Numbers within [brackets] indicate page with figures and numbers within (brackets) indicate page with photograph).

Acoridium Nees & Meyen 3
Adrorrhizon purpurascens (Thw.) Hook.f. 223
Bletilla stricta (Thunb.) Rchb. f. 219
Bracisepalum J.J. Sm. 3, 228
Bromheadia finlaysoniana (Lindl.) Miq. 218
Bulleya Schltr. 3
Calanthe rubens Ridl. 224
Camelostalix Pfitzer 15
Chelonanthera Blume 15
Chelonistele Pfitzer 3, 15, 228, 231
    amplissima (Ames & C. Schweinf.) Carr 217
    brevilamellata (J.J. Sm.) Carr 218
    ingloria (J.J. Sm.) Carr 220
    lamellulifera Carr 221
    lurida (L. Lind. & Cogn.) Pfitzer **var. lurida** 221, 224
    perakensis (Rolfe) Ridl. 222
    richardsii Carr 224
    sulphurea (Blume) Pfitzer **var. sulphurea** 217, 219, 220, 223, 225
    sulphurea (Blume) Pfitzer **var. crassifolia** (Carr) Masamune 219
    unguiculata Carr 225
Coelogyne Lindl.
    sect. *Ancipites* Pfitzer 15, 16, 17, 28, 32, 35, 57
    sect. **Bicellae** J.J. Sm. 15, 16, 17, 21, **83**, [84], 85, 228, 230
    sect. **Brachypterae** D.A. Clayton 17, 20, **57**, [58], 60, 228
    sect. *Carinatae* Pfitzer 15, 16, 17, 207
    sect. **Coelogyne** [2], 15, 16, 18, 25, **170**, [171], 202, 216, 228, 230, 231
    sect. *Cristatae* Pfitzer 15, 16, 170
    sect. **Cyathogyne** (Schltr.) D.A. Clayton 16, 18, 23, 117, [118], 228, 230, 231
    sect. **Elatae** Pfitzer 12, 15, 16, 17, 19, **28**, [28–29], 228, 230, 231, 243, 254
    sect. *Erectae* Lindl. 15
    sect. *Filiferae* Lindl. 15
    sect. **Flaccidae** Lindl. 15, 16, 17, 19, 27, 152, **207**, [209], 216, 228, 230, 243
    sect. *Flexuosae* Lindl. 15
    sect. **Fuliginosae** Lindl. 15, 16, 17, 20, **47**, [48], 57, 228, 230, 243

sect. **Fuscescentes** Pfitzer 15, 16, 17, 18, 25, **180**, [180]
sect. **Hologyne** (Pfitzer) D.A. Clayton 15, 16, 18, 24, **154**, [155], 160, 228, 230
sect. **Lawrenceanae** D.A. Clayton 18, 25, **167**, [168], 228, 230, 242
sect. **Lentiginosae** Pfitzer 15, 16, 19, 27, 176, 177, **192**, [193–194], 204, 228
sect. **Longifoliae** Pfitzer 15, 18, 22, **95**, [96-97], 104, 228, 230
sect. **Micranthae** Pradhan 16, 17, 20, **56**, [56]
sect. **Moniliformes** Carr 16, 18, 21, **85**, [86–87]
sect. **Ocellatae** Pfitzer 15, 16, 19, 26, **185**, [186], 188, 192, 230, 243
sect. **Proliferae** Lindl. 12, 15, 16, 17, 19, 37, **41**, [43], 254
sect. **Ptychogyne** (Pfitzer) D.A. Clayton 15, 16, 18, 25, **166**, [166], 228, 230
sect. **Rigidiformes** Carr 16, 18, 24, **157**, [158], 160, 165, 228, 230 231
sect. **Speciosae** Lindl. 15, 16, 17, 20, **61**, [64–66], 68, 70, 72, 82, 83, 228, 230, 243, 253
sect. **Tomentosae** Pfitzer 9, 15, 16, 18, 23, **128**, [129–131], 144, 152, 165, 228, 230, 231, 242, 243, 253
sect. **Veitchiae** D.A. Clayton 18, 25, **164**, [165], 228, 230, 231
sect. *Venustae* Pfitzer 15, 16
sect. **Verrucosae** Pfitzer 15, 16, 18, 23, 59, **119**, [120], 228, 230, 231, 242, 243
Subseries *Glumacea* Pfitzer 15
Subseries *Imbricatae* Pfitzer 15
Subseries *Nudae* Pfitzer 15
Subseries *Nudiscapae* Pfitzer 15
Subseries *Vaginatae* Pfitzer 15
acutilabium de Vogel 23, 132, **135**
advena C.S.P. Parish & Rchb.f. 217
alata A. Millar 217
alba Rchb.f. 217
albobrunnea J.J. Sm. 24, **159**
albolutea Rolfe 27, **210**, 236, 237 (301)
amplissima Ames & C. Schweinf. 217
amplissima Ames & C. Schweinf. var. *schweinfurthiana* J.J. Sm. 217
anceps Hook.f. 19, 31, **32**, 57
angustifolia Ridl. 27, 214, 217
angustifolia Wight 26, 196, 217
angustifolia A. Rich 27, 203, 217
annamensis Ridl. 19, 39, 217
annamensis Rolfe 26, 181, 217
apiculata Rchb.f. 217
arthuriana Rchb.f. 217
articulata Rchb.f. 217
arunachalensis H.J. Chowdhery & G.D. Pal 20, **50**
asperata Lindl. 8, 23, 119, 121, **122**, 219, 221, 223, 237, 239, (279)
assamica Linden & Rchb.f. 25, **181**, 217, 219, 224, (297)
balfouriana Sander 217
barbata Griff. 19, 31, **32**, (257)
beccarii Rchb.f. 20, 63, **67**, 217, 222, 238, (263)
beccarii Rchb.f. var. *micholitziana* Kraenzl. 20, 67, 217
beccarii Rchb.f. var. *tropidophora* Schltr. 20, 67, 217
bella Schltr. 21, 75, 217

Index of Scientific Names

beyrodtiana Schltr. 217
bicamerata J.J. Sm. 21, **84**, 85, (269)
biflora C.S.P. Parish ex Rchb.f. 218
bihamata J.J. Sm. 22, 94, 218
bilamellata Lindl. 22, 99, **101**, (273)
bimaculata Ridl. 25, 167, 218
birmanica Rchb.f. 218
borneensis Rolfe 22, 98, **101**
brachygyne J.J. Sm. 22, 99, **102**, (273)
brachyptera Rchb.f. 16, 20, 57, 58, **59**, 60, 61, (263)
brevifolia Lindl. 26, 191, 218
brevilamellata J.J. Sm. 218
breviscapa Lindl. 8, 26, **196**, 217, (299)
bruneiensis de Vogel 23, 135, **136**, (283)
brunnea Lindl. 26, 183, 218
buennemeyeri J.J. Sm. 23, 133, **136**
bulbocodioides Franch. 218
calcarata J.J. Sm. 21, 84, **85**
calceata Rchb.f. 218
calcicola Kerr 19, 31, **33**, (257)
caloglossa Schltr. 20, 64, **68**
camelostalix Rchb.f. 218
candoonensis Ames 22, 98, **102**
carinata Rolfe 20, 63, **68**, 71, 217, 222, 224, 225, (263)
carnea Hook.f. 22, 113, 218
carnea Rchb.f. 218
casta Ridl. 25, 174, 218
caulescens Griff. 218
celebensis J.J. Sm. 21, 64, **69**, 223, (263)
chinensis Rchb.f. 218
chlorophaea Schltr. 26, 194, **197**
chloroptera Rchb.f. 26, 195, **197**, (299)
chrysotropis Schltr. 20, 49, **51**
ciliata Boxall ex Naves 218
cinnamomea Lindl. 27, 214, 218
clarkei Kraenzl. 20, 57, 218
clemensii Ames & C. Schweinf. **var. angustifolia** Carr 24, 159, **160**
clemensii Ames & C. Schweinf. **var. clemensii** 24, 159, **160**, (293)
clemensii Ames & C. Schweinf. **var. longiscapa** Ames & C. Schweinf. 24, 159, **161**
clypeata Rchb.f. 218
compressicaulis Ames & C. Schweinf. 22, 98, **103**, (273)
conchoidea Rchb.f. 218
concinna Ridl. 25, **172**
conferta Hort 26, 189, 190, 218
confusa Ames 26, 195, **198**
contractipetala J.J. Sm. 22, 100, **104**
convallariae C.S.P. Parish & Rchb.f. 218
corniculata Rchb.f. 218

coronaria Lindl. 218
corrugata Wight 25, 179, 219
corymbosa Lindl. 26, **187**, 188, (299)
crassifolia (Carr) Masamune 219
crassiloba J.J. Sm. 21, **88**
craticulaelabris Carr 24, 159, **162**, (293)
cristata Lindl. (ii), 1, [2], 8, 15, 170, 172, **173**, 174, 231, 235–239, 241, 244, (295)
croockewittii Teysm. & Binn. 219
crotalina Rchb.f. 219
cumingii Lindl. 25, 172, **174**, 218, 221, (295)
cuneata J.J. Sm. 219
cuprea H. Wendl. & Kraenzl. **var. cuprea** 22, 100, **104**, (273)
cuprea H. Wendl. & Kraenzl. **var. planiscapa** J. J. Wood & C.L. Chan 22, 101, **105**, (273)
cymbidioides Ridl. 24, 152, 219
cymbidioides Rchb.f. 219
cynoches C.S.P. Parish & Rchb.f. 26, 183, 219
dalatensis Gagnep. 26, 181, 219
darlacensis Gagnep. 19, 39, 219
dayana Rchb.f. 24, 145, 146, 219, 237
dayana Rchb.f. var. *massangeana* Ridl. 24, 152, 219
decipiens Sander 219
decora Wall. ex Voigt 20, 53, 219
delavagi Rolfe 219
densiflora Ridl. 24, 152, 219
dichroantha Gagnep. 16, 17, 26, 77, 181, **182**
diphylla Lindl. 219
distans J.J. Sm. 23, 132, **137**
dulitensis Carr 22, 99, **105**
eberhardtii Gagnep. 16, 25, 168, **169**, 230, (295)
ecarinata C. Schweinf. 19, 42, **43**, (257)
echinolabium de Vogel 23, 133, **137**, 138, 147, 148, (283)
edelfeldtii F. Muell. & Kraenzl. 23, 122, 219
elata Lindl. 19, 28, 40, 219, 241
elata Hook. non Lindl. 19, 35, 219
elegans Rchb.f. 219
elegantula Kraenzl. 219
elmeri Ames 22, 99, **106**
endertii J.J. Sm. 22, 99, **107**, (275)
esquirolei Schltr. 27, 210, **211**
exalata Ridl. 24, 159, **162**, (293)
falcata Anders. ex Hook.f. 219
filipeda Gagnep. 19, 31, **34**, 37
fimbriata Lindl. 20, 47, 49, **51**, 221, 223, 226, 230, 238, 241, (261)
flaccida Lindl. 8, 27, 207, 208, **211**, 212, 213, 221, 236–238, 241, 244, (303)
flavida Lindl. 19, 37, 45, 219
fleuryi Gagnep. 25, 169, 170, 220
flexuosa Rolfe 25, 166, **167**, 218, (295)

foerstermannii Rchb.f. 25, 172, **175**, 220, 221, 239, (295)
formosa Schltr. 21, 64, **70**
forstenebrarum P. O'Byrne 220
fragrans Schltr. 21, 63, 69, **70**, 238, (265)
fuerstenbergiana Schltr. 21, 64, **71**, (265)
fuliginosa Lodd. ex Hook. 20, 47, 50, **52**, 55, (261)
fusca Rchb.f. 220
fuscescens Lindl. **var. brunnea** Lindl. 26, 181, **183**, 218, 219, (299)
fuscescens Lindl. **var. fuscescens** 26, 180, **182**, (297)
fuscescens Lindl. **var. integrilabia** (Pfitzer) Schltr. 26, 181, **184**, 220
fuscescens Lindl. **var. viridiflora** U. Pradhan 26, 181, **184**
fusco-lutea Teijsm. & Binn. 220
gardneriana Lindl. 220
genuflexa Ames & Schweinf. 23, 132, **138**, 224, (283)
ghatakii T.K. Paul, S.K. Basu & M. Biswas 19, 32, **34**
gibbifera J.J. Sm. 21, 87, **89**, 221, (269)
gibbosa (Blume) Rchb.f. 220
glandulosa Lindl. **var. bournei** S. Das & S.K. Jain 25, 172, **176**
glandulosa Lindl. **var. glandulosa** 25, 172, **176**, 202, 203, 222
glandulosa Lindl. **var. sathyanarayanae** S. Das & S.K. Jain 25, 172, **176**
globosa Rchb.f. 220
gongshanensis H. Li 26, 187, **188**
goweri Rchb.f. 26, 191, 220
graminifolia C.S.P. Parish & Rchb.f. 27, 215, 216, 220
grandiflora Rolfe 220
griffithii Hook.f. 19, 31, **35**, 57
guamensis Ames 21, 63, **72**, 222
hajrae Phukan 27, 210, **212**
harana J.J. Sm. 22, 87, **89**, (269)
henryi Rolfe 220
hirtella J.J. Sm. 23, 134, **139**, 223, (283)
hitendrae S. Das & S.K. Jain 26, 187, **188**
holochila P.F. Hunt & Summerh. 19, 31, **35**, (257)
hookeriana Lindl. var. *brachyglossa* Rchb.f. 220
hookeriana Lindl. 220
huettneriana Rchb f. 27, 208, **213**, (303)
humilis (J.E. Sm.) Lindl. 220
humilis (J.E. Sm.) Lindl. var. *albata* Rchb.f. 220
humilis (J.E. Sm.) Lindl. var. *tricolor* Rchb.f. 220
imbricans J.J. Sm. 23, 121, **123**
imbricata Rchb.f. 220
incrassata (Blume) Lindl. var. **incrassata** 22, 88, **90**, (271)
incrassata (Blume) Lindl. var. **sumatrana** J.J. Sm. 22, 88, **90**
incrassata (Blume) Lindl. var. **valida** J.J. Sm. 22, 88, **91**, (271)
ingloria J.J. Sm. 220
integerrima Ames 27, 210, **213**
integra Schltr. 22, 98, 102, **107**
integrilabia Schltr. 26, 184, 220

interrupta Lindl. ex Heynh. 220
javanica Lindl. 220
judithiae P. Taylor 23, 133, **139**, (283)
kaliana P.J. Cribb 23, 134, **140**, (285)
kelamensis J.J. Sm. 22, 87, **91**, (271)
kemiriensis J.J. Sm. 25, 172, **177**, (297)
khasiyana Rchb.f. 220
kinabaluensis Ames & C. Schweinf. 22, 100, **108**, (275)
kingii Hook.f. 25, 175, 220
kutaiensis J.J. Sm. 220
lacinulosa J.J. Sm. 26, 195. **198**
lactea Rchb.f. 27, 211, 221
lagenaria Lindl. 221
lamellata Rolfe 21, 73, 221
lamellulifera (Carr) Masamune 221
laotica Gagnep. 20, 51, 221
latiloba de Vogel 23, 133, **141**
lauterbachiana Kraenzl. 24, 156, 221
lawrenceana Rolfe 16, 17, 25, 77, 167, 168, **169**, 220, 228, 237, 239, 241, 242, (295)
lentiginosa Lindl. 26, 59, 192, 195, **199**, 204, (301)
leucantha W.W. Sm. 19, 31, **36**
leungiana S.Y. Hu 20, 51, 221
ligulata Teijsm. & Binn. 221
limminghei Hort. ex Gentil 221
lockii Aver. 19, 31, **37**
loheri Rolfe 26, 196, **200**
longebracteata Hook.f. 25, 174, 221
longeciliata Teijsm. & Binn. 20, 49, **53**
longibulbosa Ames & Schweinf. 23, 132, **141**, (285)
longifolia (Blume) Lindl. 22, 95, 100, **108**, (275)
longipes Lindl. 19, 42, **44**, (259)
longipes Lindl. var. *verruculata* S.C. Chen 20, 46, 221
longirachis Ames 22, 99, **109**
longipes Hook.f. 19, 45, 221
longpasiaensis J.J. Wood & C.L. Chan 22, 88, **92**
loricata Rchb.f. 221
lowii Paxt. 23, 122, 221
lurida Ames & C. Schweinf. 221
lycastoides F. Muell. & Kraenzl. 21, 62, **72**, 226, (265)
macdonaldii F. Muell. & Kraenzl. 21, 63, **73**, 221, (265)
macrobulbon Hook.f. 24, 148, 221
macroloba J.J. Sm. 22, 89, 221
macrophylla Teijsm. & Binn. 23, 122, 221
maculata Lindl. 221
maingayi Hook.f. 25, 175, 221
malintangensis J.J. Sm. 25, 172, **177**
malipoensis Z.H. Tsi 24, 155, **156**, (291)
mandarinorum Kraenzl. ex Diels 221

marmorata Rchb.f. 26, 196, **200**, 226, (301)
marthae S.E. Sierra 23, 122, **123**, (279)
massangeana Rchb.f. 24, 152, 221, 235–239
mayeriana Rchb.f. 23, 121 **124**, (279)
media Wall. ex F. Voigt 221
membranifolia Carr 21, 76, 221
merrillii Ames 26, 196, **201**, (301)
micholitziana Kraenzl. 20, 67, 222
micrantha Lindl. 20, 56, **57**, 218
miniata (Blume) Lindl. 24, 154, 155, **156**, 221, 224, 230, (293)
minutissima Kraenzl. 222
modesta J.J. Sm. 22, 112, 222
monilirachis Carr 22, 85, 88, **92**, (271)
monticola J.J. Sm. 26, 196, **201**
mooreana Sand. ex Rolfe 25, 172, **178**, 223, 237–239, 241 (297)
mossiae Rolfe em. S.K. Jain & S. Das 8, 26, 176, 177, 195, **202**, 203, 238, (301)
motleyi Rolfe ex J.J. Wood, D.A. Clayton & C.L. Chan 22, 99, **109**, (275)
moultonii J.J. Sm. 23, 132, **142**, (285)
multiflora Schltr. 23, 117, **118**, 230, (277)
muluensis J.J. Wood 23, 133, **143**, (285)
naja J.J. Sm. 22, 88, **93**, (271)
nervillosa Rchb.f. 222
nervosa A. Rich 8, 25, 172, **178**, 219, (297)
nervosa Wight 222
nigrescens P. Don ex Loudon. 222
nigro-furfuracea Guillaumin 222
nitida (Roxb.) Hook.f. 26, 191
nitida (Wall. ex D. Don) Lindl. 8, 26, 187, **189**, 218, 222, 241, (299)
obtusifolia Carr 24, 155, **157**
occulata Hook.f. **var. occulata** 26, 185, **190**
occulata Hook.f. **var. uniflora** N.P. Balakr. 26, **190**
ocellata Lindl. 26, 191, 222
ocellata Lindl. var. *boddaertiana* Rchb.f. 26, 191, 222
ocellata Lindl. var. *maxima* Rchb.f. 26, 191, 222
ochracea Lindl. 26, 189, 222, 241
odoardi Schltr. 23, 135, **143**, (285)
odoratissima Lindl. 26, 196, **203**, 217, 225, (301)
oligantha Schltr. 21, 68, 222
ovalis Lindl. 20, 50, 51, **53**, 219, 222, 238, (261)
pachybulbon Ridl. 27, 215, 222
padangensis J.J. Sm. & Schltr. 20, 49, **54**. (261)
palaelabellatum A. Gilli 222
palawanensis Ames 23, **144**
palawensis Tuyama 21, 72, 222
pallens Ridl. 20, 49, **54**, (261)
pallida Rchb.f. 222
pandurata Lindl. 8, 23, 121, 124, **125**, 126, 222, 237, 238, 242, (279)
papagena Rchb.f., *paparina* sphalm. 20, 57, 222

papillosa Ridl. ex Stapf 9, 23, 121, **125**, (279)
parishii Hook. 20, 58, 59, **60**, 238, (263)
parviflora Lindl. 222
peltastes Rchb.f. 23, 122, **126**, (279)
peltastes Rchb.f. var. *unguiculata* J.J. Sm. 23, 125, 222
pendula Summerh. ex Parry 19, 32, **37**
perakensis Rolfe 222
phaiostele Ridl. 222
pholas Rchb.f. 222
pholidotoides J.J. Sm. 23, 132, **145**, (287)
picta Schltr. 26, 181, **185**
pilosissima Planch. 222
pinniloba J.J. Sm. 223
planiscapa Carr var. **grandis** Carr 22, 100, **111**, (275)
planiscapa Carr var. **planiscapa** 22, 100, **110**
plantaginea Lindl. 24, 148, 223
platyphylla Schltr. 21, 69, 223
plicatissima Ames & Schweinf. 25, 159, **163**, (293)
pogonioides Rolfe 223
porrecta Rchb.f. 223
porrecta Rchb.f. 223
praecox (J.E. Sm.) Lindl. var. *sanguinea* Lindl. 223
praecox (J.E. Sm.) Lindl. var. *tenera* Rchb.f. 223
praecox (J.E. Sm.) Lindl. var. *wallichiana* Lindl. 223
praecox Lindl. 223
prasina Ridl. 22, 100, **112**, 222, 224, 225, (275)
primulina Barretto 20, 51, 223
prolifera Lindl. 19, 20, 34, 37, 41, 42, **44**, 46, 219 (259)
psectrantha Gagnep. 25, 178, 223
psittacina Rchb.f. 21, 74, 223
psittacina Rchb.f. var. *huttonii* Rchb.f. 21, 74, 223
pulchella Rolfe 19, 32, **38**
pulverula sensu Lamb & C.L. Chan *non* Teijsm. & Binn.: 24, 147, 223
pulverula Teijsm & Binn. 24, 134, **145**, 219, (287)
pumila Rchb.f. 223
punctulata Lindl. 1, 16, 26, 185, 188, **191**, 218, 220, 222, 223 (299)
punctulata Lindl. var. *hysterantha* Tang & Wang 26, 191, 223
purpurascens (Thw.) Hook.f. 223
pusilla Ridl. 223
pustulosa Ridl. 23, 122, 223
quadrangularis Ridl. **24, 150,** 223
quadratiloba Gagnep. 27, 210, **214**, 225
quinquelamellata Ames 22, 98, **112**, (277)
radicosa Ridl. 22, 99, **113**, 218, 225, (277)
radicosus Ridl. 223
radioferens Ames & Schweinf. 24, 133, **146**, (287)
radiosa J.J. Sm. 23, 139, 223
raizadae S.K. Jain & S. Das 19, 42, **45**, (259)

ramosii Ames 223
recurva Rchb.f. 223
reflexa J.J. Wood & C.L Chan 23, 138, 224
reichenbachiana T. Moore & Veitch 224
remediosae Ames & Quisumb. 23, 99, **114**
rhabdobulbon Schltr. 24, 134, **147**, (287)
rhizomatosa J.J. Sm. 22, 112, 224
rhizomatosa J.J. Sm. var. *quinquelobata* J.J. Sm. 22, 112, 224
rhodeana Rchb.f. 27, 214, 224
rhombophora Rchb.f. 224
richardsii (Carr) Masamune 224
ridleyana Schltr. 224
ridleyi Gagnep. 19, 39, 224
rigida C.S.P. Parish & Rchb.f. 19, 32, **38**, 225
rigidiformis Ames & Schweinf. 25, 157, 159, **163**
rochussenii de Vriese 24, 134, **148**, (289)
rossiana Rchb.f. 27, 214, 224
rubens Ridl.; Hook.f., 224
rubra Rchb.f. 224
rumphii Lindl. 21, 63, **74**, 223, (265)
rupicola Carr 24, 133, **149**, (289)
saigonensis Gagnep. 26, 181, 224
salmonicolor Rchb.f. 21, 62, **75**, 224, (265)
salmonicolor Rchb.f. var. *virescentibus* J.J. Sm. ex Dakkus 21, 75, 224
sanderae Kraenzl. ex O'Brien 19, 32, **39**, 224
sanderiana Rchb.f. 25, 172, **179**, (297)
sarawakensis Schltr. 224
sarrasinorum Kraenzl. 20, 68, 224
schilleriana Rchb.f. & Koch 27, 195, **203**
schultesii S.K. Jain & S. Das 20, 42, **46**, 221, (259)
septemcostata J.J. Sm. 21, 62, **76**, 221, (267)
siamensis Rolfe 25, 181, 224
simplex Lindl. 24, 156, 224
sparsa Rchb.f. 27, 195, **204**
speciosa (Blume) Lindl. **subsp. fimbriata** (J.J. Sm.) Gravendeel 21, 62, **77**, 224
speciosa (Blume) Lindl. **subsp. incarnata** Gravendeel 21, 62, **78**
speciosa (Blume) Lindl. **subsp. speciosa** 16, 21, 62, **76**, 224, 237, 238, 241
speciosa (Blume) Lindl. var. *alba* Hort. 21, 77, 224
speciosa (Blume) Lindl. var. *albicans* H.J. Veitch 21, 77, 224
speciosa (Blume) Lindl. var. *fimbriata* J.J. Sm. 21, 77, 224
speciosa (Blume) Lindl. var. *rubiginosa* Hort. 21, 77, 224
squamulosa J.J. Sm. 24, 133, **149**, (289)
steenisii J.J. Sm. 23, 100, **114**
steffensii Schltr. 24, 148, 224
stellaris Rchb.f. 24, 148, 224
stenobulbon Schltr. 23, 98, **115**
stenochila Hook.f. 23, 98, **115**, (277)
stenophylla Ridl. 27, 215, 225

stipitibulbum Holttum 23, 113, 225
stricta (D. Don) Schltr. 19, 28, 31, 36, **40**, 219, 230, 241, (257)
suaveolens (Lindl.) Hook.f. 27, 196, **204**
subintegra J.J. Sm. 25, 162, 225
sulphurea (Blume) Rchb.f. 225
sumatrana J.J. Sm. 24, 151, 225
susanae P.J. Cribb & B.A. Lewis 21, 63, **79**
swaniana Rolfe 24, 134, **150**, 223, (289)
taronensis Hand.-Mazz. 27, 196, **205**
tenasserimensis Seidenf. 19, 31, **40**, (257)
tenompokensis Carr 23, 99, **116**, (277)
tenuiflora Ridl. 225
tenuis Rolfe 22, 88, **94**, 218
testacea Lindl. 24, 135, **151**, 225 (289)
thailandica Seidenf. 27, 214, 225
thuniana Rchb.f., Otto & Dietr. 225
tiomanensis M.R. Henderson 21, 64, **80**, (267)
tomentosa Lindl. 24, 128, 134, **152**, 153, 219, 221, 225, 241, (291)
tomentosa Lindl. var. *? penangensis* Hook.f. 24, 153, 225
tomentosa Lindl. var. *massangeana* (Rchb.f.) Ridl. 24, 152, 225
tomiensis O' Byrne - *nom. invalid.* 21, 81, 225
tommii Gravendeel & P. O'Byrne 21, 63, **80**, 225, (267)
treutleri Hook.f. 57, 225
tricarinata Ridl. 19, 39, 225
trifida Rchb.f. 27, 203, 225
trilobulata J.J. Sm. 23, 98, **116**
trinervis Lindl. 27, 152, 210, **214**, 217, 218, 224–226, (303)
triotos Rchb.f. 225
triplicatula Rchb.f. 20, 47, 50, **55**, (261)
triptera Brongn. 225
trisaccata Griff. 225
trunicola Schltr. 20, 68, 225
tumida J.J. Sm. 23, 100, **117**
undatialata J.J. Sm. 27, 195, **206**
undulata Rchb.f. 225
undulata Wall ex Pfitzer & Kraenzl. 27, 205, 225
undulata Wall. ex F. Voigt. 225
unguiculata (Carr) Masamune 225
uniflora Lindl. 207, 225
usitana Röth & Gruss 21, 63, **81**, (269)
ustulata C.S.P. Parish & Rchb.f. 20, 42, **47**
vagans Schltr. 22, 112, 225
vanoverberghii Ames 27, 195, **206**
veitchii Rolfe 25, 164, **165**, (293)
velutina de Vogel 24, 135, **153**, (291)
ventricosa Rchb.f. 226
venusta Rolfe 24, 132, **154**, (291)
vermicularis J.J. Sm. 22, 88, **94**, (273)

verrucosa S.E.C. Sierra 23, 121, **126**, (281)
virescens Rolfe 20, 58, 59, **60**, (263)
viscosa Boxall ex Vidal 216, 226
viscosa Rchb.f., Otto & Dietr. 27, 210, **215**, 220, (303)
wallichi Hook. 226
wallichiana Lindl. 226
wettsteiniana Schltr. 27, 215, 226
whitmeei Schltr. 21, 72, 226
xanthoglossa Ridl. 21, 82, 226
xerophyta Hand.-Mazz. 20, 51, 226
xylobioides Kraenzl. 226
xyrekes Ridl. 21, 62, **82**, 226, (269)
yunnanensis Rolfe 226
zahlbrucknerae Kraenzl. 26, 200, 226
zeylanica Hook.f. 27, 196, **207**
zhenkangensis S.Chen & K.Y. Lang 19, 31, **41**
zurowetzii Carr 23, 121, **127**, (283)

Crinonia Blume 15
Cypripedium elegans 219
Dendrochilum Blume 3, 15, 222
    pumilum Rchb.f. **var. pumilum** 223
    rhombaphorum (Rchb.f.) Ames 224
Dickasonia L.O. Williams 3
    vernicosa L.O. Williams 3
Entomophobia de Vogel 3, 228
    kinabaluensis (Ames) de Vogel 3
Epidendrum caespitosum Paepp & Endl. 225
Epigeneium treutleri (Hook.f.) Ormd. 57
Eria coronaria (Lindl.) Rchb.f. 218
Geesinkorchis de Vogel 3, 228, 231
    phaiostele (Ridl.) de Vogel 222
Gynoglottis J.J. Sm. 3
    cymbidioides (Rchb.f.) J.J. Sm. 219
    palaelabellatum (A. Gilli) Garay & Kittridge 222
Hologyne Pfitzer 15, 16, 154
Ischnogyne Schltr. 3
    mandarinorum (Kraenzl.) Schltr. 3
Nabaluia Ames 3, 228
Neogyna Rchb.f. 3, 228, 231
    gardneriana (Lindl.) Rchb.f. 3
Nervila crispata (Blume) Schltr. 220
Neogyne Rchb.f. 15
Otochilus Lindl. 3, 15
    albus Lindl. 217
    fuscus Lindl. 220
    porrectus Lindl. 223
Panisea Lindl. 3, 15
    apiculata Lindl. 217

    demissa (D. Don) Pfitzer 222
    uniflora (Lindl.) Lindl. 207, 218, 219, 225
Phaius longipes (Hook.f.) Holttum 221
Pholidota (Hook.) Lindl. 3, 15, 228, 231
    advena Hook.f. 217
    articulata Lindl. 217, 220
    camelostalix Rchb.f. 218
    carnea (Blume) Lindl. 113
    chinensis Lindl. 218
    convallariae (Rchb.f.) Hook.f. 218
    gibbosa (Blume) de Vries 218, 220, 225
    globosa (Blume) Lindl. 199, 220
    imbricata Hook. 218–221, 225
    pallida Lindl. 222
    recurva Lindl. 223
    rubra Lindl. 224, 225
    ventricosa (Blume) Rchb.f. 226
Pleione D. Don 3, 15, 228, 231
    bulbocodioides (Franch.) Rolfe 218–220
    grandiflora (Rolfe) Rolfe 220
    hookeriana (Lindl.) B.S. Williams 220
    humilis (J.E. Sm.) D. Don 220
    maculata (Lindl.) Lindl. 217, 219, 221
    praecox (J.E. Sm.) D. Don 218, 223, 224, 226
    × **lagenaria** Lindl. & Paxt. 221
    yunnanensis (Rolfe) Rolfe 226
Pseudoacoridium Ames 3
Ptychogyne Pfitzer 15, 16, 166
Sigmatogyne Pfitzer 3
Tainia barbata (Lindl.) Schltr. 222

Other titles by *Natural History Publications (Borneo)*

For more information, please contact us at

Natural History Publications (Borneo) Sdn. Bhd.
A913, 9th Floor, Phase 1, Wisma Merdeka
P.O. Box 15566, 88864 Kota Kinabalu, Sabah, Malaysia
Tel: 088-233098 Fax: 088-240768 e-mail: chewlun@tm.net.my
www.nhpborneo.com

Head Hunting and the Magang Ceremony in Sabah by Peter R. Phelan

Mount Kinabalu: Borneo's Magic Mountain—an introduction to the natural history of one of the world's great natural monuments by K.M. Wong & C.L. Chan

On the Flora of Mount Kinabalu in North Borneo by O. Stapf. Reprint with an Introduction by John H. Beaman

A Contribution to the Flora and Plant Formations of Mount Kinabalu and the Highlands of British North Borneo by Lilian S. Gibbs. Reprint with with an Introduction by John H. Beaman

Discovering Sabah by Wendy Hutton (English, Chinese and Japanese editions)

Enchanted Gardens of Kinabalu: A Borneo Diary by Susan M. Phillipps

A Colour Guide to Kinabalu Park by Susan K. Jacobson

Kinabalu: The Haunted Mountain of Borneo by C.M. Enriquez (Reprint)

National Parks of Sarawak by Hans P. Hazebroek and Abang Kashim Abg. Morshidi

A Walk through the Lowland Rainforest of Sabah by Elaine J.F. Campbell

In Brunei Forests: An Introduction to the Plant Life of Brunei Darussalam (Revised edition) by K.M. Wong

The Larger Fungi of Borneo by David N. Pegler

Rafflesia of the World by Jamili Nais

Pitcher-plants of Borneo by Anthea Phillipps & Anthony Lamb

A Field Guide to the Pitcher Plants of Sabah by Charles Clarke

Nepenthes of Borneo by Charles Clarke

Nepenthes of Sumatra and Peninsular Malaysia by Charles Clarke

The Plants of Mount Kinabalu 3: Gymnosperms and Non-orchid Monocotyledons by John H. Beaman & Reed S. Beaman

The Plants of Mount Kinabalu 4: Dicotyledon Families Acanthaceae to Lythraceae by John H. Beaman, Christiane Anderson & Reed S. Beaman

Slipper Orchids of Borneo by Phillip Cribb

The Genus Paphiopedilum (Second edition) by Phillip Cribb

Orchids of Sarawak
by Teofila E. Beaman, Jeffrey J. Wood, Reed S. Beaman and John H. Beaman

Orchids of Sumatra by J.B. Comber

Dendrochilum of Borneo by J.J. Wood

Gingers of Peninsular Malaysia and Singapore by K. Larsen, H. Ibrahim, S.H. Khaw & L.G. Saw

Mosses and Liverworts of Mount Kinabalu
    by Jan P. Frahm, Wolfgang Frey, Harald Kürschner & Mario Manzel

Birds of Mount Kinabalu, Borneo by Geoffrey W.H. Davison

The Birds of Borneo (Fourth edition) by Bertram E. Smythies (Revised by Geoffrey W.H. Davison)

The Birds of Burma (Fourth edition) by Bertram E. Smythies (Revised by Bertram E. Smythies)

Swiftlets of Borneo: Builders of Edible Nests by Lim Chan Koon and Earl of Cranbrook

Proboscis Monkeys of Borneo by Elizabeth L. Bennett & Francis Gombek

The Natural History of Orang-utan by Elizabeth L. Bennett

A Field Guide to the Frogs of Borneo by Robert F. Inger & Robert B. Stuebing

A Field Guide to the Snakes of Borneo by Robert B. Stuebing & Robert F. Inger

Turtles of Borneo and Peninsular Malaysia by Lim Boo Liat & Indraneil Das

The Natural History of Amphibians and Reptiles in Sabah by Robert F. Inger & Tan Fui Lian

An Introduction to the Amphibians and Reptiles of Tropical Asia by Indraneil Das

Marine Food Fishes and Fisheries of Sabah by Chin Phui Kong

Layang Layang: A Drop in the Ocean by Nicolas Pilcher, Steve Oakley & Ghazally Ismail

Phasmids of Borneo by Philip E. Bragg

The Dragon of Kinabalu and other Borneo Stories by Owen Rutter (Reprint)

Land Below the Wind by Agnes N. Keith (Reprint)

Three Came Home by Agnes N. Keith (Reprint)

White Man Returns by Agnes N. Keith (Reprint)

Forest Life and Adventures in the Malay Archipelago by Eric Mjöberg (Reprint)

A Naturalist in Borneo by Robert W.C. Shelford (Reprint)

Twenty Years in Borneo by Charles Bruce (Reprint)

With the Wild Men of Borneo by Elizabeth Mershon (Reprint)

Kadazan Folklore (Compiled and edited by Rita Lasimbang)

An Introduction to the Traditional Costumes of Sabah (eds. Rita Lasimbang & Stella Moo-Tan)

Bahasa Malaysia titles:

Manual latihan pemuliharaan dan penyelidikan hidupan liar di lapangan oleh Alan Rabinowitz (Translated by Maryati Mohamed)

Etnobotani oleh Gary J. Martin (Translated by Maryati Mohamed)

Panduan Lapangan Katak-Katak Borneo oleh R.F. Inger dan R.B. Stuebing

Other titles available through
Natural History Publications (Borneo)

The Bamboos of Sabah by Soejatmi Dransfield

The Morphology, Anatomy, Biology and Classification of Peninsular Malaysian Bamboos by K.M. Wong

Orchids of Borneo Vol. 1 by C.L. Chan, A. Lamb, P.S. Shim & J.J. Wood

Orchids of Borneo Vol. 2 by Jaap J. Vermeulen

Orchids of Borneo Vol. 3 by Jeffrey J. Wood

Orchids of Java by J.B. Comber

Forests and Trees of Brunei Darussalam (eds. K.M. Wong & A.S. Kamariah)

A Field Guide to the Mammals of Borneo by Junaidi Payne & Charles M. Francis

Pocket Guide to the Birds of Borneo Compiled by Charles M. Francis

Kinabalu: Summit of Borneo (eds. K.M. Wong & A. Phillipps)

Ants of Sabah by Arthur Y.C. Chung

Traditional Stone and Wood Monuments of Sabah by Peter Phelan

Borneo: the Stealer of Hearts by Oscar Cooke (Reprint)

Maliau Basin Scientific Expedition (eds. Maryati Mohamed, Waidi Sinun, Ann Anton, Mohd. Noh Dalimin & Abdul-Hamid Ahmad)

Tabin Scientific Expedition (eds. Maryati Mohamed, Mahedi Andau, Mohd. Nor Dalimin & Titol Peter Malim)

Klias-Binsulok Scientific Expedition (eds. Maryati Mohamed, Mashitah Yusoff and Sining Unchi)

Traditional Cuisines of Sabah (ed. Rita Lasimbang)

Cultures, Costumes and Traditions of Sabah, Malaysia: An Introduction

Tamparuli Tamu: A Sabah Market by Tina Rimmer